Lightweight Materials from Biopolymers and Biofibers

ACS SYMPOSIUM SERIES **1175**

Lightweight Materials from Biopolymers and Biofibers

Yiqi Yang, Editor
University of Nebraska-Lincoln
Lincoln, Nebraska

Helan Xu, Editor
University of Nebraska-Lincoln
Lincoln, Nebraska

Xin Yu, Editor
Zhejiang Fashion Institute of Technology
Ningbo, Zhejiang Province, China

Sponsored by the
ACS Division of Cellulose and Renewable Materials

American Chemical Society, Washington, DC

Distributed in print by Oxford University Press

Library of Congress Cataloging-in-Publication Data

Lightweight materials from biopolymers and biofibers / Yiqi Yang, editor, University of Nebraska-Lincoln, Lincoln, Nebraska, Helan Xu, editor, University of Nebraska-Lincoln, Lincoln, Nebraska, Xin Yu, editor, Zhejiang Fashion Institute of Technology, Ningbo, Zhejiang Province, China ; sponsored by the ACS Division of Cellulose and Renewable Materials.

pages cm. -- (ACS symposium series ; 1175)

Includes bibliographical references and index.

ISBN 978-0-8412-2990-7 (alk. paper)

1. Lightweight materials. 2. Biopolymers. 3. Fibers. I. Yang, Yiqi, editor. II. Xu, Helan, editor. III. Yu, Xin (Materials researcher) IV. American Chemical Society. Cellulose and Renewable Materials Division.

TA418.9.L53L54 2014

620.1'92--dc23

2014039655

The paper used in this publication meets the minimum requirements of American National Standard for Information Sciences—Permanence of Paper for Printed Library Materials, ANSI Z39.48n1984.

Distributed in print by Oxford University Press

PRINTED IN THE UNITED STATES OF AMERICA

Foreword

The ACS Symposium Series was first published in 1974 to provide a mechanism for publishing symposia quickly in book form. The purpose of the series is to publish timely, comprehensive books developed from the ACS sponsored symposia based on current scientific research. Occasionally, books are developed from symposia sponsored by other organizations when the topic is of keen interest to the chemistry audience.

Before agreeing to publish a book, the proposed table of contents is reviewed for appropriate and comprehensive coverage and for interest to the audience. Some papers may be excluded to better focus the book; others may be added to provide comprehensiveness. When appropriate, overview or introductory chapters are added. Drafts of chapters are peer-reviewed prior to final acceptance or rejection, and manuscripts are prepared in camera-ready format.

As a rule, only original research papers and original review papers are included in the volumes. Verbatim reproductions of previous published papers are not accepted.

ACS Books Department

Contents

Indexes

Preface

Weight reduction in industrial products has been a well-received concept for quite some time. Recently, lightweight materials have become increasingly important in product design because they are capable of remarkably reducing material and energy consumption, carbon emission, and waste generation. As such, lightweight materials are used in a variety of industries, from automotive and construction to biomaterials. Currently, an overwhelming amount of lightweight materials are fabricated from plastics produced from fossil fuel. However, the soaring prices of current petro-products and depletion of crude oil has been driving substitution of petroleum-derived materials with renewable and sustainable substrates.

Biopolymers from renewable resources could be potential raw materials used to develop lightweight materials. Biopolymers, such as proteins, carbohydrates, and biosynthetic polymers, come from living organisms, including plants, animals and bacteria. Production of biopolymers substantially reduces energy consumption and carbon emission when compared with the production of petroleum-derived polymers. The future of material industries lies in developing lightweight materials made from renewable resources.

This book intends to present state-of-the-art research and technologies for design, construction, and applications of lightweight materials from biopolymers and biofibers. Though it is not possible to entirely capture the tremendous length and breadth of advancements in all relevant disciplines, we have put forth serious effort to do so, and to synergistically analyze the status and future of lightweight materials and bio-based polymers.

This collection of chapters, each one contributed by internationally recognized experts in their relevant fields, presents comprehensive coverage of the major aspects of lightweight materials. The first section of the book synergistically reviews the development of biopolymers and the manipulation of biopolymers into fibrous structures via diverse approaches. The second section mainly summarizes requirements for specific applications of bio-based lightweight materials in areas of constructions, logistics, medicine, and wastewater treatments, etc. Particular emphasis has been placed on recent advances and imminent perspectives pertaining to the development of lightweight materials, with recognition of the recent achievements in this growing field.

Yiqi Yang
yyang2@unl.edu (e-mail)
Department of Textiles, Merchandising and Fashion Design
Department of Biological Systems Engineering
Nebraska Center for Materials and Nanoscience
234, HECO Building
University of Nebraska-Lincoln
Lincoln, Nebraska 68583-0802, United States

Helan Xu
hxu14@unl.edu (e-mail)
Department of Textiles, Merchandising and Fashion Design
234, HECO Building
University of Nebraska-Lincoln
Lincoln, Nebraska 68583-0802, United States

Xin Yu
sisi_yu_xin@hotmail.com (e-mail)
International School
Zhejing Fashion Institute of Technology
495 Fenghua Rd., Zhenhai District
Ningbo, Zhejiang, China

Chapter 1

Lightweight Materials from Biofibers and Biopolymers

Danning Zhang[*,1,2,3]

[1]Department of Materials Science and Engineering, University of Delaware, Newark, Delaware 19716, United States
[2]Center for Composite Materials, University of Delaware, Newark, Delaware 19716, United States
[3]Current address: 201 Composites Manufacturing Laboratory, University of Delaware, Newark, Delaware 19716, United States
[*]E-mail: dzhang@udel.edu

Materials from natural resources are drawing growing attention in the last two decades due to the growing environmental threat and uncertainty of petroleum supply. Both academia and industries have contributed great efforts on developing lightweight materials from natural resources. This chapter briefly introduces the recent development of biofibers, biopolymers and biocomposites. Technologies, processing methods and properties optimization on these materials are included. The performance of light weight biocomposites are investigated including static mechanical properties, long term stability and functional properties. Wide range applications and future development of biobased materials are discussed.

Introduction

The growing concern of the non-renewability of petroleum resources and environmental issues results in an increasing interest in developing novel bio-based materials from renewable agricultural and natural resources. There are many advantages that bio-based materials have over the traditional ones, for example, the renewability, recyclability, sustainability, triggered biodegradability and low cost. These advantages make the importance of the bio-based materials considering the growing environmental threat and uncertainty of petroleum supply. Great efforts have been made to develop lightweight materials from abundant natural resources and agricultural byproducts, including biofibers, biopolymers and biocomposites. Technologies, processing methods and properties optimization on these materials are discussed in this chapter. The following sections are organized as follows: we start with the introduction of biofibers including types, chemical compositions, and properties. We subsequently discussed typical biopolymers from plant oil, proteins, starch and etc. Lastly, processing methods, performance and applications of biocomposites are reviewed.

Biofibers

Natural cellulose fibers such as cotton, jute, and flax, and protein fibers like wool and silk have been widely used for textiles and ropes for thousand years. As the world population increases, there is a trend for textile and other industries to search new fibers from abundant natural resources and agriculture byproducts to compete with major traditional fibers in terms of land dependency, cost and availability.

Bio-Cellulose Fibers

Bio-cellulose fibers from plant resources are the major biofibers studied. With the large variety of plant species, thousands of different cellulose fibers may be produced. Jute, flax, hemp, sisal, kenaf, ramie are common biofibers that were studied extensively and applied in industries. Abaca, oil palm, sugarcane bagasse, bamboo, pineapple leaf, coir, date palm leaf, curaua, rice straw, wheat straw and cornhusk fibers are drawing more attention and also being investigated due to their low cost, wide availability and specific properties.

Classification and Chemical Compositions

Based on the origins of the bio-cellulose fibers, they can be classified into bast, leaf, fruit, and stalk fibers, as seen in Table I with the corresponding fibers.

Table I. Classification of bio-cellulose fibers (*1–4*)

Classifications	*Bio-cellulose fibers*
Bast	Jute, flax, hemp, kenaf, ramie,
Leaf	Sisal, abaca, pineapple leaf, date palm leaf, curaua
Fruit	Coir, oil palm fruit bunch fiber, cornhusk
Stalk	wheat straw, rice straw, sugarcane bagasse, bamboo, wood

Table II. Chemical composition of some bio-cellulose fibers (*1*, *2*, *4–9*)

Biofiber	*Cellulose (wt%)*	*Hemicellulose (wt%)*	*Lignin (wt%)*	*Wax (wt%)*	*Pectin (wt%)*
Jute	61-71.5	13.6-20.4	12-13	0.5	0.2
Flax	71	18.6-20.6	2.2	1.5-1.7	2.3
Hemp	68-74.4	15-22.4	3.7-10	0.8	0.9
Kenaf	72	20.3-21.5	9-19	-	3-5
Ramie	68.6-76.2	13.1-16.7	0.6-0.7	0.3	1.9
Sisal	60-78	10-14.2	8-12	2	10
Abaca	56-63	20-25	7-9	3	1
Pineapple leaf	70-82	-	5-12.7	-	-
Date palm leaf	46	28	20	-	-
Curaua	73.6	9.9	7.5	-	-
Coir	32-43	0.15-0.25	40-45	-	3-4
Oil palm	42.7-65	17.1-33.5	13.2-25.3	-	-
Wheat straw	38-48.8	15-35.4	12-23	-	8
Cornhusk	42.3	41	12.6	-	-
Rice straw	64	23	8-19	8	-
Sugarcane bagasse	55.2	16.8	25.3	-	-
Bamboo	26-43	30	21-31	-	-

Dry bio-cellulose fibers are mainly consisting of cellulose, hemicellulose, lignin and small amount of pectin, wax, and ash (silica). The former three contribute to the mechanical and physical properties of cellulose fibers with different composition and molecular structures. The chemical compositions of the biofibers vary from species, parts of the plants and also age of the plants. Typical range of chemical components of some important bio-cellulose fibers are shown in Table II.

Table III. Physical and mechanical properties of some bio-cellulose fibers (*1, 2, 4, 5, 7, 9–16*)

Fiber	*Density (g/cm³)*	*Diameter (μm)*	*Tensile strength (MPa)*	*Young's Modulus (GPa)*	*Strain to failure (%)*
Jute	1.3-1.5	25-250	393-800	13-26.5	1.5-1.8
Flax	1.5	40-600	345-1500	27-39	2.7-3.2
Hemp	1.5	25-250	550-900	38-70	1.6-4
Kenaf	1.5-1.6	260-400	350-930	40-53	1.6
Ramie	1.5-1.6	34-49	400-938	24.5-128	1.3-3.8
Sisal	1.5	50-200	468-640	9.4-22	2-7
Abaca	1.5	-	400	12	3-10
Pineapple leaf	1.4	20-80	400-1627	34.5-82.5	1.6
Date palm leaf	0.9-1.2	100-1000	233	9	2-19
Curaua	1.4	500-1150	500-1150	11.8	3.7-4.3
Coir	1.2-1.5	10-460	175	4-6	15-51.4
Oil palm	0.7-1.6	150-500	50-400	0.5-9	4-18
Bagasse	1.25	-	290	17	-
Bamboo	0.6-1.1	-	140-230	11-17	-
Wheat straw	1.6	94	273	13	2.7
Rice straw	-	-	449	26	2.2
Cornhusk	-	20	351	9.1	15.3
E-glass	2.5	15-25	2000-3500	70-73	2.5-3.7

Physical and Mechanical Properties

Different from synthetic fibers, biofibers exhibit large variation on their properties due to their complex structure, chemical composition, dimensions, defects, maturity and extracting methods to obtain the fibers. The physical and mechanical properties of biofibers comparing with man-mad fibers are shown in Table III. The strength and stiffness can be related to the fiber diameter, fiber length, aspect ratio (length/diameter ratio), the angle between fiber axis and microfibrils and crystallinity of cellulose. The properties also depend on the form of the fibers, testing methods and environmental conditions. Generally, the strength and stiffness of biocellulosic fibers are lower than glass fiber. The major drawback of bio-cellulose fiber is their hydrophilic nature and moisture absorption, which can introduce weak wetting and bonding between fiber and matrix, and also increase the moisture absorption of composites when used as reinforcements for composite materials. The moisture regain of the bio-cellulose fibers can be more than 10% (*1*, *12*, *14*). Dimensional instability of composites in humid environments will also be an issue for biobased lightweight composites

The properties of biofibers are subject to degradation under biological, UV light, thermal and moisture conditions due to their lignocellulosic nature. The crystalline cellulose can be weakened by biodegradation, resulting in reduced strength of the fibers. Lignin is much more sensitive to the photochemical degradation than cellulose. Hemicellulose and cellulose degrade faster than lignin when subject to heat. Moisture absorption mainly occurs in hemicellulose and non-crystalline cellulose regions (*4*, *17–19*).

Bio-Protein Fibers

Bio-protein fibers could provide additional resources for the alternatives of synthetic fibers. However, the only fibers (excluding the regenerated fibers) being studied so far are from chicken feather which is a major by-product of the poultry processing industry with abundant resources (*20–25*). The density of chicken feathers is 0.89g/cm^3, much lower than almost all cellulose fibers and synthetic fibers (*20*). Containing about 90% of protein, chicken feathers mainly consist of quills (rachis) and feather fibers (barbs); each are about 50% in weight, and separated commercially. The presence of honeycomb shaped air pockets in chicken barbs and quills shows the potential acoustic application of chicken feathers (*26*, *27*). The fine and flexible barbs show strength of 180 MPa and modulus of 4.7GPa (*26*). The raw quill is stiff, long and thick, and can be grinded and used as reinforcement of polymer composites (*27*). With an improved web form matrix, whole chicken feather can be well distributed in polymer matrix when applied in composite materials, reducing the process of feather processing (*21*). A recent study produced nanoparticles from chicken feather through enzymatic hydrolysis and ultrasonic treatment, showing the potential applications of chicken feathers on the reinforcement of nanocomposites and adsorbents (*28*). Biothermoplastics were obtained from chicken feathers through acrylate grafting and hydrolyzing and citric acid crosslinking for various applications (*29*, *30*).

Biopolymers

Biopolymers derived from natural resources can be thermoplastics and thermosets. They are playing important roles on replacing petroleum based polymers for packaging, coating, biomedical materials and composite materials. Based on the chemical transformation process and the origin, bio-thermoplastics are generally categrized into: (i) the agro-polymers that are directly extracted from biomass, mainly polysaccharides such as thermoplastic starch (TPS); (ii) biopolyesters synthesized with biobased monomers, such as polylactide acid (PLA) polymerized with lactic acid from sugar fermenting; and (iii) polymer generated by microorganism or modified bacteria, such as polyhydroxyalkanoate (PHA) (*31–35*). Vegetable oils including soybean oil, linseed oil and corn oil are the major resources for various biobased thermosetting resins (*36*). Plant proteins from soybean, wheat and corn are investigated and made into both thermoset and thermoplastic polymers (*37*, *38*).

Using biobased polymers could potentially help us shift our dependent on non-renewable petroleum resources to sustainable natural resources, and also reduces the cost of many products. Many biopolymers are also biodegradable and recyclable, providing environmentally friendly products. Improvements on the performance, manufacturing process, and recycling techniques, and reducing the total cost of the products are currently in progress.

Thermoplastic Starch (TPS)

Plants like wheat, rice, corn, potato, oats and peas are the major resources of starch in the form of starch granule. Starch contains linear and branched amylopectin. Native starch has to be plasticized with water, glycerol, sorbitols, glycols and so on, and then processed under temperature and shear to produce thermoplastic polymers. Type and content of plasticizers, and ratio of amylose/amylopectin in the raw starch affect glass transition temperature and mechanical properties of thermoplastic starch (Table IV), and consequently led to development of the different products for different applications. Various techniques can be used to process starch, such as solution casting, extrusion, injection molding, compression molding, similar to those applied for synthetic thermoplastics. Although using cheap and green raw materials, thermoplastic starch suffers from weak mechanical properties compared to synthetic polymers, very poor moisture resistance and long term stability (*4*, *39*). Chemical modification such as esterification by acetylation or hydroxylation can successfully improve the water resistance of TPS, but at the expense of toxic byproduct and higher production cost. Blending starch with other biodegradable polymers such as PLA and PHAs is another way to improve the properties of starch (*40*, *41*). Cellulose fibers reinforced starch based biocomposites, and nanocomposites with phyllosilicates and polysaccharide nanofillers generally provide improved moisture resistance, mechanical and thermal properties (*42–44*). Functionalized nanocomposites were developed through reinforcing starch with nanofillers such as carbon nanotubes, graphite, metal oxides, metalloid oxides, and hydroxyapatite for potential applications such as stimulators of bone cells, photovoltaic solar cells, gas sensors, photodiodes, clinical orthopaedics and scaffolds (*43*).

Polylactide Acid (PLA)

Polylactic acid is one of the most thoroughly investigated and widely used biopolymers. It is an aliphatic polyester made through ring-opening polymerization of lactide, or condensation polymerization of lactic acid. PLA can be amorphous or semicrystalline with different monomers (*31*, *36*, *45*, *46*). It has comparable mechanical properties to polyetherlene terephthalate (PET) and polypropylene (PP) (mechanical properties are shown in Table IV. Due to its relative brittleness which can results low impact performance when using as matrix of polymeric composites, PLA were toughened with polyethyleneglocols (PEGs), organophilic modified montmorillonite (MMT), linear low density polyethylene (LLDPE), or rubber blends (*47–50*). A recent work showed that a small loading (1-2wt%) of methyl monofunctional and tetra-functional silicon-oxygen silicon can significantly enhance the elongation at break of PLA without scarifying tensile strength (*51*). PLA is a sustainable biodegradable hydrophobic polymer and eventually decomposed into water and CO_2 returning to the biomass, which make it a very good matrix candidate for green composites.

Table IV. Mechanical properties of major biopolymers (*12, 33, 42, 48, 50, 52, 53*)

Polymer	*Density (g/cm³)*	*Tensile strength (MPa)*	*Tensile modulus (GPa)*	*Strain to failure (%)*
TPS*	1.35-1.39	3.3-21.4	0.04-1.1	3-104
PLA	1.26	21-72	3.6	2.4
P3HB	1.28	40	3.5	0.4-7
P(3HB-*co*-3HV)	1.22-1.25	23-40	3.5	1.6-20
P(4HB)	1.22	104	-	1000
Polystyrene (PS)	1.05	25-69	3.5	3
PP	0.85-0.95	26-41.4	0.95-1.7	40

* Plasticized with 10% - 24% glycerol and 9% - 12% water.

Polyhydroxyalkanoates (PHAs)

Polyhydroxyalkanoates is a family of linear polyesters containing short chain length (SCL) (3-5 carbon atoms) and medium chain length (MCL) (6-18 carbon atoms) hydroxyacids. Poly (3-hydroxybutyrate) (P3HB) which is one of the most common PHAs is stiff and brittle due to its high crystallinity. As the chain length of monomer increases, PHAs are becoming increasingly flexible, resulting in relatively low crystallinity, strength and melting point. The brittleness significantly affected the impact resistance of P3HB. The copolymer of Poly(3-hydroxybutyrate-*co*-3-hydroxyvalerate) (P(3HB-*co*-3HV)) was found to have enhanced toughness compared to P3HB. P(4HB) have much higher strength and stain to failure as seen in Table IV. Techniques on controlling the chemical composition, molecular weight, microstructure, crystallization to control the thermal and mechanical properties of PHAs were extensively studied. Tensile properties can be improved through physical drawing of the polymer during the processing; nucleating agent can help to control the crystallization process; rubbery additives and plasticizers can significantly improve the impact properties of PHAs. PHAs blended with miscible molecules, copolymers, fibers, and nano particles can achieve enhanced properties as well. The chemical and mechanical properties of PHAs were comprehensively reviewed in (*54*). Besides the mechanical properties, PHAs are piezoelectric, hydrophobic, have good air stability and UV resistance. Due to the biodegradability and interesting physical and mechanical properties, wide range of applications in medicine, agriculture, tissue engineering, composite materials and nanocomposites were developed for PHAs (*55*). Despite the interesting properties and potential applications of PHAs, production of PHAs is not cost effective at an industrial scale. Mix culture production was proposed as a strategy to reduce the production cost over the pure culture method and the catalyst synthesis (*54*, *56*). Industrial and agricultural fermented wastes such as sugar molasses, sludge from wastewater, bio-oil from fast-pyrolysis of chicken beds and so on can be used as the feedstock to accumulate PHAs to improve the production with relatively low cost (*56–60*).

Thermosets from Vegetable Oil

Vegetable oils from soybean, palm trees, linseeds, sunflowers, castors and olives consist of triglycerides structured of various fatty acids. The distribution of fatty acids varies in different vegetable species. Most fatty acids in the vegetable oils have long straight chains and unsaturated double bonds which is ready to polymerize. Tung oil naturally contains highly unsaturated and conjugated carbon carbon double bonds, increasing the activity of polymerization. Other conjugated oils can be prepared artificially with catalysts (*36*). Various polymerization techniques are involved to produce thermosets from vegetable oil, such as free radical, cationic, ring opening metathesis, and condensation polymerization (*61*, *62*). The carbon carbon double bonds also react with acrylates to increase the reactivity. Acrylic epoxidized soybean oil (AESO) was synthesized after introduce the acrylates to the epoxidized soybean oil, which is commercially available (*63*). Blended with styrene (ST), AESO-ST thermosets was formed and used as matrix of polymeric composites with good properties. (*64*) The triglyceride ester groups in the vegetable oil can also been modified (*65*). After transesterification of soybean oil or linseed oil and glycerol and reacting with maleic anhydride (MA), the corresponding maleat half easters were obtained such as soybean oil monoglyceride (SOMG). These monomers can then copolymerize with ST, providing promising properties to replace some conventional petroleum based thermosets. Vegetable oil based polyols can react with diisocyanates to form polyurethane elastomers. Polyesters can be formed through condensation polymerization, for instance, from SOMG reacting with an anhydride, and Nahar seed oil monoglyceride reacting with phthalic and/or maleic anhydride (*66*). Although successful commercial products are available, these polymerization methods involve expensive reactants and catalysts. Recently, click chemistry reactions emerged to obtain thermosets from vegetable oils (*67*). A thermal catalyst and solvent free click chemistry approach was developed by introducing the azide groups into vegetable oils. Various fully cross-linked biopolymers were obtained from different vegetable oils (*68*, *69*). The thermal and mechanical properties of the polymer were systematically correlated with cross-linking density. Green composites can be produced with those thermosets resins reinforced with natural cellulose fibers (*70–72*).

Biopolymers from Plant Proteins

Plant proteins are inexpensive natural resources with abundant availability. However, industrial applications of natural proteins were limited by their brittleness and poor water resistance. Soy protein is one of the major raw materials to make bio-protein resins. Three purity levels of soy proteins are commercially available: defatted soy flour (53%), soy protein concentrate (72%) and soyprotein isolate (90%). In order to improve the mechanical properties and hydrophobicity of the films made from soy proteins, cross-linkers such as glutaraldehyde (GA), formaldehyde, glyoxal, and natural genipin (Gen) (*37*, *73–75*). A novel green approach utilized the hydrogen peroxide oxidized sugars which originally exist in the soy flour to cross-link the rest of the soy protein was developed (*76*). The thermoset protein obtained from this approach reinforced with natural fibers can provide fully biodegradable green composites.

Thermoplastics from proteins can be obtained through adding plasticizers such as water, glycerol, sorbitol or fatty acid (*74*, *77–80*). Chemical modifications and physical blending synthetic polymers were also applied to enhance the performance of natural protein products (*81*, *82*). In terms of processing, thermoplastic protein films were fabricated through solution cast, compression molding or injection molding. It is found that compression molding provide films with better properties, lower cost and a more environmentally friendly procedure than solvent casting (*83*). Processing temperature and amount of plasticizers can significantly affect the mechanical properties of films (*84*, *85*) Crosslinkers can also been added to protein thermoplastics to improve the mechanical properties (*86*, *87*). The processing and properties of thermoplastics developed from soy proteins, wheat gluten, zein and some lesser known cereal crops were reviewed recently (*38*). It is seen that the tensile strength and modulus of most protein thermoplastic developed are not comparable to synthetic polymers. New techniques on improving the performance and processing will be the key for natural protein polymers to successfully replace synthetic polymers.

Biocomposites

Biocomposites formed by natural fibers and polymer matrices are lightweight and low cost, providing comparable specific properties to some conventional fiber composites. The polymer matrices can be petroleum-based such as PP, PE, and epoxy, and bio-based polymers as introduced in the previous session. Both thermosets and thermoplastics can be used as matrices. Thermosets polymers usually provide better strength, modulus, solvent resistance and creep resistant due to their 3D cross linking structure. There is a growing interest and trend on using thermoplastic matrices due to their better impact resistance, easier forming methods, and the recyclability.

Reinforcement Architecture

Besides the species, chemical composition and microstructures of the fibers, the fiber geometry (length and aspect ratio), orientation, packing arrangement, and fiber volume content are important to the processing and mechanical properties of biocomposites. Longer fibers and larger fiber aspect ratio provie better load sustaining effect, therefore higher strength of composites. More fibers aligning along the loading direction also improve the strength of the composites. The natural fibers can be made into monofilaments, non-woven mats, rovings, yarns, and fabrics to reinforce the polymers. Different forms of the fiber preforms provide different fiber packing, alignment, volume content to the composites, as well as different suitable processing techniques.

Processing Techniques

Generally, the processing methods for biocomposites are similar to those developed for conventional composite materials. Based on the type of resin used, different manufacturing processing techniques can be applied as seen in Table V. (*1*, *53*) These techniques have been well developed and applied commercially for composite manufacturing. Some processing solutions may developed to improve the property of biocomposites.

Conventional injection molding or compression molding usually involves melt mix of thermoplastic resins and natural fibers, which can result in fiber damage due to the high shear force and limited fiber volume content. Fiber and matrix commingling was developed to overcome the high shear force. Banana fiber PP composites was initially manufactured through randomly mixing the short banana fibers and PP fibers following by compression molding (*88*). It is found that higher fiber volume content can be achieved. The use of solvent that required in mixing the polymer to the reinforcing fibers also significantly reduced. At the same time, a uniform distribution of fiber and matrix was obtained. A novel approach that winding the jute fiber and PP yarns onto a tooling substrate was designed, providing a more productive and cost effective fiber and matrix mixing. Using of fiber yarns can improve the fiber volume content, fiber alignment, as well as fiber packing, potentially improved the strength of the composites (*89–91*).

Long switchgrass (SG) stems reinforced PP light weight composites with density of $0.47 g/cm^3$ was developed through compression molding. PP webs – non-woven web with long pp fibers bonded by heat treatment was used as the matrix. In this case, the large size SG stems can be well distributed in the PP web, increasing homogeneity of the composites. PP webs containing SG stems can be stacked layer by layers to form composites under compression molding. The regular SG stem were also splited into 2 and 4 parts in order to increase the aspect ratio and fiber matrix adhesion area. It is found that the composites with split SG stem have 52% lower tensile strength, but 53% higher Young's modulus and 56% higher impact resistance compared with jute-PP composites (*92*). The raw fiber splitting and PP web were also applied on a long wheat straw PP composite manufacturing, and higher tensile strength and stiffness with lower impact resistance was observed (*93*). Non-woven web formed matrices were also applied to make hop bines PP composite, chicken quill PP composites, bamboo strips PP composites (*27*, *94*, *95*).

Kenaf fiber reinforced polylactide biocomposites were fabricated by non-woven carding followed by hot-pressing. In this process, the kenaf fibers and PLA fibers were uniformly blended and needle punched into non-woven webs (*96*). Although the mechanical properties improved after reinforcing with natural fibers, this non-woven process can introduce damage to the fibers. In a later work, air laying technique was adopted to produce flax fibers and PLA fibers mixed non-woven webs avoiding potential fiber damage caused by needle punching for non-woven composites (*97*).

Table V. Processing methods for biocomposites

Thermoplastic polymers	*Thermoset polymers*
• Compounding	• Extrusion
• Injection molding	• Compression molding
• Long fiber thermoplastic – direct (LFT-D)	• Resin transfer molding (RTM)
• Vacuum assisted RTM – VARTM	• Sheet molding compound
• Compression molding	• Pultrusion

Micro-braiding was developed to braid the reinforcement and matrix fibers/yarns together to improve the impregnation of thermoplastic resin into the fibers (*98*). This technique was applied to unidirectional hemp fiber/PLA composites and bamboo-rayon PLA composites followed by compression molding (*99*, *100*). With different types of micro-braiding, different level of fiber volume content can be achieved.

Pultrusion is a well-developed and commercially available manufacturing method for thermoset composites, usually used to produce straight parts with constant cross-sections. It provides continuous mass production with low lost, and potentially high mechanical properties of parts due to the continuous and aligned fibers. Recenlty, pultrusion was successfully applied on thermoplastic biocomposites. With the braded commingled fiber matrix preform, jute fiber/PLA tubular part was fabricated through pultrusion (*101*). The effect of molding temperature and speed to the mechanical properties of part was also studied. Thermoplastic pultrusion have the potential to provide economically production of high performance biocomposites and large components.

More processing details of biocomposites were reviewed in several papers (*1*, *102*). The fiber or preform type, molding temperature, processing pressure and time can all affect the performance of the final parts, and researches have been conducting to find the optimizing processing conditions for certain types of biocomposites. Besides improving the quality of products, reducing the processing cost, increasing the productivity, utilizing the green energy will be the main focus for biocomposite manufacturing, especially for large components.

Fiber Modifications for Performance Enhancement

Besides the properties of raw materials and manufacturing methods, fiber/matrix interface is an important factor affecting the performance of composites interface. Strong interface provides effective load transferring from matrix to the fibers which is the main load carrying component in composites. The hydrophility and moisture absorption of natural cellulose fibers significantly reduce the adhesion between fiber and matrix. Extensive research was conducted on modifying the natural fibers (mainly surface modification) to improve the fiber/matrix interfacial strength. Physical and chemical modifications were investigated. Physical techniques such as plasma treatment and corona treatment mainly rely on fiber surface structure (roughness) to increase mechanical interlock between fiber and matrix. Chemical modifications introduce functional groups to the surface of the fibers to improve the chemical bonding between fiber and matrix, which is more effective and reliable than physical methods. The major techniques are silane treatment, alkaline treatment, etherification, acetylation, maleated coupling, stearic acid treatment, benzylation, TDI (toluene-2, 4-diisocyanate) treatment, peroxide treatment, anhydride treatment, permanganate treatment, isocyanate treatment, and enzyme treatment (*1*, *12*, *103–105*). Using enzyme is a new trend since it is a more environmentally friendly approach than using other chemicals. a Pectinase treated hemp fiber reinforced PP composites showed higher modulus, tensile and flexural strength than untreated composites (*106*). Jute fabrics were treated with pectinase, laccase, cellulose and xylanase enzyme solutions before reinforcing polyester (*107*). A mixture of lipase, protease and amylase-xylanase enzymes was used to modify wheat husk, rye husk and soft wood fibers for natural fiber PP and PLA composite (*108*). In these works, improved tensile and flexural properties of biocomposites were obtained. The diameters of natural fiber were smaller so that the aspect ratio was larger. Different enzyme targets different chemical composition of the natural fibers, providing a more controlled preparation for different resin systems. Additionally, enzyme treatment could reduce the cost for fiber preparations while enhancing the performance of biocomposites.

Performance of Biocomposites

In order to develop the applications of biocomposites, understanding the performance of biocomposites is crucial. Tensile, flexural, and impact properties are the basic mechanical performance of biocomposites that are extensively investigated in the last two decades, and were reviewed in papers (*1*, *2*, *4*, *10*, *12*, *13*, *31*, *34*, *53*, *102*, *104*, *109*, *110*). Fatigue and creep behaviors that are critical for long term applications are also studied. Besides the mechanical properties, properties such as acoustic insulation and fire retardant of biocomposites are also explored, leading to wide potential applications.

Most of the biocomposites studied contain single type natural fibers. For example, kenaf fiber/polyester biocomposites was fabricated through hand lay-up with fibers treated using propionic and succinic anhydride (*111*). The tensile, flexural and low velocity impact properties were tested and shown in Figure 1.

It is seen that the chemical treatment can significantly increase the mechanical properties of the composites. Through the thermogravimetric analysis (TGA), it is showed that better thermal stability was obtained in the modified biocomposites than the untreated ones. The tensile and notch impact properties of PLA and P(3HB-*co*-3HV) reinforced with man-made cellulose, jute and abaca fibers after injection molding were reported to be better than natural fiber reinforced PP composites (*112*).

Hybrid natural fiber/synthetic fiber composites were developed. With 20%wt glass, 10wt% coir fibers with length of 15 mm, the maximum flexural strength of 63MPa was obtained for this hybrid epoxy composites. Increasing the natural fiber loading and length increases the moisture absorption of the composites (*113*). Jute/glass hybrid epoxy composites was fabricated via resin infusion under flexible tooling (RFFT) with jute plain weave fabrics sandwiched with glass satin weave fabrics. Compared with pure jute/epoxy composites, the added glass fibers increased the bending and drop-weight impact properties significantly. The increase of tensile strength depends on the position and amount of glass fabrics. The water penetration to the jute fibers was decelerated by the glass woven outer layers (*114*).

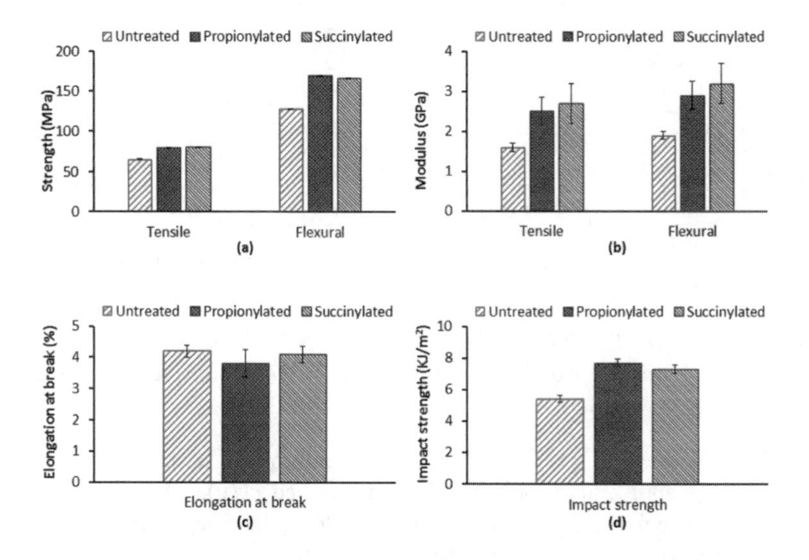

Figure 1. (a) Tensile and flexural strength, (b)tensil and flexural modulus, (c) elongation at break and (d) impact strength of Kenaf/polyester composites (111).

Fatigue behaviors of jute, hemp and flax reinforced unsaturated polyester composites comparing to E-glass/polyester composites (GFRPs) were comprehensively studied with the effects of fiber type/qulity, fiber volume fraction, fiber orientation, and loading stress ratios (*115*). Tension-tension, tension-compression, and compression-compression fatigue testing were performed. It is found that the fiber type, volume fraction and orientation affect the tensile properties and fatigue loading capacities significantly. However, from the S-N curves of the materials, they have little impact on the fatigue strength coefficient. Higher fatigue strength coefficient was observed with increasing loading stress ratios. Although lower absolute fatigue performance, lower fatigue strength degradation rates were observed for the studied biocomposites than GFRPs. Tension-Tension fatigue behavior of flax fiber/epoxy composites with different stacking sequence was studied. Better fatigue resistance was found from the composites with higher static strength. Transverse cracking around fiber and matrix interface and fiber pull-out were observed on the fracture surfaces, indicating the fiber/matrix adhesion can be a main issue for performance improvement (*116*). IR thermography technique was used to monitor and predict the fatigue life of flax/epoxy composites. Increasing temperature of flax/epoxy composites during the stepwise loading of fatigue tests was recorded with good repeatability and small deviations (*117*). NaOH-clay treated sisal fibers was found to improve the fatigue life of sisal/epoxy and sisal/PP composite; lower fiber volume fraction and water absorption result in short fatigue life (as seen in Figure 2) (*118*).

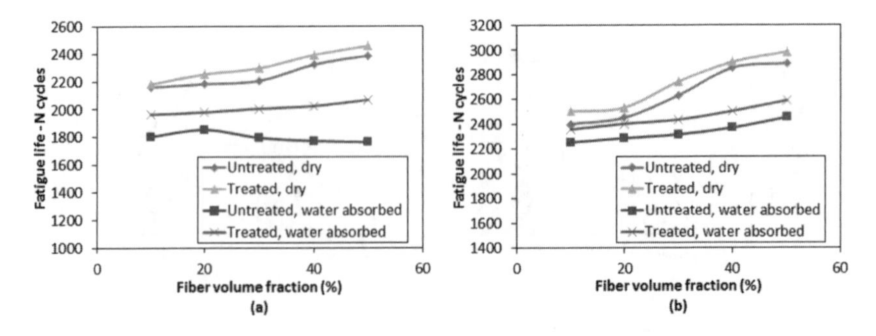

Figure 2. Fatigue life of (a)sisal/epoxy and (b)sisal/PP composites under stress ratio of 10%, frequency of 1Hz and stress level of 90% of tensile strength (118).

Flexural creep behavior of hemp fiber reinforced unsaturated polyester composites manufactured though RTM was studied (*119*). With 21% volume content of hemp fibers, 27% and 43% of failure load, the creep strain as a function of time was recorded. No significant deformation occurred with a slightly increase of initial strain then flat-out under 27% load for 750hrs; while an initial increase in the early stage followed by a linear increase of creep strain under 43% load was observed, indicating long term instability under relatively high loading conditions. Similar stress dependent behavior was observed on injection molded flax/starch composite with fiber weight content of 40% (*120*).

The sound absorption property of light weight long switchgrass/PP, wheat straw/PP and hop bine fibers/PP composites were studied and compared with jute/PP composites (*92*, *93*, *95*). It is found that outer bark of bop bine fibers/PP and wheatstraw/PP composites shows better sound absorption properties than jute/PP composites in the sound range of 1.6-3.0kHz and 0.3-2.2Hz, respectively.

Swtchgrass/PP composites show similar sound absorption behavior as jute/PP composites. Sound properties of natural sandwich composites – bamboo/vinyl ester skin with balsa wood core, cotton/vinyl ester skin with pine wood core, cotton/vinyl ester skin with synthetic foam core were compared to carbon/epoxy skin with synthetic foam core. It is found that the coincidence frequency is doubled when using biocomposite face sheets and balsa core, and tripled with biocomposites face sheets with a synthetic core. The noise radiation was also reduced by using natural materials for the sandwich beams (*121*).

Fire retardant property of industrial hemp fibers reinforced unsaturated polyester composites manufactured from sheet molding compound method was investigated. A flame retardancy containing aluminum trihydrate was added into the resin system, which prolonged the ignition delay time and reduced the peak heat release value from 361kW/m^2 to 176 kW/m^2. With the fire retardant additives, the fire performance of natural fiber composites is competitive with other building materials (*122*). The hemp fiber in the polyester resin act like a thermally resistant char layer. Increasing fiber volume provides a more effective thermally insulating char layer. The decomposition of hemicellulose of hemp fibers may result in less ignitable gases at the surface of the material and delay the ignition (*123*). More detailed flammability of the biofibers, polymer matrices, and composites are summarized in Ref. (*124*, *125*).

Applications

Light weight biocomposites and biopolymers have wide range applications. Biopolymers such as starch and PLA are already applied in packaging industry. A major area of using biocomposites is the automotive industry. The automotive vehicles in the world now have already been using considerable biocomposite components, such as door panels, seat backs, cargo area floors, windshields, and business tables and so on. Biocomposites are also used as furniture, indoor structures and building materials due to their light weight, fire retardant and sound absorption properties. Biodegradable and biocompatible composites are used as biomaterials, for instance the interference screws and tenodesis for ligament reconstruction. Biocomposites with comparable mechanical properties with glass fiber reinforced composites can be used in the form of tubes, sandwich panels, panels, and other shapes as required. A case study on making a 3.5m composite rotor blade with flax fiber/polyester was conducted recently. It is found that the flax blade is 10% lighter than the identical constructed E-glass; and the flax blade can sustain the load under normal operation and worst condition. However, the flexural stiffness of flax blades is only about half of that of the E-glass blade. Therefore, flax still cannot replace E-glass on this large load bearing component (*126*). The addition of nanoparticles/whiskers provides more functionalities to the biopolymer or biocomposites, extending their application on larger area such as electrical materials, sensors, and solar cells.

Summary and Perspectives

Biofibers, biopolymers and biocomposites have been significantly developed over the last two decades because of their significant advantages: renewable resources, biodegradability, low cost, light weight and high specific strength. Massive techniques were developed and applied on: (i) processing the biocomposites; (ii) improving the properties of biopolymers and biocomposites; (iii) enhancing the fiber compatability to the matrix; (iv) controlling the final material properties by modifying the raw materials, composite microstructure, and processing. Many applications were also developed and commercialized for biocomposites in industries. The use and demand of biobased materials will definitely keep increasing in the future.

For more commercialized applications and productions, further improvement and research are required on many aspects: (i) improving extracting techniques of biofibers to reduce the damages to the fibers; (ii) reducing the scattering of the fiber properties; (iii) exploring biofibers and biopolymers with better properties; (iv) developing low cost, green, and cleaner techniques for the processing and treatment of the materials; (v) involving more green energy; (vi) improving the mechanical properties; and (vii) developing multi-functionality of biocomposites. In addition, the understanding of mechanics can be beneficial to simulate and predict the behavior of the biocomposites. The long term properties, lift cycle assessment, biodegradability and recycling of biobased materials require further investigations. Establishing the reliable system of raw material properties – processing – product properties will provide efficient product development cycles.

References

1. Faruk, O.; Bledzki, A. K.; Fink, H.-P.; Sain, M. *Prog. Polym. Sci.* **2012**, *37*, 1552–1596.
2. Shinoj, S.; Visvanathan, R.; Panigrahi, S.; Kochubabu, M. *Ind. Crops Prod.* **2011**, *33*, 7–22.
3. Gutiérrez, M. C.; De Paoli, M.-A.; Felisberti, M. I. *Ind. Crops Prod.* **2014**, *52*, 363–372.
4. Mohanty, a. K.; Misra, M.; Hinrichsen, G. *Macromol. Mater. Eng.* **2000**, *276-277*, 1–24.
5. AL-Oqla, F. M.; Sapuan, S. M. *J. Clean. Prod.* **2014**, *66*, 347–354.
6. Hoareau, W.; Trindade, W. G.; Siegmund, B.; Castellan, A.; Frollini, E. *Polym. Degrad. Stab.* **2004**, *86*, 567–576.
7. Khiari, R.; Mhenni, M. F.; Belgacem, M. N.; Mauret, E. *Bioresour. Technol.* **2010**, *101*, 775–780.
8. Martí-Ferrer, F.; Vilaplana, F.; Ribes-Greus, A.; Benedito-Borrás, A.; Sanz-Box, C. *J. Appl. Polym. Sci.* **2006**, *99*, 1823–1831.
9. Huda, S.; Yang, Y. *Macromol. Mater. Eng.* **2008**, *293*, 235–243.
10. Mishra, S.; Mohanty, A. K.; Drzal, L. T.; Misra, M.; Hinrichsen, G. *Macromol. Mater. Eng.* **2004**, *289*, 955–974.
11. Yılmaz, N. D. *J. Text. Inst.* **2013**, *104*, 396–406.

12. Avella, M.; Malinconico, M.; Buzarovska, A.; Grozdanov, A.; Gentile, G.; Errico, M. E. *Polym. Compos.* **2007**, *28*, 98–107.
13. Saheb, D. N.; Jog, J. P. *Adv. Polym. Technol.* **1999**, *18*, 351–363.
14. Reddy, N.; Yang, Y. *J. Agric. Food Chem.* **2007**, *55*, 8570–8575.
15. Reddy, N.; Yang, Y. *Green Chem.* **2005**, *7*, 190–195.
16. Reddy, N.; Yang, Y. *J. Agric. Food Chem.* **2006**, *54*, 8077–8081.
17. Faruk, O.; Bledzki, A. K.; Fink, H.-P.; Sain, M. *Macromol. Mater. Eng.* **2014**, *299*, 9–26.
18. Nishino, T. *Green Composites: Polymer Composites and the Environment*; Baillie, C., Ed.; Woodhead Publishing: Cambridge, U.K., 2004; pp 49–67.
19. Azwa, Z. N.; Yousif, B. F.; Manalo, a. C.; Karunasena, W. *Mater. Des.* **2013**, *47*, 424–442.
20. Barone, J. R. *Composites, Part A* **2005**, *36*, 1518–1524.
21. Reddy, N.; Yang, Y. *J. Appl. Polym. Sci.* **2010**, *116*, 3668–3675.
22. Wrześniewska-Tosik, K.; Adamiec, J. *Fibres Text. East. Eur.* **2007**, *15*, 106–112.
23. Popescu, C.; Höcker, H. *Chem. Soc. Rev.* **2007**, *36*, 1282–1291.
24. Hong, C. K.; Wool, R. P. *J. Appl. Polym. Sci.* **2005**, *95*, 1524–1538.
25. Zhan, M.; Wool, R. P. *J. Appl. Polym. Sci.* **2013**, *128*, 997–1003.
26. Reddy, N.; Yang, Y. *J. Polym. Environ.* **2007**, *15*, 81–87.
27. Huda, S.; Yang, Y. *Compos. Sci. Technol.* **2008**, *68*, 790–798.
28. Eslahi, N.; Hemmatinejad, N.; Dadashian, F. *Part. Sci. Technol.* **2014**, *32*, 242–250.
29. Reddy, N.; Jiang, Q.; Jin, E.; Shi, Z.; Hou, X.; Yang, Y. *Colloids Surf., B* **2013**, *110*, 51–58.
30. Reddy, N.; Chen, L.; Yang, Y. *Mater. Sci. Eng., C* **2013**, *33*, 1203–1208.
31. Mukherjee, T.; Kao, N. *J. Polym. Environ.* **2011**, *19*, 714–725.
32. Halley, P. J.; Dorgan, J. R. *MRS Bull.* **2011**, *36*, 687–691.
33. Asokan, P.; Firdoous, M.; Sonal, W. 2012, 30, 254–261.
34. John, M. J.; Thomas, S. *Carbohydr. Polym.* **2008**, *71*, 343–364.
35. Chodak, I. *Monomers, Polymers and Composites from Renewable Resources*; Belgacem, M., Gandini, A., Eds.; Elsevier: Oxford, U.K., 2008; pp 451–477.
36. Quirino, R. L.; Garrison, T. F.; Kessler, M. R. *Green Chem.* **2014**, *16*, 1700–1715.
37. González, A.; Strumia, M. C.; Alvarez Igarzabal, C. I. *J. Food Eng.* **2011**, *106*, 331–338.
38. Reddy, N.; Yang, Y. *J. Appl. Polym. Sci.* **2013**, *130*, 729–738.
39. Zhang, Y.; Rempel, C.; Liu, Q. *Crit. Rev. Food Sci. Nutr.* **2014**, *54*, 1353–1370.
40. Wang, X.-L.; Yang, K.-K.; Wang, Y.-Z. *J. Macromol. Sci., Part C: Polym. Rev.* **2003**, *43*, 385–409.
41. Yu, L.; Dean, K.; Li, L. *Prog. Polym. Sci.* **2006**, *31*, 576–602.
42. Avérous, L.; Halley, P. J. *Biofuels, Bioprod. Biorefining* **2009**, *3*, 329–343.
43. Xie, F.; Pollet, E.; Halley, P. J.; Avérous, L. *Prog. Polym. Sci.* **2013**, *38*, 1590–1628.
44. Bodirlau, R.; Teaca, C.-A.; Spiridon, I. *Composites, Part B* **2013**, *44*, 575–583.

45. Albertsson, A.; Varma, I. K. *Adv. Polym. Sci* **2002**, *157*, 1–40.
46. Bajpai, P. K.; Singh, I.; Madaan, J. *J. Thermoplast. Compos. Mater.* **2012**, *27*, 52–81.
47. Baiardo, M.; Frisoni, G.; Scandola, M.; Rimelen, M.; Lips, D.; Ruffieux, K.; Wintermantel, E. *J. Appl. Polym. Sci.* **2003**, *90*, 1731–1738.
48. Balakrishnan, H.; Hassan, A.; Imran, M.; Wahit, M. U. *Polym. Plast. Technol. Eng.* **2012**, *51*, 175–192.
49. Balakrishnan, H.; Hassan, A.; Wahit, M. U.; Yussuf, a. a.; Razak, S. B. A. *Mater. Des.* **2010**, *31*, 3289–3298.
50. Anderson, K.; Schreck, K.; Hillmyer, M. *Polym. Rev.* **2008**, *48*, 85–108.
51. Shi, X.; Chen, Z.; Yang, Y. *Eur. Polym. J.* **2014**, *50*, 243–248.
52. Van de Velde, K.; Kiekens, P. *Polym. Test.* **2002**, *21*, 433–442.
53. Yan, L.; Chouw, N.; Jayaraman, K. *Composites, Part B* **2014**, *56*, 296–317.
54. Laycock, B.; Halley, P.; Pratt, S.; Werker, A.; Lant, P. *Prog. Polym. Sci.* **2014**, *39*, 397–442.
55. Philip, S.; Keshavarz, T.; Roy, I. *J. Chem. Technol. Biotechnol.* **2007**, *82*, 233–247.
56. Queirós, D.; Rossetti, S.; Serafim, L. S. *Bioresour. Technol.* **2014**, *157*, 197–205.
57. Anjali, M.; Sukumar, C.; Kanakalakshmi, a.; Shanthi, K. *Compos. Interfaces* **2014**, *21*, 111–119.
58. Morgan-Sagastume, F.; Valentino, F.; Hjort, M.; Cirne, D.; Karabegovic, L.; Gerardin, F.; Johansson, P.; Karlsson, a; Magnusson, P.; Alexandersson, T.; Bengtsson, S.; Majone, M.; Werker, a. *Water Sci. Technol.* **2014**, *69*, 177–184.
59. Albuquerque, M. G. E.; Torres, C. a V; Reis, M. a M. *Water Res.* **2010**, *44*, 3419–3433.
60. Moita, R.; Lemos, P. C. *J. Biotechnol.* **2012**, *157*, 578–583.
61. Xia, Y.; Larock, R. C. *Green Chem.* **2010**, *12*, 1893–1909.
62. Miao, S.; Wang, P.; Su, Z.; Zhang, S. *Acta Biomater.* **2014**, *10*, 1692–1704.
63. Lu, J.; Khot, S.; Wool, R. P. *Polymer* **2005**, *46*, 71–80.
64. La Scala, J.; Wool, R. P. *Polymer* **2005**, *46*, 61–69.
65. Khot, S. N.; Lascala, J. J.; Can, E.; Morye, S. S.; Williams, G. I.; Palmese, G. R.; Kusefoglu, S. H.; Wool, R. P. *J. Appl. Polym. Sci.* **2001**, *82*, 703–723.
66. Dutta, N.; Karak, N.; Dolui, S. *Prog. Org. Coat.* **2004**, *49*, 146–152.
67. Lligadas, G.; Ronda, J. C.; Galià, M.; Cádiz, V. *J. Polym. Sci., Part A: Polym. Chem.* **2013**, *51*, 2111–2124.
68. Hong, J.; Shah, B. K.; Petrović, Z. S. *Eur. J. Lipid Sci. Technol.* **2013**, *115*, 55–60.
69. Hong, J.; Luo, Q.; Wan, X.; Petrović, Z. S.; Shah, B. K. *Biomacromolecules* **2012**, *13*, 261–266.
70. Hong, C. K.; Wool, R. P. *J. Appl. Polym. Sci.* **2005**, *95*, 1524–1538.
71. Quirino, R. L.; Larock, R. C. *J. Appl. Polym. Sci.* **2011**, *121*, 2039–2049.
72. Liu, Z.; Erhan, S. Z.; Akin, D. E.; Barton, F. E. *J. Agric. Food Chem.* **2006**, *54*, 2134–2137.
73. Chabba, S.; Netravali, A. N. *J. Mater. Sci.* **2005**, *40*, 6263–6273.
74. Kim, J. T.; Netravali, A. N. *J. Agric. Food Chem.* **2010**, *58*, 5400–5407.

75. Chabba, S.; Matthews, G. F.; Netravali, a. N. *Green Chem.* **2005**, *7*, 576–581.
76. Ghosh Dastidar, T.; Netravali, A. N. *Green Chem.* **2013**, *15*, 3243–3251.
77. Reddy, N.; Yang, Y. *Polym. Int.* **2011**, *60*, 711–716.
78. Kunanopparat, T.; Menut, P.; Morel, M.-H.; Guilbert, S. *Composites, Part A* **2008**, *39*, 1787–1792.
79. Kunanopparat, T.; Menut, P.; Morel, M.-H.; Guilbert, S. *Composites, Part A* **2008**, *39*, 777–785.
80. Reddy, N.; Yang, Y. *Biomass Bioenergy* **2011**, *35*, 3496–3503.
81. Zhang, X. *Aust. J. Chem.* **2014**, *67*, 6–10.
82. Diao, C.; Dowding, T.; Hemsri, S.; Parnas, R. S. *Composites, Part A* **2014**, *58*, 90–97.
83. Mangavel, C.; Rossignol, N.; Perronnet, a; Barbot, J.; Popineau, Y.; Guéguen, J. *Biomacromolecules* **2004**, *5*, 1596–1601.
84. Chen, L.; Reddy, N.; Wu, X.; Yang, Y. *Ind. Crops Prod.* **2012**, *35*, 70–76.
85. Sun, S.; Song, Y.; Zheng, Q. *Food Hydrocolloids* **2007**, *21*, 1005–1013.
86. Wihodo, M.; Moraru, C. I. *J. Food Eng.* **2013**, *114*, 292–302.
87. Song, Y.; Zheng, Q.; Liu, C. *Ind. Crops Prod.* **2008**, *28*, 56–62.
88. Paul, S. A.; Joseph, K.; Mathew, G.; Pothen, L. A.; Thomas, S. *Polym. Compos.* **2010**, *31*, 816–824.
89. George, G.; Tomlal Jose, E.; Jayanarayanan, K.; Nagarajan, E. R.; Skrifvars, M.; Joseph, K. *Composites, Part A* **2012**, *43*, 219–230.
90. George, G.; Tomlal Jose, E.; Åkesson, D.; Skrifvars, M.; Nagarajan, E. R.; Joseph, K. *Composites, Part A* **2012**, *43*, 893–902.
91. George, G.; Joseph, K.; Nagarajan, E. R.; Tomlal Jose, E.; Skrifvars, M. *Composites, Part A* **2013**, *48*, 110–120.
92. Zou, Y.; Xu, H.; Yang, Y. *J. Polym. Environ.* **2010**, *18*, 464–473.
93. Zou, Y.; Huda, S.; Yang, Y. *Bioresour. Technol.* **2010**, *101*, 2026–2033.
94. Huda, S.; Reddy, N.; Yang, Y. *Composites, Part B* **2012**, *43*, 1658–1664.
95. Zou, Y.; Reddy, N.; Yang, Y. *J. Appl. Polym. Sci.* **2010**, *116*, 2366–2373.
96. Lee, B.-H.; Kim, H.-S.; Lee, S.; Kim, H.-J.; Dorgan, J. R. *Compos. Sci. Technol.* **2009**, *69*, 2573–2579.
97. Alimuzzaman, S.; Gong, R. H.; Akonda, M. *Polym. Compos.* **2013**, *34*, 1611–1619.
98. Sakaguchi, M.; Nakai, a.; Hamada, H.; Takeda, N. *Compos. Sci. Technol.* **2000**, *60*, 717–722.
99. Kobayashi, S.; Takada, K. *Composites, Part A* **2013**, *46*, 173–179.
100. Kobayashi, S.; Takada, K.; Song, D. *Adv. Compos. Mater.* **2012**, *21*, 79–90.
101. Memon, A.; Nakai, A. *Adv. Mech. Eng.* **2013**, *2013*, 1–8.
102. Fowler, P. A.; Hughes, J. M.; Elias, R. M. *J. Sci. Food Agric.* **2006**, *86*, 1781–1789.
103. Mohanty, A. K.; Misra, M.; Drzal, L. T. *Compos. Interfaces* **2001**, *8*, 313–343.
104. La Mantia, F. P.; Morreale, M. *Composites, Part A* **2011**, *42*, 579–588.
105. Kalia, S.; Kaith, B. S.; Kaur, I. *Polym. Eng. Sci.* **2009**, *49*, 1253–1272.
106. Saleem, Z.; Rennebaum, H.; Pudel, F.; Grimm, E. *Compos. Sci. Technol.* **2008**, *68*, 471–476.

107. Karaduman, Y.; Gokcan, D.; Onal, L. *J. Compos. Mater.* **2012**, *47*, 1293–1302.
108. Mamun, A. a.; Bledzki, A. K. *Compos. Sci. Technol.* **2013**, *78*, 10–17.
109. Tripathi, G.; Choudhury, P; Basu, B. *Mater. Technol.* **2010**, *25*, 158–176.
110. Thakur, V. K.; Thakur, M. K.; Gupta, R. K. *Int. J. Polym. Anal. Charact.* **2014**, *19*, 256–271.
111. Khalil, H. a.; Suraya, N.; Atiqah, N.; Jawaid, M.; Hassan, a. *J. Compos. Mater.* **2012**, *47*, 3343–3350.
112. Bledzki, a. K.; Jaszkiewicz, a. *Compos. Sci. Technol.* **2010**, *70*, 1687–1696.
113. Bhagat, V. K.; Biswas, S.; Dehury, J. *Polym. Compos.* **2014**, *35*, 925–930.
114. Pandita, S. D.; Yuan, X.; Manan, M. a.; Lau, C. H.; Subramanian, A. S.; Wei, J. *J. Reinf. Plast. Compos.* **2013**, *33*, 14–25.
115. Shah, D. U.; Schubel, P. J.; Clifford, M. J.; Licence, P. *Compos. Sci. Technol.* **2013**, *74*, 139–149.
116. Liang, S.; Gning, P.-B.; Guillaumat, L. *Int. J. Fatigue* **2014**, *63*, 36–45.
117. El Sawi, I.; Fawaz, Z.; Zitoune, R.; Bougherara, H. *J. Mater. Sci.* **2013**, *49*, 2338–2346.
118. Kanny, K.; T.P., M. *Compos. Interfaces* **2013**, *20*, 783–797.
119. Rouison, D.; Sain, M.; Couturier, M. *Compos. Sci. Technol.* **2006**, *66*, 895–906.
120. Varna, J.; Spārniņš, E.; Joffe, R.; Nättinen, K.; Lampinen, J. *Mech. Time-Dependent Mater.* **2011**, *16*, 47–70.
121. Sargianis, J. J.; Kim, H.-I.; Andres, E.; Suhr, J. *Compos. Struct.* **2013**, *96*, 538–544.
122. Hapuarachchi, T. D.; Ren, G.; Fan, M.; Hogg, P. J.; Peijs, T. *Appl. Compos. Mater.* **2007**, *14*, 251–264.
123. Naughton, A.; Fan, M.; Bregulla, J. *Composites, Part B* **2014**, *60*, 546–554.
124. Kozłowski, R.; Władyka-przybylak, M. *Polym. Adv. Technol.* **2008**, *19*, 446–453.
125. Chapple, S.; Anandjiwala, R. *J. Thermoplast. Compos. Mater.* **2010**, *23*, 871–893.
126. Shah, D. U.; Schubel, P. J.; Clifford, M. J. *Composites, Part B* **2013**, *52*, 172–181.

Chapter 2

Proteins and Protein-Based Fibers

Yan Vivian Li*

**Department of Design and Merchandising,
College of Health & Human Science, Colorado State University,
Fort Collins, Colorado 80523, United States
*E-mail: Yan.Li@colostate.edu**

Protein-based fibers are generated from many protein sources including plants, insects and animals. The use of natural protein fibers is historical, while man-made regenerated protein fibers have been produced since 1950s and their development remains constant innovation. Protein-based fibers become important in the development of lightweight materials because they offer not only light weight but also biodegradability, excellent moisture and temperature regulation, resiliency and possibly exceptional mechanical properties. This chapter discusses the fiber structures, properties and performance of both conventional and advanced protein-based fibers. Advanced nanofibers and nanocomposites made from regenerated proteins and other polymers exhibit great potential to make new lightweight functional materials for textiles, health and medical, energy and engineering applications.

Introduction

Fibers composed of proteins are protein-based fibers. Some used commonly are silkworm silk and merino wool. Others are less known as fiber materials, for example, feathers and wheat. In the fiber formation, amino acids in the proteins are polymerized through condensation polymerization to form repeating polyamide units with various substituent on the carbon atoms. The properties of protein-based fibers depend on the substituent on the carbon atoms in the polyamide units as well as the microstructures of proteins. The fibers can be formed by natural protein sources including plants (e.g. zein, soy bean and wheat), insects (e.g. silkworms, spiders and ants) and animals (e.g. sheep, alpaca and angora). Fiber properties are usually varied depending on protein sources. In general, protein-based fibers have excellent moisture absorbency and transport characteristics, moderate strength, resiliency and elasticity. These superior fiber properties lead to wide uses of protein-based fibers in the applications of textiles, medicals, energy and sustainability.

Protein-based fibers have become important in developing innovative lightweight materials. The use of natural protein fibers is historical, while man-made regenerated protein fibers were commercially produced since 1950s (*1*). In terms of medical applications, lightweight hollow materials are widely used, due to their capability to allow cells penetrate, grow and communicate. It is easier to maintain the functions of the extracellular matrices (ECMs) with proteins than with carbohydrates or synthetic polymers as the natural ECMs are composed of collagen, and hence proteins are preferred. Man-made regenerated protein fibers make lightweight biofibers, which is proven promising. For example, spider silk fibers have been wet spun using transgenic goat milk proteins. The regenerated spider silk fibers exhibit lightweight as well as super mechanical properties, which make them preferable to many applications in reinforcing materials, lightweight textiles and other industrial uses. This chapter discusses protein-based fibers from different resources, the general chemistry and microstructures of the fibers, and innovative protein-based biofibers and their applications.

Insect Protein-Based Fibers

Insects such as silkworm, spider and ants naturally produce fibers with luster, soft hand, light weight and excellent mechanical properties. These protein-based biofibers show high value in fiber market. For example, silk is an agricultural commodity at premium price, although the production volume is less than 1 percent of the market for natural textile fibers. Insect biofibers become important when the weight of fibers is particularly considered in the applications such as lightweight body armor and other military used textiles. Silk has a long history of be used as a fiber material due to the luster and soft hand. Recently, superior mechanical properties and low specific gravity of spider silk have been discovered, resulting in many new development of natural and man-made spider silk fibers. Silk fibers from silkworm, natural and man-made regenerated spider silk fibers are particularly discussed in this section.

Silkworm Silk

Silkworm which is primarily native in China produces silk fibers. Silk is the only natural fiber that is a filament. The density of silk fibers is 1.34 g/cm^3. Silk comes from the cocoon of the silkworm and requires a great deal of handling and processing (*2*). Silkworm silk consists of two main proteins: sericin and fibroin.

The fibroin composed of amino acids makes up the primary structure of the silk, beta pleated sheets (see Figure 1). The sericin is a sticky protein and glues the two fibroin together. There is hydrogen bonding formed between silk polymer chains and between the beta sheets, resulting a well-connected network in the silk microstructure. Other small amino acids such as sercine, glycine and alacine allow tight molecular packing in silk. Therefore, the silk fibers are strong and resistant to breaking (*3*).

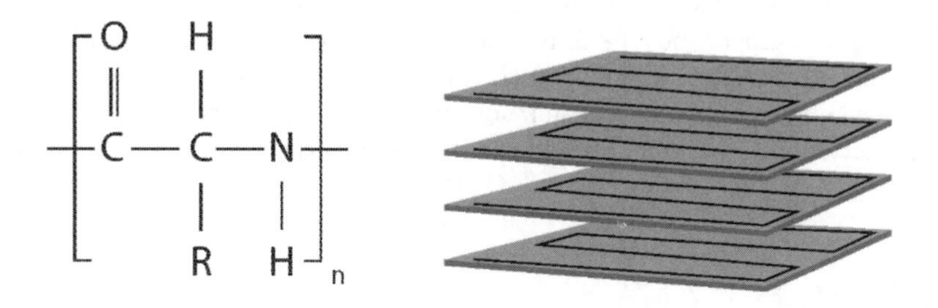

Figure 1. (Left) A repeated unit of amino acid is shown as the primary chemical composition of the silk. (Right) An illustration of the beta-pleated sheets is shown as the primary microstructure of the silk.

Protein Structures and Fiber Properties

Silkworm silk fibers show triangular cross sections with rounded corners. The beta pleated sheets composed of an amino acid repeat sequence with some variations are found in the silk fibroin, resulting flat surfaces of silk fibers. The flat surfaces reflect light at many angles and hence give silk a natural shine appearance. The beta-configuration also provides silk fibers smooth and soft hand. Silk makes strong fibers due to linear and beta-configuration in the microstructure. The linear and beta-configuration structures make molecular packing easy, resulting 65-70% of crystalline regions in silk. They also promote the formation of hydrogen bonds in a regular manner. High crystallinity and hydrogen bonds are responsible for high strength of silk fibers. Dilute organic acids show little effect on silk at room temperature, but when concentrated, the dissolution of fibroin may take place. On the other hand, alkaline solutions cause the silk fiber to swell because the alkali molecules can hydrolyze the peptide bonds on silk polymer chains. Silk is sensitive to light. Prolonged exposure to sunlight can cause partially spotted color change due to photo degradation by the UV radiation of the sun. The resistance of silk to environment is low mainly due to no covalent cross-link in the silk polymer.

Innovative Development of Silk Fibers

Other than being used in textile industry for thousand years, recently silkworm silk provides important clinical repair options because of its biodegradblity and biocompatibility (*4*). New characteristic properties of silk have been reported recently. Reed and Viney (*5*) found that even under microwave radiation, the silk fibers were maintained well in tensile properties without significant deterioration. They concluded silk fiber is preferably used in composites as a reinforcing fiber in some severe conditions. Liu et al. (*6*) discovered a thermally induced increase in energy transport capacity of silkworm silk, suggesting potential application as biosensors. Chen et al. (*7*) demonstrated fabrication of bioinspired bead-on-string silkworm silk with superhydrophilicity in addition to mechanical properties. Their results support novel silk fiber applications such as the control of fluidic transport (*8*), drug delivery (*9*), particle sorting (*10*) and sensor devices (*11*). With the development of biomedical and biotechnological engineering, silk finds more and more applications in implantation, artificial organs (*12*), biosensors and drug delivery (*13*).

Spider Silk

Spiders produce silk fibers to make webs or other structures, which function as nets to catch other animals, or as nests or cocoons for protection of their offspring. A single spider can produce up to seven different types of silk for their different ecological uses (*14*). Spiders produce many types of silk in different conformation including glues and fibers to meet the specification and requirement for all ecological uses, such as structural support and protective construction. Some spider silk can absorb energy effectively, whereas others transmit vibration efficiently (*15*).

Protein Structures and Fiber Properties

Spider silk fibers are composed of fibril bundles. The fibrile primarily consists of two repetitive alanine and glycine, which make crystalline and amorphous regions in spider silk, respectively (*16*). On a secondary structure level, the short side chained alanine is primarily found in the crystalline segments of the silk fiber, while the glycine is mainly discovered in the amorphous matrix. The rigid semi-crystalline segments are distributed in the strained elastic semi-amorphous regions and connected one and another, creating the interplay of these two molecular structures (See Figure 2) (*17*). Such an interlocking molecular structure determines excellent elasticity and makes the spider fiber extremely resistant to rupture (*18*). Various compounds other than protein are present in spider silk too and enhance the fiber's properties. Non-protein ingredients found in spider silk include sugars, lipids, ions and pigments that are considered as a protection layer in the fiber structure (*19*).

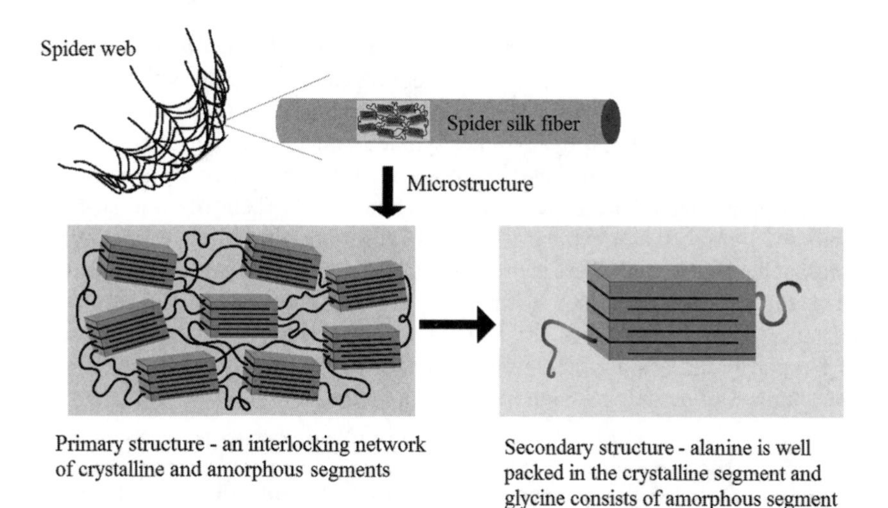

Figure 2. A spider silk fiber within a spider web is shown and a zoom-in microstructure review is inserted. A representative scheme illustrates the primary structure of spider silk. The entanglement of crystalline and semi-amorphous segments promotes exceptional mechanical properties of spider silk fibers. On a secondary structure, short chained alanine is highly packed, resulting crystalline segments with high mechanical properties. Glycine is loosely interlocked among packed crystalline segments, resulting semi-amorphous segments (17). (Figure adapted from http://en.wikipedia.org/wiki/Spider_silk).

Innovation of Spider Silk Fibers

Most spider silks exhibit exceptional mechanical properties. They show high tensile strength and excellent elongation (extensibility), which enables a silk fiber to absorb a lot of energy before breaking. Compared with high-performance synthetic fibers, spider silk fibers exhibit lightweight, high-strength, high-elasticity and excellent resilience (which implies the capability to store energy) (see Table 1).

These superior mechanical properties make spider silk fibers attractive for lightweight textiles applications such as protective clothing. The interest of using spider silk in protective applications recently grows fast. The author leads a research team at Colorado State University, currently studying the possibility of using spider silk fibers on protective clothing for firefighters. Besides the lightweight property, protective clothing for firefighters has requirements on fiber thermal properties and moisture absorption for the purpose of protection in severe working conditions. Spider dragline silk was found to have exceptionally high thermal conductivity (416 $Wm^{-1}K^{-1}$) that surpasses most synthetic fibers (0.25 $Wm^{-1}K^{-1}$ for Nylon, 90 $Wm^{-1}K^{-1}$ for Nomex and 205 $Wm^{-1}K^{-1}$ for aluminum, respectively, in Table 1) due to the highly oriented crystalline domains in molecular structure (*20*). The surprising behavior provides a new opportunity for spider silk fibers to be used on protective clothing. On the other hand, resilience is the ability of the fiber to recover after it has been deformed by compression, an indicator of fiber flexibility. The excellent resilience of spider silk fibers (150 MJ/m^3 in Table 1) potentially offers a protection to a new degree to firefighters, which cannot be achieved by any other synthetic fibers. Response of spider silk fibers to moisture/water absorption is also critical for their strength, stability and thermal properties of protective clothing. Spider silk fibers become stiff with increased humidity and exhibit a stiffness reduction with rising temperature at constant humidity (*21*). There is little known so far about the spider silk behaviors when the silk adsorbs water, particularly used as fibers relevant to protective clothing. The ongoing work at Colorado State University focuses on the study of thermal properties and moisture adsorption properties of spider silk for the development of new lightweight material applications.

Table 1. Comparison of average of mechanical properties of spider silk fibers and synthetic fibers

	Density (g/cm^3)	*Breaking Elongation (%)*	*Tensile Strength (GPa)*	*Resilience (MJ/m^3)*
Spider silk	1.3	40	2.0	150
Kevlar	1.1	2.4	3.6	50
Nomex	0.6	20	0.1	12
Nylon	1.4	90	0.9	80
Steel	7.8	0.6	3.0	6

Regenerated Spider Silk Fibers

Although spider silk has been recognized as super light and super strong materials, the mass production of spider silk is still not practical and very challenging. Normally, getting enough spider silk requires large numbers of spiders. However, spiders tend to be territorial, so when the researchers tried to set up spider farms (like silkworm farms), the spiders killed each other. Therefore, research has focused on man-made regenerated spider silk, which replicates the complex structure in natural spider silk. One promising method is to put the spiders' dragline silk gene into goats in such a way that the goats would only make the protein in their milk (*22*). The goat milk is collected and then purified into spider silk protein with significantly high quantities (*23*). The purified spider silk protein is later spun into fiber filaments using wet spinning. This process has so far not been sufficient to completely replicate the superior properties of native spider silk (*24*). Fibers were also regenerated using spider silk proteins from other biological resources including bacteria *E. coli* (*25*), mammalian cells (*26*) and transgenic plants (*27*). A strategy developed toward commercial mass production of spider silk is to use spider silk proteins in varying host organisms and to produce recombinant spider silk proteins. The recombinant spider silk fibers and non-woven meshes have been developed using innovative processing techniques including biomimetic spinning, wet-spinning or electrospinning. Synthetic spider silk was recently used to design an artificial tendon/graft at Utah State University (*28*). This is an attempt of using regenerated spider silk fibers to make artificial organs in the human body. The results explain that spider silk has the potential to enhance strength and mobility of Achilles tendon after repair by allowing for early movement of the damaged tendon, which prevents scar formation and promotes neo-tissue development (*29*).

Animal Protein-Based Fibers

Animal protein-based fibers are naturally derived from animal hair, fur and feathers. Merino wool, alpaca fiber, cashmere and mohair from angora goats are very popular in textiles. Specialty fibers from angora rabbits, camel, llama, chicken feathers also exist, however, are rarely found in mass production. The fibers discussed in this section include wool and chicken feathers in developing lightweight materials.

Wool

Wool is the fiber from the fleece of domesticated sheep. It is a natural, protein, multicellular, staple fiber. Wool fibers vary in length between 2 – 38 cm, depending on the breed of the sheep and the part of the animal from where the fibers were removed. The diameters of the wool fibers also vary, giving that fine wool is 15 μm in diameter and coarse wool is 50 μm in diameter. The density of wool fiber is 1.31 g/cm^3 (*30*), marking wool as a light weight fiber.

Protein Structures and Fiber Properties

The protein of the wool fiber is keratin composed of amino acids in polypeptide chains (*30*). The wool keratin is a linear polymer with some short side groups, normally exhibiting a helical molecular configuration. In the linear chain structure, there are also cross-linkages called cystine or sulphur linkages, ion-to-ion bonds called salt bridges and hydrogen bonds (*31*). The cross-linkages allow the molecular chains to restore deformation, providing the resilience of the wool fibers (*3*). The hydrogen bonding between the oxygen and hydrogen atoms of alternate spirals of the helix, attributing to fiber strength.

The fiber micro-structure consists of three components: cuticles, cortex and fibrils. The cuticle is an outer layer of the fiber, which features scales. These scales are responsible for the felting shrinkage of untreated wool textiles, as a consequence of the difference of friction in the "with-scale" (*32*) and "against-scale" directions (*33*). The outermost layer of the cuticle has many microscopic pores which permit the wool to transport moisture. The middle layer called cortex is the bulk content of the fiber consisting of millions of long and narrow fibrils. Each fibril is about 100-200 nm in diameters and of indeterminate length. These fibrils are held together by a protein matrix. Two distinct sections are characterized in the cortex, known as the ortho-cortex and the para-cortex, due to different level of cystine content. Cystine contains amino acid which is capable to form cross-linkages. A higher cystine content is found in the para-cortex, resulting in greater chemical stability and molecular order. This difference also leads the spiral form of the fiber, the spontaneous curling and twisting of wool (*3*). The center of the fiber is hollow, which is call the medulla. The hollow structure is attributable for excellent insulating power of the wool fiber (*30*).

Wool is a weak natural fiber because it has large amorphous area lack of molecular packing (*34*). The fiber is weaker when wet because moisture weakens the hydrogen bonding and salt linkage. However, the scale structure of the wool fiber imparts excellent abrasion resistance, resulting fiber durability (*3*). Wool is hydrophilic and contains various amounts of absorbed water depending on the conditions (*35*). However, water absorption is usually prevented by the wool fiber due to the protection by the scales, interfacial surface tension, uniform distribution of pores and low bulk density. Once the moisture seeps between the scales, the high degree of capillarity within the fiber will cause water absorption (*36*). Most of the moisture is absorbed into the spongy matrix, then causing the rupture of hydrogen bonds and leading to swelling of the fibers. Wool is easily attacked by alkalis, because alkaline solutions can hydrolyze peptides as well as open the disulphide cross-links of wool and hence damage the fiber. Wool is more resistant to acids. Strong acids hydrolyze the peptide groups in the wool but have no interaction with the cross-linking in the polymers (*34*). Exposure to sunlight and weather tends to yellow white or dull colored wool fibers. The ultraviolet radiation of sunlight causes the peptide and disulphide bonds to sever.

Innovative Development of Wool Fibers

Wool has become more important when both lightweight and warmth are considered in athletic and outdoor textiles. In the last decade, commercial wool fabrics such as Icebreaker® and SmartWool® were developed, focusing on lightweight, insulation and moisture-wicking performance. Icebreaker products originated thermal underwear made from 100% pure New Zealand merino wool were first developed in 1994. Icebreaker collections include Superfine Journeys lightweight Travel from warm to hot conditions and all season wear; City Lightweight Urban Wear; Icebreaker GT stand alone and insulation layers for active sports such as skiing and snowboarding; Icebreaker GT Running, Road Cycling and Mountain biking lines with Lycra; Bodyfit Active Base Layers for outdoor sports; wind resistant Outer Layers; and Nature underwear for women and Beast underwear for men. SmartWool makes textile products primarily from treated merino wool. This treatment makes the wool itch-free and resistant to shrinking. SmartWool is also claimed to have moisture-wicking performance and odor-reducing, anti-microbial properties; it is thus marketed primarily as performance apparel. Moreover, these superior properties of wool fibers are also preferable in other high-performance fiber applications such as multifunctional protective clothing. There are new development of wool fibers.

For examples, silver nanoparticles have been used to color merino wool fibers and imparted antimicrobial and antistatic properties to the fibers, resulting a novel silver nanoparticle-wool composite material (*37*). Wool has also found uses in reinforcing composites. For examples, wool fibers were introduced in an earthen materials with the improvements in strength and crack resistance (*38*).

Chicken Feathers

Chicken feathers are approximately composed of 91% protein (keratin), 1% lipids and 8% water (*39*). Chicken feathers are agricultural byproducts that are low in cost and essentially a renewable source of protein fibers (*40–43*). Chicken feathers have a density of 0.8 g/cm^3, which is much lower than that of cotton and wool (see Figure 3). The feather barbs show honeycomb structures, resulting in unique properties including low density, excellent compressibility and resiliency, ability to dampen sound, warmth retention and distinctive morphological structure of feather barbs. The unique honeycomb structures made feacher not only light but also acting as air and heat insulators. Therefore, feathers are suitable for lightweight material applications that also require sound adsorption properties. Feathers are not only very light but also strong because they generally must withstand the aerodynamic force generated during flight (*44*). The lightweight property makes feathers possible for many applications such as light and warm textiles, reinforcing materials and tissue engineering. Xu et al. developed a de-cross-linking method and disentangle the keratin from chicken feathers into linear and aligned molecules (*45*). The modified keratin was readily electrospun into scaffolds with ultrafine fibers oriented randomly in 3D dimention. The 3D ultrafine fibrous composed of pure keratin scaffolds are promising materials for cartilage tissue engineering. Reddy and Yang have developed lightweight composites using chicken feathers blended with cotton fibers (*40*). The composites can provide unique properties to products which have low cost, lightweight, ability to dampen sound and warmth retention. It is still difficult to process feathers as the common protein fibers such as wool and silk due to the complex structure of the feathers. A potentially green process was recently developed to fabricate pure karetin fibers from chicken feathers by Xu and Yang (*46*). The regenerated fibers were successfully produced as linear keratin with preserved backbones that could be untangled and aligned in a controlled manner. Further research is necessary to understand the behavior and contribution of chicken feather barbs to the processability and properties of various products.

Figure 3. A photograph of chicken feather.

Plant Protein-Based Fibers

Fibers made from regenerated plant proteins began to arouse interest in the middle of the 20th century. They were used as a wool or silk substitute in 1960s due to the high cost and high demand of wool and silk at the time. The regenerated fibers are made from oilseed peanut proteins (*47*), from corn zein proteins (*48*) and from soybean proteins (*49*). These plant protein fibers are usually soft, lustrous, resilient and thermally resistant. The fibers discussed in this section are made from zein, soybean and wheat gluten proteins.

Zein

Zein (prolamine) is a class of prolamine, the protein dissolves in aqueous alcohols, found in corn. It is known for its solubility in binary solvents (*50*). Historically, zein has been commercially used for many products, including coatings, fibers, inks, molded articles, adhesives and binders (*1*, *51*).

Protein Structures and Fiber Properties

Biologically, zein is a mixture of proteins varying in molecular size and solubility. These proteins can be separated by differential solubility and their related structures into four distinct types: α, β, γ and δ (*52*). α-Zein is by far the most abundant, accounting for approximate 70% of the total, and can be extracted using only aqueous alcohol (*53*). The other types of zeins (β, γ and δ) are thought to contribute to gelling. α-Zein is the major zein found in commercial zein primarily because of the solvent used and the material from which zein is extracted. Commercial zein is not extracted from whole corn but from corn gluten meal. Zein fibers were commercialized under the trade name, "Vicara", first in 1951 (*48*). In the early spinning methods, a zein solution can be extruded either into air (dry spinning) or into water or some other coagulation medium (wet spinning). Two types of zein solutions were used to produce fibers. The first solution called for zein to be dissolved in aqueous ethanol or similar organic solvent mixtures containing water, methanol, diethylene glycol, ethylene glycol monoethyl ether, or diacetone (*48*). An alternative method required for zein to be dispersed into aqueous solutions of formaldehyde without organic solvents. After the fibers were spun and produced, a curing treatment was required to bake the fibers at 60-90 °C for 8-10 hr (*51*). The resulting fibers had excellent water resistance and satisfactory wet strength, generally superior to the artificial fibers in 1930s. Spun zein fibers were stretched in coagulation medium before they were cured and dried so that the molecular orientation was further improved in favor of high tensile strength of the fiber. However, the production of zein fibers did continue and dropped significantly by 1960 mainly due to the development of cheaper synthetic materials. The process using formaldehyde could become environmental issues and pose health risk to the personnel.

Innovative Development of Zein Fibers

Although there is no commercial production of Vicara any more, the interest in producing fiber from zein still remains and the development of innovative zein fibers continues. It remains, mainly considering increased demand for true 100% biodegradable fibers. Recently, zein fibers have again been produced in the lab by using electrospinning, where research will be performed for zein fibers to re-enter the fiber market (*54*, *55*). Torres-Giner et al. electro-spun ultrathin zein fibers embedded with nanoclays. The nanoclays were oriented along the fiber axis and increased the thermal properties of zein fibers (*56*). The hybrid fibers were incorporated in poly(lactic acid) films via compression molding, resulting hybrid composites with improved mechanical and barrier properties and sustained release properties (*57*). Coaxial electrospinning process was also used to develop ultrathin zein fibers containing functional components including ibuprofen, chitosan and tannin for medical applications. Zein nanofibers/nanocomposites have shown promising medical applications in implantation, scaffolds, drug delivery, wound dressing and surgical meshes (*58*, *59*). Cai et al. developed a novel electrospun scaffolds from zein and illustrated that the structure of the electrospun zein scaffolds could more closely mimic the 3D randomly oriented fibrous architechtures in many native extracellular matrics (*60*).

Soybean

Soybean is a protein-rich plant containing 40% protein with minimum saturated fat in comparison to milk (3.2%), corn (10%) and peanuts (25%). Except for food use, soybean proteins are used in many industrial applications including adhesives, emulsions, cleansing materials, pharmaceuticals, inks, plastics and also textiles fibers. Soybean protein fibers were first patented in 1960 by Aarons and later the first commercial production began in China in 2000s. Raw material for spinning textile fibers is obtained from soybean remaining flakes after the extraction of oils and other fatty substances (*49*).

Protein Structure and Fiber Properties

Soybean proteins contain 18 amino acids beneficial to the human body and added anti-bacterial elements. There are about 23% of acidic amino acids, 25% alkaline amino acids and about 30% of neutral amino acids. Critical ingredients in soybean protein as a raw material for producing fibers are globulins consisting of ß-conglycinin and glycinin (*61*). Subunits in the protein are non-covalently associated into trimeric proteins by hydrophobic interactions and hydrogen bonding without any disulphide bonds (*62*).

Pristine soybean fibers exhibit a cream color and can be dyed using acid and active dyes. Especially the active dye contributes fine color and luster, good sunlight resistance and perspiration fastness to the fibers. Soybean protein fabrics have soft, smooth and light hand, which is comparable with that of fabrics made from silk blended with cashmere. The commercially available soybean fiber is a manufactured regenerated protein fiber wet-spun from soybean protein blended with poly (vinyl alcohol) (PVA) (*63*). The fiber cross-section is a kidney bean shape (in a diameter of 20 μm) and there are longitudinal striations on the fiber surface parallel to the axis, varying in length and depth (*64*). The fabric has the same moisture absorption as that of cotton and better moisture transmission than that of cotton. The content of PVA in soyprotein fibers may lead to environmental degradation problems. In addition, large quantity of toxic crosslinkers, such as formaldehyde or glutaraldehyde was used in productions, posing hazard to environment and personnels.

Innovative Development of Soybean Protein Fibers

The 100% soybean fibers without any treatment have a tendency to be weak. However, a number of treatments were developed to enhance the tensile properties of soybean fibers, such as treating fibers with nitrous acid. The introduction of PVA was considered as a competitive method to improve the tensile properties of soybean fibers efficiently and cost-effectively (*63*). Soybean is a competitive production material for fibers in the textile industry since it is abundant, protein-rich and cost-effective. The possibilities that a plant protein can be modified by molecular genetic techniques, provide the opportunity to improve the properties of the fiber in specific applications. Xu et al. recently demonstrated that tissue engineering could be benefitted from biological properties of soybean protein. They electrospun soybean protein into intrinsically water-stable scaffolds that well supported uniform distrubtion and adipogenic differentiation of adipose derived mesenchymal stem cells (*65*). The invention of soybean protein fibers contributes to the protection of resources, the care of the environment and the consideration of the global sustainability.

Wheat Gluten

Wheat gluten is a cheap ($0.5 per pound), abundant (500,000 tons per year) and renewable source for producing protein fibers. Wheat gluten consists of protein, starch and lipids. The chemical composition of wheat gluten is highly complex and heterogeneous. The wheat gluten proteins have good stability to water and heat, excellent elasticity and easy degradability. These properties are preferable for forming fibers. Reddy and Yang (*66*) developed 100% wheat gluten fibers using wet spinning followed by drawing and annealing. The fibers had breaking tenacity of about 115 MPa, breaking elongation of 23% and a Young's modulus of 5 GPa. The mechanical properties were similar to those of wool and then better than those of 100% soybean and zein fibers. Good stability to weak acidic and weak alkaline conditions at high temperature was exhibited by the wheat gluten fibers. Fiber applications provide an opportunity for high value addition and also offer a large market for consumption of wheat gluten.

Nanofibers were successfully electrospun from wheat gluten. The nanofibers mats were composed of highly heterogeneous flat ribbon-like fibers and a core-shell structure (*67*). The fibrous mats from wheat gluten show promising applications as biomaterials for tissue engineering and drug delivery (*68*). Inexpensive and biodegradable composites were developed from wheat gluten matrix and jute or coconut fiber as natural reinforcing materials (*69–71*). The biocomposites had better flexural and tensile properties than similar polypropylene composites reinforced with jute fibers (*69*). Xu and Yang (*72*) studied the drug release properties of wheat gluten fibers. The results showed that the high affinity, low drug loading concentration and high activation energy for diffusion lead to lower initial burst and more constant drug release.

Outlook of Protein-Based Fibers for Lightweight Materials

The demand of lightweight materials is constantly increasing in many industries. Protein-based fibers are important in developing these lightweight materials. Proteins offer fibers biodegradability, antimicrobial properties, sustainability and other functional properties. Conventional protein-based fibers including wool and silkworm silk can be chemically treated to obtain enhanced mechanical strength and be used in reinforcing materials. Therefore, they find new applications in medical textiles such as implantation and surgical meshes. On the other hand, new regenerated protein fibers from spider silk, chicken feather, zein, soybean and wheat gluten have been recently developed. Nanofibers and nanocomposites made from the regenerated proteins and other polymers provide new lightweight functional materials potentially for many medical applications such as drug delivery, scaffolds, wound dressings and biosensors. In a summary, natural and man-made regenerated protein-based fibers have great potential to make biodegradable, renewable, sustainable lightweight materials.

References

1. Sturken, O. U.S. Patent 2,178,924, 1939.
2. Rudall, K. M. Comparative Biochemistry V4: A Comprehensive Treatise. In *Silk and Other Cocoon Proteins*; Florkin, M., Ed.; Academic Press, Inc: London, 1962; pp 397−431.
3. Gohl, E. P. G.; Vilensky, L. D. *Textile Science: An Explanation of Fiber Properties*; Guilford Publications: Melourne, Australia, 1983; pp 68−88.
4. Altman, G. H.; Diaz, F.; Jakuba, C.; Calabro, T.; Horan, R. L.; Chen, J.; Lu, H.; Richmond, J.; Kaplan, D. L. *Biomaterials* **2003**, *24*, 401–416.
5. Reed, E. J.; Viney, C. *J. Mater. Res.* **2014**, *29*, 833–842.
6. Liu, G.; Xu, S.; Cao, T.-T.; Lin, H.; Tang, X.; Zhang, Y.-Q.; Wang, X. *Biopolymers* **2014**, *101*, 1029–1037.
7. Chen, Y.; He, J.; Wang, L.; Xue, Y.; Zheng, Y.; Jiang, L. *J. Mater. Chem. A* **2014**, *2*, 1230–1234.
8. Winkleman, A.; Gotesman, G.; Yoffe, A.; Naaman, R. *Nano Lett.* **2008**, *8*, 1241–1245.
9. Kim, M. S.; Khang, G.; Lee, H. B. *Prog. Polym. Sci.* **2008**, *33*, 138–164.
10. Chen, Y.; Wang, L.; Xue, Y.; Jiang, L.; Zheng, Y. *Sci. Rep.* **2013**, *3*, 2927–2935.

11. Kang, E.; Jeong, G. S.; Choi, Y. Y.; Lee, K. H.; Khademhosseini, A.; Lee, S.-H. *Nat. Mater.* **2011**, *10*, 877–883.
12. Hu, X.; Kaplan, D.; Cebe, P. *Macromolecules* **2006**, *39*, 6161–6170.
13. Wong Po Foo, C.; Kaplan, D. L. *Adv. Drug Delivery Rev.* **2002**, *54*, 1131–1143.
14. Foelix, R. *Biology of Spiders*; Oxford University Press: Oxford, NY, 2010; pp 432–440.
15. Liu, Y.; Sponner, A.; Porter, D.; Vollrath, F. *Biomacromolecules* **2007**, *9*, 116–121.
16. Simmons, A. H.; Michal, C. A.; Jelinski, L. W. *Science* **1996**, *271*, 84–87.
17. van Beek, J. D.; Hess, S.; Vollrath, F.; Meier, B. H. *Proc. Natl. Acad. Sci.* **2002**, *99*, 10266–10271.
18. Papadopoulos, P.; Ene, R.; Weidner, I.; Kremer, F. *Macromol. Rapid Commun.* **2009**, *30*, 851–857.
19. Porter, D.; Vollrath, F.; Shao, Z. *Eur. Phys. J. E: Soft Matter Biol. Phys.* **2005**, *16*, 199–206.
20. Huang, X.; Liu, G.; Wang, X. *Adv. Mater.* **2012**, *24*, 1482–1486.
21. Elices, M.; Guinea, G.; Pérez-Rigueiro, J.; Plaza, G. *JOM* **2005**, *57*, 60–66.
22. Lewis, R. V. *Chem. Rev.* **2006**, *106*, 3762–3774.
23. Lewis, R. V.; Hinman, M.; Kothakota, S.; Fournier, M. J. *Protein Expression Purif.* **1996**, *7*, 400–406.
24. Scheibel, T. *Microb. Cell Fact.* **2004**, *3*, 14–24.
25. Xia, X.-X.; Qian, Z.-G.; Ki, C. S.; Park, Y. H.; Kaplan, D. L.; Lee, S. Y. *Proc. Natl. Acad. Sci.* **2010**, *107*, 14059–14063.
26. Lazaris, A.; Arcidiacono, S.; Huang, Y.; Zhou, J.-F.; Duguay, F.; Chretien, N.; Welsh, E. A.; Soares, J. W.; Karatzas, C. N. *Science* **2002**, *295*, 472–476.
27. Scheller, J.; Guhrs, K.-H.; Grosse, F.; Conrad, U. *Nat. Biotechnol.* **2001**, *19*, 573–577.
28. Rodriguez, F.; Thomas, H.; Hengge, N. Designing an Artificial Tendon/Graft Derived from Silkworm Silk and Synthetic Spider Silk with Respect to Structure, Mechanical Properties, Biocompatibility, and Attachment. Student Showcase, Paper 102, 2014. Utah State University. http://digitalcommons.usu.edu/student_showcase/102.
29. James, R.; Kesturu, G.; Balian, G.; Chhabra, A. B. *J. Hand Surg.* **2008**, *33*, 102–112.
30. Kadolph, S. J. *Textiles*, 11th ed.; Prentice Hall: Upper Saddle River, NJ, 2010; pp 79–106.
31. Collier, B. J.; Bide, M. J.; Tortora, P. G. *Understanding Textile*, 7th ed.; Prentice Hall: Upper Saddle River, NJ, 2008; pp 64–70.
32. Silva, C. J. S. M.; Zhang, Q.; Shen, J.; Cavaco-Paulo, A. *Enzyme Microb. Technol.* **2006**, *39*, 634–640.
33. Cortez, J.; Bonner, P. L. R.; Griffin, M. *Enzyme Microb. Technol.* **2004**, *34*, 64–72.
34. Corbman, B. P. *Textiles: Fiber to Fabric*; McGraw-Hill Professional: New York, 1983; pp 69–73.
35. Cook, J. R.; Fleischfresser, B. E. *Text. Res. J.* **1990**, *60*, 77–85.

36. Ito, H.; Muraoka, Y.; Umehara, R.; Shibata, Y.; Miyamoto, T. *Text. Res. J.* **1994**, *64*, 440–444.
37. Kelly, F. M.; Johnston, J. H. *ACS Appl. Mater. Interfaces* **2011**, *3*, 1083–1092.
38. Aymerich, F.; Fenu, L.; Meloni, P. *Constr. Build. Mater.* **2012**, *27*, 66–72.
39. Lederer, R. J. *Wilson Bull.* **1972**, *84*, 193–197.
40. Reddy, N.; Yang, Y. *J. Polym. Environ.* **2007**, *15*, 81–87.
41. Cheng, S.; Lau, K.-t.; Liu, T.; Zhao, Y.; Lam, P.-M.; Yin, Y. *Composites, Part B* **2009**, *40*, 650–654.
42. Zhan, M.; Wool, R. P.; Xiao, J. Q. *Composites, Part A* **2011**, *42*, 229–233.
43. Barone, J. R.; Schmidt, W. F. *Compos. Sci. Technol.* **2005**, *65*, 173–181.
44. Bonser, R.; Purslow, P. *J. Exp. Biol.* **1995**, *198*, 1029–1033.
45. Xu, H.; Cai, S.; Xu, L.; Yang, Y. *Langmuir* **2014**, *30*, 8461–8470.
46. Xu, H.; Yang, Y. *ACS Sustainable Chem. Eng.* **2014**, *2*, 1404–1410.
47. Fletcher, D. L.; Ahmed, E. M. *Peanut Sci.* **1977**, *4*, 17–21.
48. Lawton, J. W. *Cereal Chem.* **2002**, *79*, 1–18.
49. Boyer, R. A. *Ind. Eng. Chem.* **1940**, *32*, 1549–1551.
50. Wilson, C. M. Proteins of the Kernel. In *Corn: Chemistry and Technology*; Watson, S. A., Ramstad, P. E., Eds.; American Association of Cereal Chemists: St. Paul, MN, 1987; pp 273−310.
51. Coleman, R. E. U.S. Patent 2,236,521, 1941.
52. Coleman, C. E.; Larkins, B. A. The Prolamins of Maize. In *Seed Proteins*; Shewry, P. R., Casey, R., Eds.; Kluwer Academic Publishers: The Netherlands, 1999; pp 109−139.
53. Thompson, G. A.; Larkins, B. A. *BioEssays* **1989**, *10*, 108–113.
54. Miyoshi, T.; Toyohara, K.; Minematsu, H. *Polym. Int.* **2005**, *54*, 1187–1190.
55. Selling, G. W.; Biswas, A.; Patel, A.; Walls, D. J.; Dunlap, C.; Wei, Y. *Macromol. Chem. Phys.* **2007**, *208*, 1002–1010.
56. Torres-Giner, S.; Lagaron, J. M. *J. Appl. Polym. Sci.* **2010**, *118*, 778–789.
57. Torres-Giner, S.; Martinez-Abad, A.; Lagaron, J. M. *J. Appl. Polym. Sci.* **2014**, *131*, 9270–9276.
58. Jiang, Q.; Reddy, N.; Yang, Y. *Acta Biomater.* **2010**, *6*, 4042–4051.
59. Torres-Giner, S.; Ocio, M. J.; Lagaron, J. M. *Carbohydr. Polym.* **2009**, *77*, 261–266.
60. Cai, S.; Xu, H.; Jiang, Q.; Yang, Y. *Langmuir* **2013**, *29*, 2311–2318.
61. Pearson, A. M. Soy Proteins. In *Developments in Soy Proteins, Vol. 2*; Elsevier Applied Science Publishers: New York, 1984; pp 461−494.
62. Kinsella, J. E.; S., D.; German, B. Physicochemical and Functional Properties of Oilseed Proteins with Emphasis on Soy Proteins. In *New Protein Foods Vol.5: Seed Storage Properties*; Altschul, A. M., Ed.; Academic Press: New York, 1985; pp 108−180.
63. Zhang, Y.; Ghasemzadeh, S.; Kotliar, A. M.; Kumar, S.; Presnell, S.; Williams, L. D. *J. Appl. Polym. Sci.* **1999**, *71*, 11–19.
64. Vynias, D.; Carr, C. M. *J. Appl. Polym. Sci.* **2008**, *109*, 3590–3595.
65. Xu, H.; Cai, S.; Sellers, A.; Yang, Y. *RSC Advances* **2014**, *4*, 15451–15457.
66. Reddy, N.; Yang, Y. *Biomacromolecules* **2007**, *8*, 638–643.
67. Dong, J.; Asandei, A. D.; Parnas, R. S. *Polymer* **2010**, *51*, 3164–3172.

68. Reddy, N.; Yang, Y. *Trends Biotechnol.* **2011**, *29*, 490–498.
69. Reddy, N.; Yang, Y. *Polym. Int.* **2011**, *60*, 711–716.
70. Hemsri, S.; Grieco, K.; Asandei, A. D.; Parnas, R. S. *Composites, Part A* **2012**, *43*, 1160–1168.
71. Xu, W.; Yang, Y. *J. Appl. Polym. Sci.* **2010**, *116*, 708–717.
72. Xu, H.; Cai, S.; Sellers, A.; Yang, Y. *J. Biotechnol.* **2014**, *184*, 179–186.

Chapter 3

Synthetic Fibers from Renewable Resources

Li Shen[*]

Group Energy and Resources,
Copernicus Institute of Sustainable Development,
Utrecht University, Heidelberg 2, 3584CS Utrecht, The Netherlands
[*]Tel: +31-30253-7600. Fax: +31-30-253-7601. E-mail: l.shen@uu.nl.

Synthetic fibers from renewable resources account for only a small fraction in the world fiber market. Due to the concerns of limited fossil resources, the environment and climate change, materials made from renewable resources have attracted much attention in the past decades. In the meantime, biobased polymers have experienced a renaissance. Many traditional synthetic polymers, such as PET, PTT and PA have been, or can be potentially made from renewable resources. The historical use of biomass for material production shows that biobased polymers are neither fictional nor new. If emerging biobased polymers, such as PLA, biobased PET, biobased PTT and biobased PA, succeed in following this example, they could possibly replace their petrochemical counterparts in large quantities in the future.

Introduction

The production of fibers has undergone dramatic changes in the last century. Prior to the industrial revolution in the 19th century, natural materials such as cotton, wool and silk had been used for thousands of years. In the first decades of the twentieth century, cotton accounted for 70% of all textile raw materials in the world. It was not until the 1930s that the first man-made fiber, viscose, became one of the principal fibers. Figure 1 shows the global fiber production in the past one hundred years. Before World War II, one of the most important motivations for the research and development of man-made fibers was to find the alternatives of cotton. After World War II the production of man-made cellulosic kept increasing, until in the 1960s synthetic fibers "swept" the whole textile market.

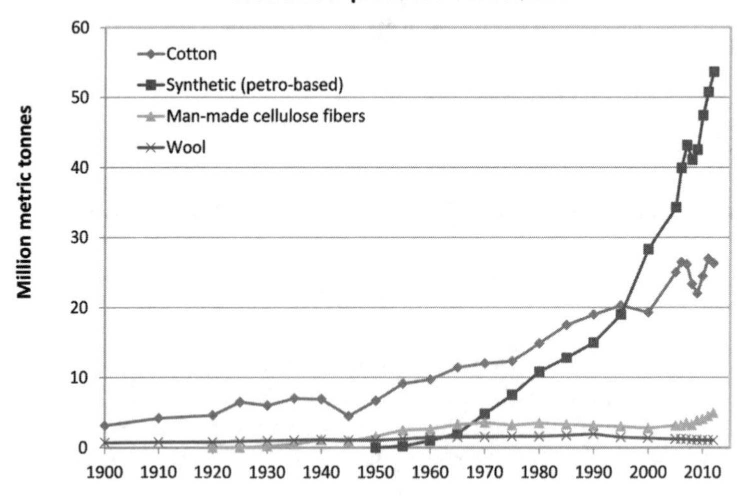

Figure 1. World fiber production 1920 – 2005 (1, 2).

Man-made fibers can be made from inorganic origin, petrochemical feedstocks and renewable resources. Within man-made fibers from renewable resources, two categories can be distinguished:

1) man-made fibers by transforming natural polymers (e.g. viscose), and
2) synthetic fibers based on emerging biobased polymers which are equivalent to the petroleum-based counterparts (see Figure 2).

Man-made cellulose fibers are produced via the transformation of natural polymers, for example, regenerated cellulose or cellulose esters. The most important man-made cellulose fiber is viscose fiber, which has been produced at industrial scale since the 1930s. Today approximately 3 Mt (million metric tonnes) of viscose fiber are produced per year. In the meantime, emerging biobased polymers have gained much attention due to the concern of environment, limited fossil fuels and climate change. Some of these biobased polymers are novel polymers, e.g. PLA (polylactic acid or polylactide); others are chemically identical with their petroleum-based counterparts, e.g. biobased PET (polyethylene terephthalate) and biobased PA (polyamide). Today, PLA has been produced at industrial scale and are used for textile fiber. Partially biobased PET polymer has been used to make beverage bottles. Biobased PAs are still in the research and development phase.

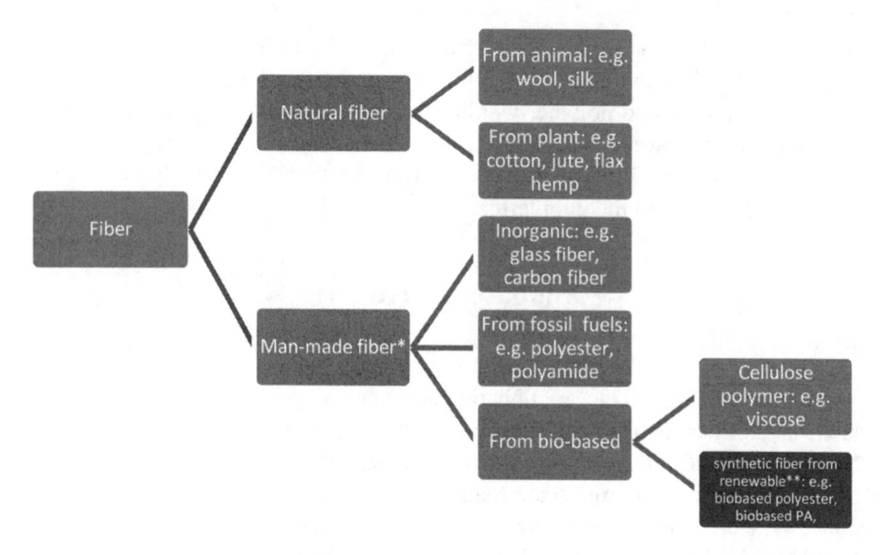

*Figure 2. Classification of fibers based on polymer origin. *Generic fiber names according to BISFA (3). **Fibers covered in this chapter.*

Biomass Resources for Emerging Biobased Polymers

Polymers abound in nature. Wood, leaves, fruits, seeds and animal furs all contain natural polymers. Biobased polymers have been used for food, furniture and clothing for thousands of years. Every year about 170 billion tonnes (1 billion = 10^9) of biomass are produced by nature, of which only 3.5% (6 billion tonnes) are utilised by mankind. Most of these 6 billion tonnes are used for food, about one third is for energy, paper, furniture and construction, and only 5% (300 million tonnes) are consumed for other non-food purposes such as chemicals and clothing (*4*).

Like biofuels, biobased polymers can be produced from first or second generation biomass. First generation biomass originates from sugar crops such as sugarcane and sugar beet, from starch crops such as corn, wheat and tapioca, or from animal fats and vegetable oils. Platform chemicals, such as ethanol and lactic acid, can be produced by directly fermenting sugar or starch via enzymes and microorganisms. Second generation biomass refers to non-food crops (e.g. switch grass), agricultural and forest residues (e.g. stems and husks), or industrial/municipal waste (e.g. woodchips and municipal waste water streams). Second generation technology aims to use cellulose, lignocellulose or lignin as the feedstock instead of sugar or starch, to produce biofuels and biochemicals. Pre-treatment (e.g. hydrolysis) is necessary to obtain fermentable sugar from lingo-cellulosic feedstock.

Today nearly all emerging biobased polymers are produced from first generation crops. Much R&D effort has been focused on cellulosic feedstock; but so far no commercial product is available. Emerging biobased polymers still have a very small share in the work market. The global production capacity was 1.2 Mt in 2011 (*5*), which is about 0.5% of total plastic production. It is projected that by 2020 the world capacity of emerging biobased polymers will increase to 3.5 Mt (*6*). Approximately million hectares of arable land would be required if all of these biobased polymer would be produced from first generation crops; this land use is, however, less than 0.3% of the arable land in Europe or 0.06% worldwide (*7*). As a consequence, no interference with food supply needs to be feared for the short to medium term.

Biobased Polyesters

PLA

PLA (see Figure 3) is an aliphatic polyester produced via polymerization of lactic acid or lactide, which are sugar fermentation products. With the start of the NatureWork LLC's manufacturing plant in 2002, PLA became the first biobased plastic produced on a large scale (name plate capacity 150 kt p.a. in 2009). In 2007, the world's largest lactic acid producer Corbion/PURAC started to produce lactide, which is a precursor of PLA, for technical applications (capacity 75 kt p.a. lactide in 2008, plant located in Thailand).

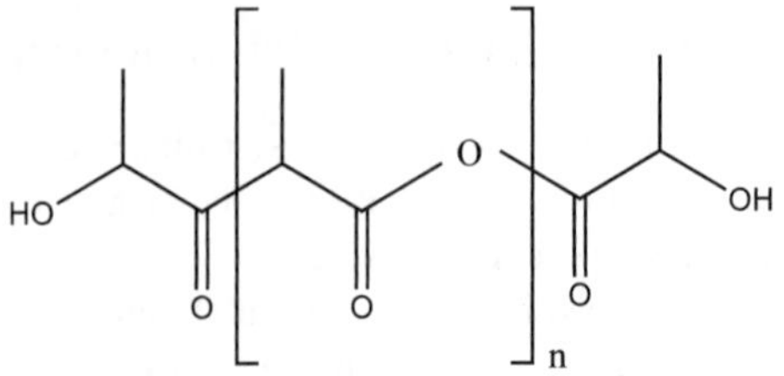

Figure 3. PLA molecule.

Lactic acid, 2-hydroxypropionic acid, is the simplest hydroxycarboxylic acid with an asymmetrical carbon atom. Lactic acid may be produced by anaerobic fermentation of carbon substrates, either pure (e.g. glucose and sucrose) or impure (e.g. starch). NatureWorks' PLA is produced from corn and PURAC's lactides are produced from cane sugar, potato starch and tapioca starch. In the future, it is expected that cellulosic biomass can be used to produce PLA.

Lactic acid produced by fermentation is optically active; specific production of either L (+) or D (–) lactic acid can be achieved by using an appropriate lactobacillus (*8*). Polymerization of L-lactide results in PLLA and polymerization of D-lactide results in PDLA. The majority of current commercial PLA is poly (meso-lactide), which is a mix of L-lactide (> 95%) and D-lactide (<5%).

Poly (meso-lactide) exhibits no stereochemical structure. It is highly amorphous, does not rotate polarized light and is optically inactive. It has a relatively low glass transition temperature (T_g = 55-60 °C), low vicat softening point and low heat deflection temperature. End-products made from this PLA are not suitable for applications which do not requiring high temperatures (similar to PET).

The recently developed heat-resistant PLA is based on stereocomplex technology. Stereocomplex formation between PLLA and PDLA occurs when L-lactide unit sequences and D-lactide unit sequences coexist in one system (*9*). Melt-blending PLLA and PDLA with a D/L ratio of 1:1 produces sc-PLA crystals with a melting temperature (T_m) of 210-240 °C, which is about 30-60 °C higher than the T_m of homo-crystalline PLLA.

The current applications of PLA cover a wide range, e.g. packaging (cups, bottles, films and container), textiles (shirts, furniture textiles), nonwovens (diapers), agricultural mulch films (usually blended with thermoplastic starch) and cutlery. Stereocomplex PLA is potentially suitable for melt-spun fibers and biaxially stretched films.

Biobased PET

Polyethylene terephthalate (PET, Figure 4) was first commercialized in the 1940s and has been used since then for the production of synthetic fibers and for film applications. In the mid-1970s important technological breakthrough was made in bottle blow moulding process. PET bottle market segment has since then increased steadily and is still continuing to grow. Fully biobased PET has not yet been announced by any producers so far. However, PET has been partly produced from biobased feedstock.

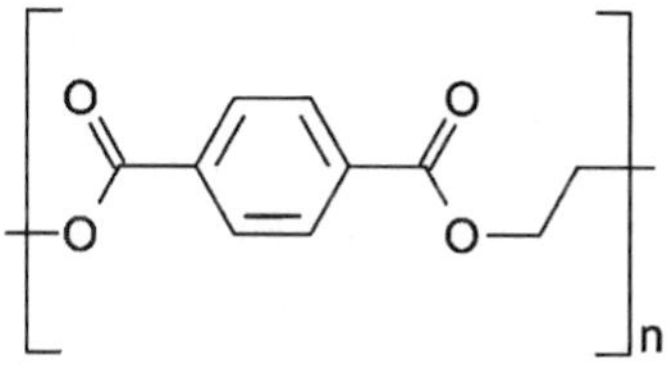

Figure 4. PET molecule.

PET can either be produced from dimethylterephthalate (DMT) and ethylene glycol via transesterification, or from purified terephthalic acid (PTA) and ethylene glycol via esterification. The transesterfication reaction of DMT and ethylene glycol is a catalytic process, followed by polycondensation. The direct esterification of PTA with ethylene glycol is nowadays the most applied process for PET production. Both DMT and PTA are derived from oxidized paraxylene.

Ethylene glycol is formed via direct oxidation of ethylene followed by thermal hydrolysis into ethylene glycol. Biobased ethylene glycol is produced from biobased ethylene, which is an ethanol dehydration product. Braskem, Dow and Solvay are the current major producers of biobased ethylene. Furthermore, ethylene glycol can also be derived from sorbitol based on hydrogenolysis (*10*).

There are several possibilities to produce PTA from biobased resources. It could potentially be made using biobased xylene produced via biomass pyrolysis. Here, cellulosic or lingo-cellulose biomass such as wood chips can be used as starting material. Alternatively, it may also be replaced by 2.5-furandicarboxylic acid (FDCA) and limonene (*10*). A third way to produce biobased PTA is via the conversion of limonene to p-cymene using zeolites and the subsequent oxidation to terephthalic acid (*11*). This technology needs, however, further development and does not offer an economically viable production route to bulk aromatic chemicals for the short to mid-term.

The current partially biobased PET contains approximately 30% of biobased content, which is the contribution from biobased ethylene glycol. Partially biobased PET is chemically identical with petrochemical PET and its properties are therefore also essentially identical. Partially biobased PET is not biodegradable.

Globally about 35 million tonnes of PET are used for fiber production, which is 65% of the total PET production (*12*). The remaining 35% are used for packaging applications including bottles (8 million tonnes) and films (2 million tonnes) (*12*). Of the 2.8 million tonnes of PET used for packaging purposes in Western Europe in 2004, 76% were bottles, 11% containers and 13% films (*12*). For partially biobased PET, the global capacity in 2011 has reached 450 kt p.a. (*5*) It is predicted that the market will grow strongly in the future; by 2020 the capacity of biobased PET will reach 5 million tonne p.a. (*5*)

PTT from Biobased PDO

Poly (trimethylene terephthalate) (PTT, Figure 5) is a linear aromatic polyester produced by polycondensation of 1,3-propanediol (trimethylene glycol, 3G, or 1,3-PDO) with either PTA or DMT. According to the conventional production route both monomers – the diacid and the diol component - are derived from petrochemical feedstocks. The production of biobased PDO has been developed and commercialized by the joint venture DuPont Tate & Lyle LLC. DuPont's biobased PDO (Bio-PDO™) is produced by aerobic fermentation of glucose from corn starch. The yield and productivity are relatively high with the aerobic process, opening the way for bulk production of biobased PTT. In 2006, the joint venture expanded the Bio-PDO™ production capacity to 45,000 t.p.a. (metric tonnes per year) in Loudon, Tennessee. Apart from PTT, other acronyms used for the same polymer are PTMT (polytrimethylene terephthalate) and PPT (polypropylene terephthalate).

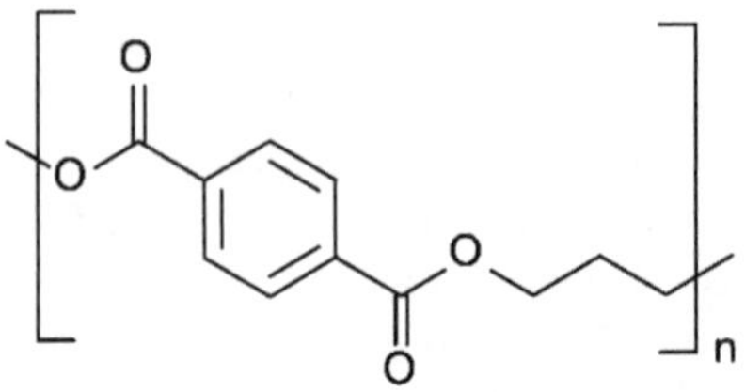

O
O
O
O
n

Figure 5. Polytrimethylene terephthalate (PTT) molecule.

As an engineering thermoplastic, PTT has a very desirable property, combining the rigidity, strength and heat resistance of polyethylene terephthalate (PET), with the good processability of polybutylene terephthalate (PBT). PTT may be used to produce fibers for carpets and industrial textiles where it has the good resiliency and wearability of nylon, as well as the dyeability, static resistance and chemical resistance of PET. As a staple or filament fiber spun into yarn for apparel, its property set includes good stretch recovery, softness and dyeability. When blended with other resins it can improve strength, flexibility, and barrier properties in moulding and extrusion applications.

From Biomass to 1,3-Propandiol

Biobased PDO is produced industrially via fermentation of glucose. There is a fermentation pathway in nature which consists two steps: naturally occurring yeasts first ferment glucose to glycerol, then microbes ferment glycerol to 1,3-propanediol. In the patented bioprocess developed by DuPont with Genencor, glucose derived from wet-milled corn is metabolized by genetically engineered microorganism *E. coli*. This microorganism converts glucose to 1,3-propanediol in a single step (Figure 6). The microorganism is placed in a fermentation tank with water and glucose, along with vitamins, minerals and oxygen. After the organism ingests the glucose, it produces the three-carbon molecule 1,3-PDO. The PDO is then separated from the fermentation broth by filtration, and concentrated by evaporation, followed by purification by distillation. 1,3-PDO appears as clear, slightly viscous liquid.

It is also possible to produce PDO by fermentation of glycerol (*13*). Glycerol is a by-product from biodiesel production. The increase in biodiesel production in recent years has led to a dramatic drop in glycerol prices, making glycerol a potentially attractive starting material for PDO and other chemicals (e.g. epichlorohydrin and propylene glycol).

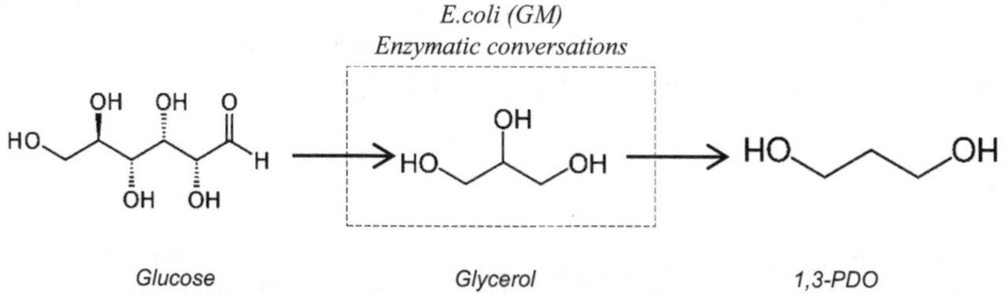

Figure 6. Fermentation route to PDO.

Propylene glycols (including both 1,2-propanediol and 1,3-propanediol) can be synthesized through thermo-chemical conversion of glycerol (see Figure 7) (*14*). In the thermo-chemical pathway patented by Celanese in 1987, aqueous glycerol solution is converted into propylene glycols (both 1,2-propanediol and 1,3-propanediol) at 200 °C and 300 bar (forming 1,3-propanediol and 1,2-propanediol at 20% and 23% yield, respectively (*15*). Among the propylene glycols, only 1,3-propanediol is suitable for making a semi-crystalline condensation polymer.

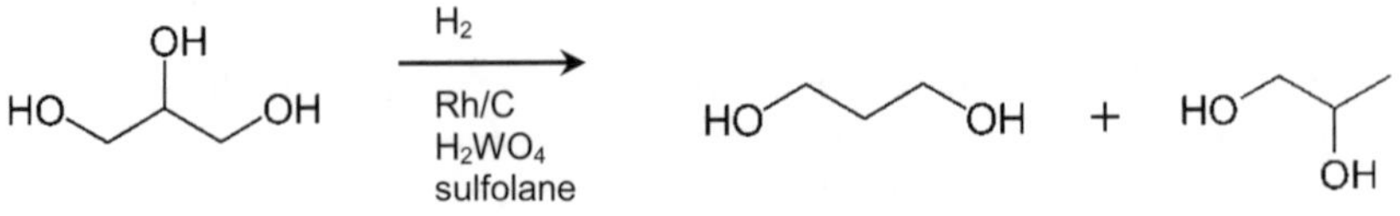

Figure 7. Conversion of glycerol to propylene glycols via the thermo-chemical route (16).

From Biobased 1,3-PDO to PTT

Like PET, PTT can be produced either by transesterification of DMT with PDO, or by the esterification route, starting with PTA and PDO. The polymerization can be a continuous process and is similar to the production of PET. In the first stage of polymerization, low molecular weight polyester is produced in the presence of excess PDO, with water (in the case of PTA) or methanol (in the case of DMT) being removed. In the second stage, polycondensation, chain growth occurs by removal of PDO and remaining water/methanol. As chain termination can occur at any time (due to the presence of a monofunctional acid or hydroxyl compound), both monomers must be very pure. As the reaction proceeds, removal of traces of PDO becomes increasingly difficult. This is compensated for by having a series of reactors operating under progressively higher temperatures and lower pressures. In a final step, the highly viscous molten polymer is blended with additives in a static mixer and then palletised (*17*).

Nylon

Nylon is a generic name for a family of long-chain polyamides (PA). They are thermoplastics which have recurring amide groups [-CONH-]. Worldwide in 2012, about 2.7 million tonnes polyamides are used for engineering coumpounds; this accounts for approximately 40% of the global engineering plastics production (*18*). Important commercial polyamides include:

- PA 6 (polycaprolactam) - made by the polycondensation of ε-caprolactam [$CH_2(CH_2)_4NHCO$];
- PA 66 (polyhexamethylene adipamide) - made by condensing hexamethylenediamine [$H_2N(CH_2)_6NH_2$] with adipic acid [$COOH(CH_2)_4COOH$];
- PA 46 (polytetramethylene adipiamide) - made by condensing tetramethylenediamine with adipic acid [$COOH(CH_2)_4COOH$];
- PA 69 (polyhexamethylene azelaamide) - made by condensing hexamethylenediamine [$H_2N(CH_2)_6NH_2$] with azelaic acid [$COOH(CH_2)_7COOH$];
- PA 610 - made by condensing hexamethylenediamine with sebacic acid [$COOH(CH_2)_8COOH$];
- PA 612 - made from hexamethylenediamine and a 12-carbon dibasic acid;
- PA11 - produced by polycondensation of the monomer 11-aminoundecanoic acid [$NH_2CH_2(CH_2)_9COOH$];
- PA12 - made by the polymerization of laurolactam [$CH_2(CH_2)_{10}CO$] or cyclododecalactam, with 11 methylene units between the linking -NH-co- groups in the polymer chain.
- PA 6T/66 - i.e. Hexamethyleneadipamide-hexamethylene terephthalamide copolyamid, polymer with 1,6-hexanediamine and hexanedioic acid
- PA 6T/6 ε-Caprolactam-hexamethyleneterephthalamide copolyamide, polymer with hexahydro-2H-axepine-2-one and 1,6-hexanediamine

Polyamides are generally synthesized from diamines and dibasic (dicarboxylic) acids, amino acids or lactams. Where two types of reactive monomer are required, the polymerization is said to be an AABB type; where one suffices, an AB type. A and B stand for the functional groups $–NH_2$ and –COOH, respectively. The different polyamide types are identified by numbers denoting the number of carbon atoms in the monomers (diamine first for the AABB type). In commercial manufacture, polyamides are in general directly prepared from (1) dicarboxylic acid and diamines, (2) ε-aminoacids, or (3) lactams.

Table I shows the biobased or partially biobased polyamides that are commercially available today and the potential polyamides that can be produced from biobased raw materials in the future. Commercially available biobased polyamides are PA 11 from castor oil (monomer 11-aminoundecanoic acid) and PA 610 which is partially biobased (sebacic acid from castor oil). This selection includes biobased polyamides that are commercialized already today and some polyamides that may be produced from biobased feedstocks in future.

Adipic acid can be potentially produced via fermentation of sugar. If this process will ultimately be successful at industrial scale, it will serve as a basis for the manufacture of partially biobased PA 66. Azelaic acid, one of the monomers to produce PA 69, can be obtained from oleic acid, which is abundant in certain vegetable oils. ε-caprolactam is the monomer to produce PA 6. R&D is ongoing to arrive at an industrially attractive route having lysine production by fermentation as its first step (*17*, *19–21*). However, this route is not yet economically viable compared to the conventional route (*13*, *22*).

Table I. Commercially available biobased polyamides and potential biobased polyamides

Polyamide	*Monomers*	*Raw material*	*Commercialization status*	*Tradename (Company)*
11	11-aminoundecanoic acid	Castor oil	Commercial product	Rilsan® PA 11 (Arkema)
610	Hexamethylenediamine	Butadiene, propene	Commercial product	Ultramid® (BASF) Amilan® (Toray)
	Sebacic acid	Castor oil		
66	Hexamethylenediamine	Butadiene, propene	R&D	
	Adipic acid	Glucose		
69	Hexamethylenediamine	Butadiene, propene	R&D	
	Azelaic acid	Oleic acid		
6	ε-Caprolactam	Glucose	R&D	
46	Tetramethylenediamine	acrylonitrile and HCN	unknown	
	Adipic acid	Glucose		
36	Dimer acid	Oleic and linoleic acids	unknown	

PA11 from Castor Oil

11-aminoundecanoic acid, produced from castor oil, is the monomer of PA11 Castor oil is transesterified with methanol to produce methyl ricinoleate (with glycerol as the by-product). The methyl ricinoleate then goes through a pyrolysis process (500°C) and is converted into methyl-10-undecylenate and heptaldehyde. The methyl-10-undecylenate is hydrolysed, and the resultant undecyenic acid is treated with hydrogen bromide in the presence of peroxides to yield 11-bromoundecanoic acid. This compound is then reacted with ammonia and 11-aminoundecanoic acid is obtained (Figure 8) (*23*, *24*).

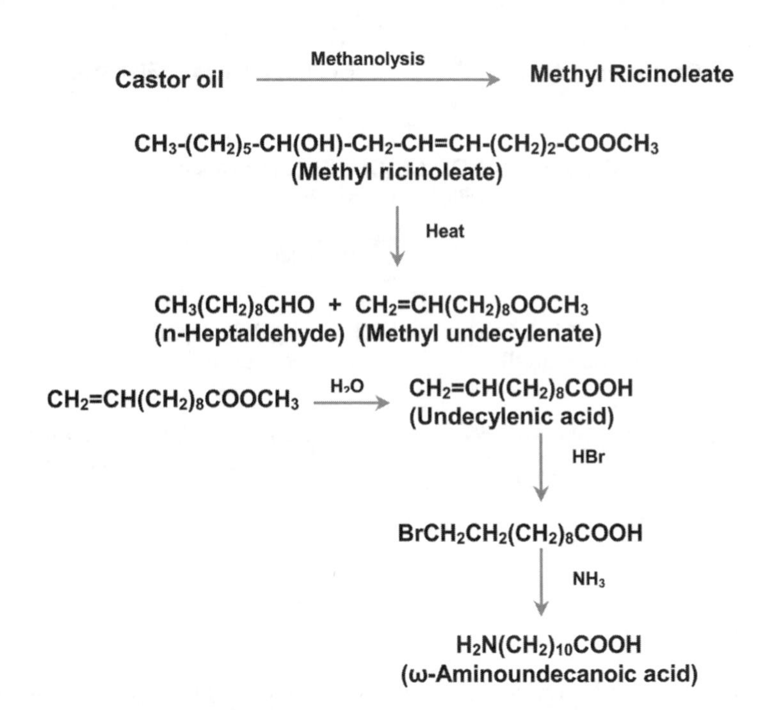

Figure 8. Production of ω-aminoundecanoic acid from castor oil. Reproduced with permission from reference (23). Copyright (2006) Elsevier.

PA 610 from Castor Oil

The monomers of PA610 are hexamethylenediamine and sebacic acid. Currently hexamethylenediamine can only be produced from petrochemical routes, i.e. from butadiene or propylene; sebacic acid can be obtained from castor oil. Therefore PA610 has approximately 60% of biobased carbon content.

In the production of sebacic acid, castor oil is first heated to a high temperature (about 180-270°C) with alkali (e.g. NaOH), resulting in saponification of the castor oil to ricinoleic acid and glycerol. Ricinoleic acid is then cleaved to give 2-octanol and sebacic acid (see Figure 9). Although the sebacic acid yields are low, this route still has been found to be cost competitive (*23*).

$$\text{Castor oil} + \text{NaOH} \xrightarrow[\text{NaOH}]{180\text{-}270^{\circ}\text{C}} \text{Ricinoleic acid} + \text{Glycerol}$$

$CH_3(CH_2)_5CH(OH)CH_2CH{=}CH(CH_2)_7COOH$
(Ricinoleic acid)

↓ NaOH 130°C

$HOOC(CH_2)_8COOH$ + $CH_3(CH_2)_5CH(OH)CH_3$
(sebacic acid) (capryl alcohol)

Figure 9. Production of sebacic acid from castor oil. Reproduced with permission from reference (23). Copyright (2006) Elsevier.

PA 66 from Biobased Adipic Acid

The monomer of PA66 is adipic acid. Conventionally adipic acid is made from petrochemical cyclohexane. Most large scale production uses nitric acid oxidation of cyclohexanol, cyclohexanone or a mixture of the two (*25*). In the biobased route to adipic acid, *E. coli* bacteria first convert sugar to 3-dehydroxyshikimate, which is then converted into cis, cis-muconic acid. Cis, cis-muconic acid is then hydrogenated to adipic acid under high pressure (50 psi) and the presence of a catalyst (see Figure 10) (*26*). Finally the production of nylon 66 from adipic acid and diamine is performed in a conventional step polymerization by means of a carbonyl addition/elimination reaction (*25*).

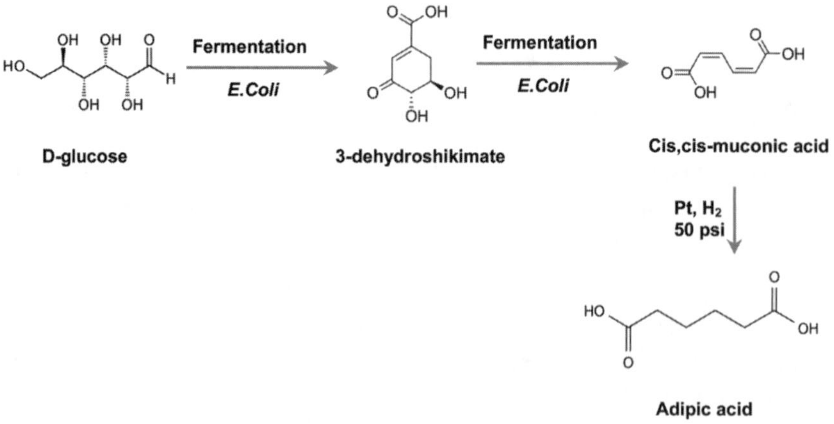

D-glucose —(Fermentation, E.Coli)→ 3-dehydroshikimate —(Fermentation, E.Coli)→ Cis,cis-muconic acid —(Pt, H_2, 50 psi)→ Adipic acid

Figure 10. Biotechnological production of adipic acid (26).

PA 69 from Biobased Azelaic Acid

Azelaic acid (nonanedioic acid, see Figure 11), the diacid monomer of PA69, is produced by a chemical synthesis pathway from oleic acid. Oleic acid is a monounsaturated 18-carbon fatty acid which is found in most animal fats and vegetable oils (e.g. in olive oil, palm berry oil). Azelaic acid used to be prepared by oxidation of oleic acid with potassium permanganate, but it is now produced by oxidative cleavage of oleic acid with chromic acid or by ozonolysis (*27*). The polymerization of azelaic acid and diamine to PA69 is a conventional step polymerization. It is very similar to that for PA66; however, the process conditions differ due to different melt viscosities and melting points (*28*).

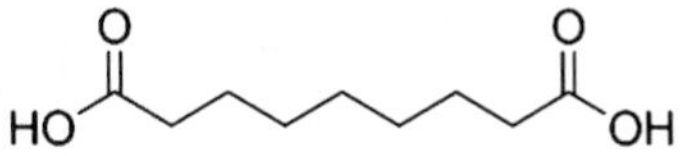

Figure 11. Azelaic acid.

PA6 from Biobased Caprolactam

The monomer of PA6 is ε-caprolactam. The commercial processes of caprolactam production are based on benzene or toluene from BTX (benzene, toluene and xylene) stream. Large scale industrial processes use cyclohexanone, cyclohexane or toluene as starting materials. In 1899 the first (biobased) caprolactam was synthesized by the cyclization of ε-aminocaproic acid (*29*). The recent studies reported that it may be produced in future by fermentation from glucose, possibly via the precursor lysine (see Figure 12) (*30*). Michigan University patented a chemical route to produce caprolacatam from lysine (*19*). PA6 follows from the ring opening polymerization of caprolactam (*31*).

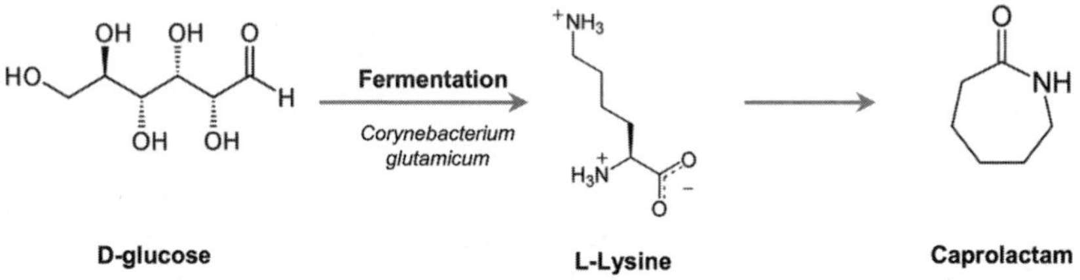

Figure 12. Biobased caprolactam via Lysine (19, 30).

Properties of Synthetic Fibers from Renewable Resources

Table II listed some mechanical, thermal and water retention properties of staple fibers reviewed in this chapter. PET, PTT and PAs are partially biobased; they are chemically identical with the petrochemical counterparts and are not biodegradable.

Table II. Selected mechanical, thermal and water retention properties of staple fibers (unmodified)

Fiber	*Density (g/cm^3)*	*Elongation (%)*	*Tenacity [a] (cN/tex)*	*Water retention (%)*	*Melting point (°C)*
PLA (*32*)	1.24	10-70	n/a [b]	n/a	155-170
PET (*33*)	1.36-1.41	30-55	25-40	3-5	250-260
PTT	1.35 (*34*)	80-90 (*35*)	27-30 (*35*)	n/a	225 (*34*)
PA 11 (*36*)	1.04	15-40	40-70	2.9	190
PA 610 (*37*)	1.08	n/a	n/a	n/a	225
PA 6 (*36*)	1.14	15-80	30-90	9-15	214-220
PA 66 (*36*)	1.14	15-80	35-90	9-15	255-260

[a] Tenacity is expressed in relative to the fineness (1 tex = 1 gram per 1000 metres). Figures for tenacity are based on both fiber fineness (tex) and cross-sectional area of the sample. [b] n/a = data not available or not applicable.

Conclusions and Outlook

The historical use of natural polymers demonstrates that biobased materials are neither fictional nor new. Instead, for many decades, they have become an industrial reality on a million-tonne scale. This shows that the production of biobased products at very large scale is not unprecedented and that related challenges, for instance related to supply chain, can be mastered. If the emerging biobased polymers, such as PLA, biobased PET, biobased PTT and biobased PA, succeed in following this example, they could possibly replace their petrochemical counterparts in large quantities in the future.

Today, the volume of synthetic fibers made from biobased resources are still small compared to that of synthetic fibers made from petrochemical feedstock. However, biobased materials have demonstrated an emerging, dynamic and very active research and development area. Several factors clearly speak for biobased materials, including limited and uncertain supply of fossil fuels, economic viability, environmental considerations (e.g. greenhouse gas abatement) and innovation offering new opportunities and rejuvenation in all steps from fundamental chemistry research to final product development and waste management. Challenges that need to be successfully addressed in the next decades are their relatively high cost for production and processing and the need to minimize agricultural land use and forests, in order to avoid competition with food production and adverse effects on biodiversity and other environmental impacts.

References

1. Shen, L.; Patel, M. K. *Lenzinger Ber.* **2010**, *88*, 1–59.
2. World Man-Made Fibres Production. CIRFS. http://www.cirfs.org/KeyStatistics/WorldManMadeFibresProduction.aspx (accessed March 27, 2014).
3. Generic Fibre Names with Their Codes. BISFA. http://www.bisfa.org/GENERICFIBRENAMES/Existingnames.aspx (accessed July 13, 2014).
4. Thoen, J.; Busch, R. In *Biorefineries − Industrial Processes and Products: Status Quo and Future Directions*; Kamm, B.; Gruber, P. R., Kamm, M., Eds.; Wiley-VCH: Weinheim, Germany, 2006; Vol. 2, pp 347–366.
5. Bioplastics: Facts and Figures. European Bioplastics. www.european-bioplastics.org (accessed January 1, 2014).
6. Shen, L.; Worrell, E.; Patel, M. *Biofuels, Bioprod. Biorefin.* **2010**, *4*, 25–40.
7. Shen, L.; Haufe, J.; Patel, M. K. *Product Overview and Market Projection of Emerging Bio-Based Plastics*; Copernicus Institute of Sustainable Development, Utrecht University: Utrecht, The Netherlands, 2009; pp 227.
8. Chahal, S. P.; Starr, J. N. In *Ullmann's Encyclopaedia of Industrial Chemicals*; Wiley-VCH Verlag GmbH & Co. KGaA: Berlin, Germany, 2006; online version.
9. Tsuji, H. *Macromol. Biosci.* **2005**, *5*, 569–597.
10. Werpy, T.; Petersen, G. *Top Value Added Chemicals from Biomass Volume I: Results of Screening for Potential Candidates from Sugars and Synthesis Gas*; PNNL and NREL; Richland, WA and Golden, CO, 2004; pp 76.
11. Rauter, A. P.; Palma, F. B.; Justino, J.; Araújo, M. E.; Pina dos Santos, S. *Natural Products in the New Millennium: Prospects and Industrial Application*; Springer Netherlands: Dordrecht, The Netherlands, 2003; pp 505.
12. Glenz, W. W. *Kunststoffe* **2007**, *10/2007*, 76–80.
13. Patel, M.; Crank, M.; Dornburg, V.; Hermann, B.; Roes, L.; Hysing, B.; Overbeek van, L.; Terragni, F.; Recchia, E. *Medium and Long-Term Opportunities and Risks of the Biotechnological Production of Bulk Chemicals from Renewable Resources - The BREW Project*; Utrecht University, Utrecht, The Netherlands, 2005; pp 452.
14. Van Haveren, J.; Scott, E. L.; Sanders, J. *Biofuels, Bioprod. Biorefin.* **2008**, *2*, 41–57.
15. Tessie, M. C. U.S. Patent 4,642,394, 1987.
16. Chaminand, J.; Djakovithc, L.; Gallezot, P.; Marion, P.; Pinel, C.; Rosier, C. *Green Chem.* **2004**, *6*, 359–361.
17. Chen, G.-Q.; Patel, M. K. *Chem. Rev.* **2011**, *112*, 2082–2099.
18. Bienmuller, M.; Joachimi, D.; Klein, A.; Munker, M. *Kunststoffe Int.* **2013**, *10/2013*, 36–43.
19. Frost, J. W. WO Patent 2005123669, 2005.
20. Kimura, E.; Asakura, Y.; Uehara, A.; Inoue, S.; Kawahara, Y.; Yoshihara, Y.; Nakamatsu, T. EP Patent 1293560, 2003.
21. Matsuzaki, Y.; Nakamura, J.; Hashiguchi, K. U.S. Patent 20050014236 A1, 2004.

22. Hermann, B. G. *Opportunities for Biomaterials: Economic, Environmental and Policy Aspects along Their Life Cycles*; Ph.D. Dissertation, Utrecht University: Utrecht, The Netherlands, 2010; pp 185.
23. Ogunniyi, D. S. *Bioresour. Technol.* **2006**, *97*, 1086–1091.
24. *Chemicals Screening Information Dataset (SIDS) for High Volume Chemicals*; United Nation Environment Programme: Chatelaine, Switzerland, 2002.
25. Musser, M. T. In *Ullmann's Encyclopedia of Industrial Chemicals*; Wiley-VCH Verlag GmbH & Co. KGaA: Weinheim, Berlin, Germany, 2000; online version.
26. Niu, W.; Draths, K. M.; Frost, J. W. *Biotechnol. Prog.* **2002**, *18*, 201–211.
27. Zahardis, J.; Petrucci, G. A. *Atmos. Chem. Phys.* **2007**, *7*, 1237–1274.
28. Kohan, M. I.; Mestemacher, S. A.; Pagilagan, R. U.; Redmond, K. In *Ullmann's Encyclopaedia of Industrial Chemistry*; Wiley-VCH Verlag GmbH & Co. KGaA: Berlin, Germany, 2003; online version.
29. Ritz, J.; Fuchs, H.; Kieczka, H.; Moran, W. C. In *Ullmann's Encyclopaedia of Industrial Chemicals*; John Wiley & Sons, Inc.: Weinheim, Berlin, Germany, 2011; online version.
30. Pfefferle, W.; Möckel, B.; Bathe, B.; Marx, A. In *Microbial Production of l-Amino Acids SE - 3*; Faurie, R., Thommel, J., Bathe, B., Debabov, V. G., Huebner, S., Ikeda, M., Kimura, E., Marx, A., Möckel, B., Mueller, U., Pfefferle, W., Eds.; Advances in Biochemical Engineering/Biotechnology; Springer Berlin, Heidelberg, Germany, 2003; Vol. 79, pp 59–112.
31. Herzog, B.; Kohan, M. I.; Mestemacher, S. A.; Pagilagan, R. U.; Redmond, K. In *Ullmann's Encyclopedia of Industrial Chemicals*; Wiley-VCH Verlag GmbH & Co. KGaA: Berlin, Germany, 2013; online version.
32. Ingeo Biopolymer 6202D Technical Data Sheet. NatureWorks LLC. http://www.natureworksllc.com/~/media/Technical_Resources/Technical_Data_Sheets/TechnicalDataSheet_6202D_fiber-melt-spinning_pdf.pdf (accessed July 17, 2014).
33. Frank, E.; Bauch, V.; Schultze-Gebhardt, F.; Herlinger, K.-H. In *Ullmann's Encyclopaedia of Industrial Chemicals*; Wiley-VCH GmbH & Co. KGaA: Weinheim, Berlin, Germany, 2011; online version.
34. Hwo, C.; Forschner, T.; Lowtan, R.; Gwyn, D.; Cristea, B. In *Future-Pak 98 Conference*; Shell Chemicals: Chicago, 1998; pp 10–12.
35. Casey, P. K.; Dangayach, K.; Oliveri, L. H.; Shiffler, D. A. CA Patent 2402533 C, 2001.
36. Estes, L. L.; Helmut, S.; Berg, H.; Wolf, K.-H.; Kausch, M.; Schroer, H.; Pellegrini, A.; Oliveri, P.; Schoene, W.; Noaj, A.; Suling, C.; Menault, J.; Osugi, T.; Morimoto, O.; Frankenburg, P. E. In *Ullmann's Encyclopedia of Industrial Chemistry*; Wiley-VCH Verlag GmbH & Co. KGaA: Weinheim, Germany, 2007; online version.
37. Basic Physical Properties. Toray Nylon Resin Amilan CM 2001. Toray. http://product.toray.com/download/pdf/en_plastics/classification_0027.pdf (accessed July 18, 2014).

Chapter 4

Lightweight Materials Prepared from Vegetable Oils and Their Derivatives

Jian Hong*

Kansas Polymer Research Center, Pittsburg State University, 1701 South Broadway, Pittsburg, Kansas 66762, United States
***E-mail: jhong@pittstate.edu**

Both vegetable oils and their derivatives are good resources for preparing synthetic polymers to replace petroleum-based polymers for lightweight material. Vegetable oils are bio-resources that have large availability and renewability. Vegetable oil-based polymers, ranging from elastomers to rigid hard plastic, have been developed via different synthetic approaches. The main focus of vegetable oil-based polymers includes copolymers of vegetable oils with vinyl monomers, epoxy resins from epoxidized vegetable oils, polyurethanes from vegetable oil-based polyol and polyisocyanates, polyesters from vegetable oil derivatives, and other thermosetting materials from functionalized vegetable oils. These polymers have comparable or better thermal and mechanical properties than petroleum-based polymers.

Introduction

Synthetic polymer, as a kind of lightweight material resource, is mainly prepared from petroleum chemicals. Due to conventional energy and resource crisis, more and more synthetic polymers based on sustainable sources are being developed. Vegetable oils, have attracted considerable attention as raw materials for polymer synthesis, are annually renewable and environmentally benign sources. Moreover, synthetic polymers from vegetable oils possess comparable or better thermal and mechanical properties than petroleum-based polymers.

Triesters of glycerol (triglycerides or triacylglycerols, Figure 1) with saturated or unsaturated long-chain acids (fatty acids) are main constituents of vegetable oils. There are hundreds of fatty acids in nature, while five particular fatty acids contribute to most vegetable oils in general (*1*). The most common saturated fatty acids are palmitic (C16:0) and stearic (C18:0), which are 16- and 18-carbon acids, respectively, without double bond. The major unsaturated fatty acids are oleic (C18:1), linoleic (C18:2), and linolenic (C18:3), which are 18-carbon acids with one, two and three double bonds, respectively. Castor and tung oils are exceptional. The main hydrolysis component of castor oil is ricinoleic acid, an 18-carbon acid containing one double bond and one hydroxyl group. Tung oil contains more than 80% of eleostearic acid, a conjugated fatty acid with three double bonds. The structures of the main fatty acids are shown in Figure 2.

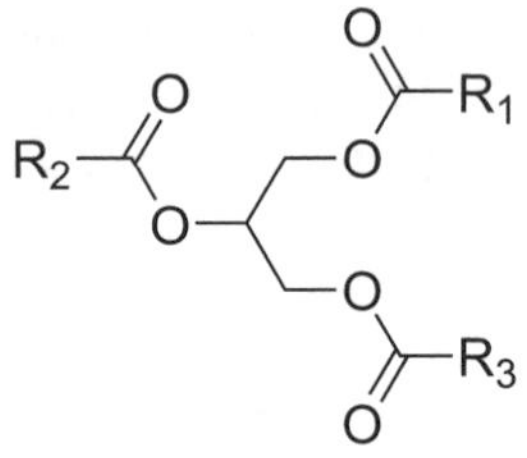

Figure 1. The structure of triglyceride (R1-R3: fatty acid chains).

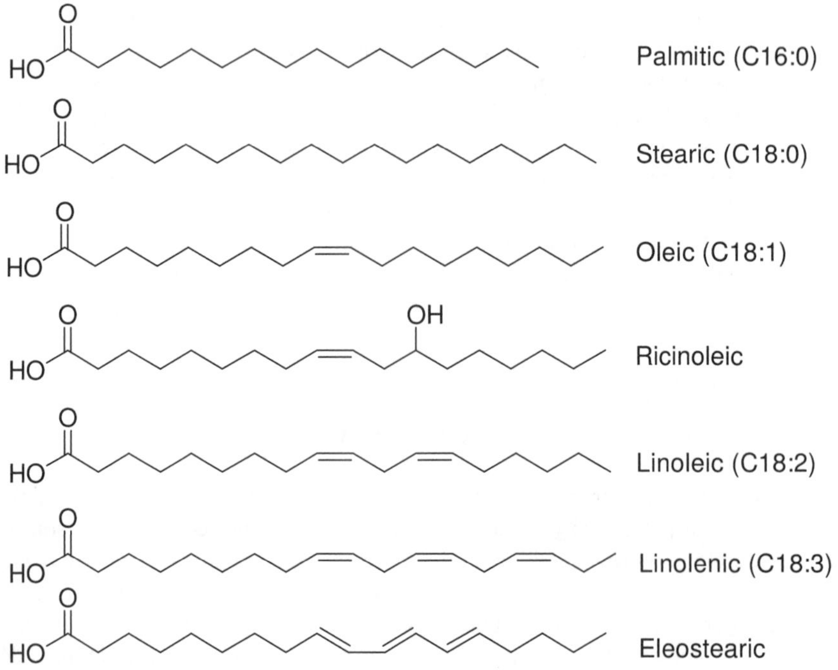

Figure 2. The structures of main fatty acids.

Every oil has a characteristic profile of fatty acids. Different vegetable oils have various numbers of functionality (double bond). Those that contain high content of palmitic and stearic acids, such as coconut oil and palm kernel oil, have functionality as low as two. While those that contain high content of linoleic and linolenic acids, such as linseed oil, have functionality as high as seven.

With multiple functional groups, vegetable oils are good resources for preparing crosslinked polymers via two approaches. The first approach is by direct polymerization through double bonds with or without other vinyl monomers. The second approach is by introducing new polymerizable group through double bonds in chains, such as epoxidation, hydroformylation, reaction with maleic anhydride, ring-opening of epoxidized oils with hydrogen active compounds.

Hydrolysis or transesterification of oils can produce glycerol and fatty acids or fatty esters. Glycerol is an important industry raw material and can be used to prepare polymers but will not be discussed here. Fatty acids and fatty esters can not only build up polymeric structures, but also be converted to tailor made monomers by further chemical modification (*2*).

Polymers Prepared via Polymerization of Vegetable Oils

Although vegetable oils contain sufficient functional groups to form polymers, it is hard to obtain polymers via direct polymerization. It is because the functional groups (double bonds) are in the middle of long chains, and thus have low reactivity. Even under high temperature conditions, products from direct radical or cationic polymerization are soft due to low degree of polymerization and cannot be used as engineering materials. However, tung oil is an exception. Tung oil is widely used in coating and painting because it has high content of conjugated fatty acids which can undergo oxidative polymerization or oxidative free radical initiation at room temperature. Therefore, converting double bonds in fatty acids to conjugated ones by catalyzed isomerization is one method to derive polymers from vegetable oils.

Copolymerization of vegetable oils and other vinyl monomers can also overcome the low reactivity of vegetable oil. Larock group (*3*) prepared polymeric materials via thermal copolymerization of tung oil, styrene (ST), and divinylbenzene (DVB). The products were light yellow and transparent materials with glossy surfaces, exhibiting glass transition temperatures (T_gs) of -2 to 116 °C, compressive moduli of 0.02-1.12 GPa and compressive strengths of 8-144 MPa. These materials were thermally stable below 300 °C and thermally degraded with a maximum degradation rate at 493-506 °C. The same group (*4–6*) also obtained a series of thermosetting materials via radical or cationic copolymerization of natural or conjugated soybean and linseed oils with vinyl monomers. Copolymers ranged from elastomers to rigid and tough plastics, depending on stoichiometry, type of vegetable oils and co-monomers. For examples, LSS45-St32-DVB15, which was prepared from 45 wt% low-saturation soybean oil (LLS), 32 wt% ST, 15 wt% DVB, and 8 wt% Norway fish oil (NFO) ethyl esters -modified boron trifluoride diethyl etherate (BFE) initiator (5 wt% NFO plus 3 wt% BFE), exhibited tensile strength of 6.0 MPa and breaking elongation of 64.1%; CLS45-ST32-DVB15, which was prepared from 45 wt% conjugated low-saturation soybean oil (CLS), 32 wt% ST, 15 wt% DVB, and same initiator, exhibited tensile strength of 11.5 MPa and breaking elongation of 40.5%.

Conjugated soybean oil has also been copolymerized with acrylonitrile and dicyclopentadiene (*7*). These polymeric materials have multiple T_gs distributed in the range of -56.7 to -43.6 °C, 1.1 to 1.5 °C, and 47.3 to72.7 °C.

Flame retardant materials can be obtained by cationic copolymerization of soybean oil with ST, DVB, and 4-trimethylsilylstyrene (*8*). The initial decomposition temperature (10% weight loss) of these flame retardant materials can reach up to 354 °C under nitrogen and 377 °C in air.

Epoxy Resin

Epoxy resins are pre-polymers or polymers containing epoxy groups. Epoxy resins prepared from epoxidized vegetable oils (EVOs) and fatty acids have been the most frequently studied in the research field of vegetable oil-based polymers.

There are several methods for vegetable oil epoxidation (Figure 3). The conventional method was *in situ* formation of peracids using strong acid as a catalyst (*9*). However, strong acids, such as sulfuric acid and nitric acid, give rise to side reactions and can present difficulties during removal from the product. To minimize such drawbacks, another method using ion exchange resin as catalyst was developed (*10*). It not only avoided use of strong acids but also facilitated high conversion of double bonds to epoxide (> 90%). Another method, using metal as catalysts has also been developed (*11*). However, due to the high cost of metal catalysts, this method was only used on the laboratory scale. At last, another method described was based on lipase-catalysis (*12*), which was environmentally friendly. As catalyst, lipase was very expensive and had limited applications.

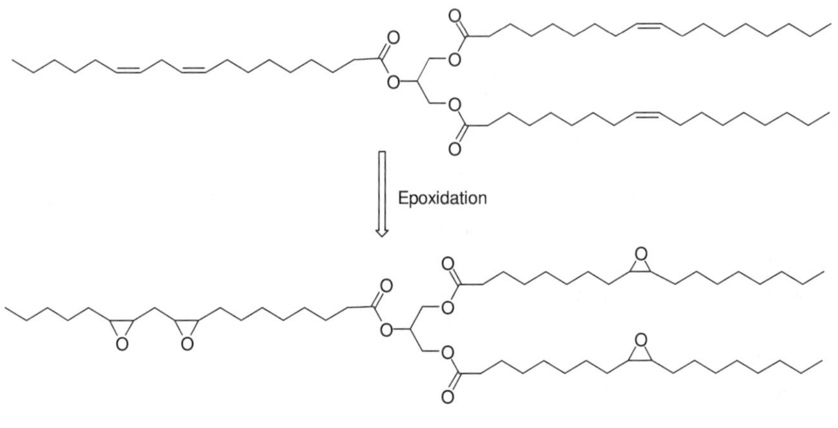

Epoxidation

Figure 3. Synthesis of epoxidized vegetable oils.

EVOs are important intermediators to produce vegetable oil derivatives, which will be discussed in detail. Besides, EVOs can be polymerized directly to form polyether in the presence of acid catalysts such as Lewis acids (*13*, *14*) or superacids (*15*, *16*). Epoxidized soybean oil (ESBO) has been polymerized in the presence of BFE. Polymerized epoxidized soybean oil (PESBO) were typically highly crosslinked networks and had T_g ranging from -6 to -48 °C. All ESBO polymers were thermally stable under the temperature of 220 °C (*14*). Hydrolysis of PESBO can lead to polyacids, which can be used to cure epoxy resins.

EVOs can be cured to form thermoset polymers with anhydrides using amine as catalyst (*17*, *18*). For example, cyclic acid anhydride cured ESBOs had T_g in the range of -16 to 65 °C (*18*). It was found that as the higher the anhydride/epoxy ratio and amount of epoxy groups were, so were the T_g and hardness. Moreover, ESBO can partially replace traditional petroleum-based epoxy, diglycidyl ether of bisphenol A (DGEBA) (*19*). With 40 wt % of ESBO replacement of DGEBA, impact strength of methyltetrahydrophthalic anhydride cured epoxy resin increased 38%.

When triglycol was replaced by sucrose, EVOs became epoxidized sucrose esters of fatty acids (ESEFA) which had a rigid core of sucrose and eight epoxidized fatty acid chains (*20*). Anhydride cured ESEFAs had T_g in the range of 48-104 °C, depending on epoxy amount and ratio of anhydride/epoxy. Tensile strength of anhydride cured ESEFAs (20-45 MPa) were several times that of ESBO-based epoxy resins (10 MPa).

Polyurethane (PU)

PU is a vital raw material for lightweight products that play important roles in our daily life, from footwear and skateboard to construction and automotive materials. Conventionally, PU is made via reaction of polyols and isocyanates. Polyether and polyester polyols currently predominate in the polyol market. However, vegetable oil-based polyols have attracted more and more interest. In the last two decades, scientists have developed polyols from various vegetable oils and many companies have started producing vegetable oil-based polyols (*21*).

As mentioned before, castor oil can be directly used as polyol since it contains hydroxyl groups. Elastomer of castor oil-based PU has two T_gs, -1.6 and 62.7 °C (*22*). It was found that castor oil-based PU with aliphatic diisocyanate had better thermal stability than aromatic diisocyanate (*23*). Decomposition of castor oil cured with isophorone diisocyanate (IPDI) in air began at 286 °C, while that with toluene diisocyanate (TDI) began at 261 °C. Under optimum synthesis conditions, elastomers of interpenetrating polymer networks (IPNs), (polyether-castor oil)PU/polyacrylonitrile (PAN) and (polybutadiene-castor oil)PU/polystyrene (PS), possessed tensile strength over 13 and 11 MPa and ultimate elongation over 240% and 270%, respectively (*24*).

Except for castor oil, other vegetable oils need two to three steps to be converted into polyols. Figure 4 shows four methods to obtain polyols from vegetable oils. Method 1, by reaction of epoxidized vegetable oils with a nucleophile, is the most popular choice (*25–28*). The polyols obtained by this method have lower reactivity because of their secondary hydroxyl groups. Method 2-4 produce polyols (*29–32*) with primary hydroxyl groups which have higher reactivity. It can also be seen in Figure 4 that polyols prepared from Methods 3 and 4 have terminal alcohols and lower molecular weights, while those from Methods 1 and 2 have dangling chains with eight or nine carbon atoms. Dangling chains would result in imperfections in the network structures, acting as plasticizers that reduce rigidity and improve flexibility of the polymer.

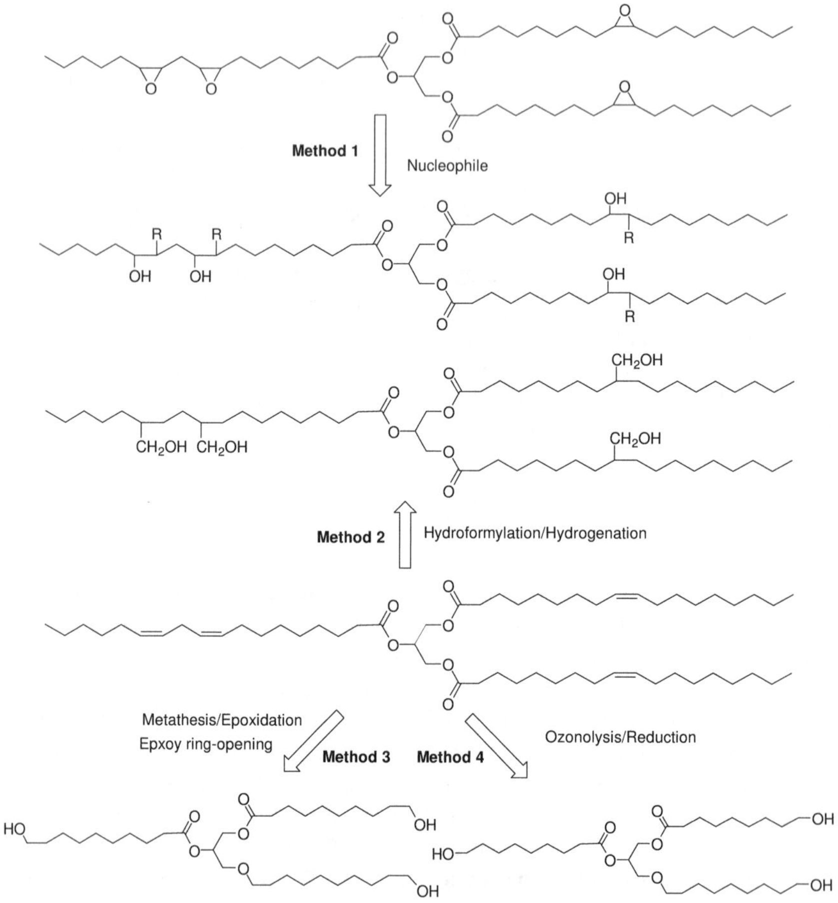

Figure 4. Vegetable oil-based polyols synthesis methods.

It is well known that the chemical structures of monomers play a vital effect on the properties of polymers. PU based on soybean polyols prepared from Method 1 exhibited strengths of 40-50 MPa (*33*). Petrovic et al. (*34*) studied the effect of NCO/OH molar ratio on PU based on soybean polyols prepared from Method 1 using methanol as the nucleophile. The resulting PUs were glassy with NCO/OH ratios of 1.05–0.8, while rubbery with NCO/OH ratios of 0.7-0.4. As NCO/OH ratio decreased from 1.05 to 0.4, density of PU decreased from 1.104 to 1.064 g/cm^3, tensile strength decreased from 47.3 to 0.3 MPa, while breaking elongation increased from 7% to 232%. Effect of different vegetable oils has also been studied (*35*). Low functionality oil-based polyol, such as mid-oleic sunflower with functionality of 3.74, resulted in PU with tensile strength of 14.8 MPa and breaking elongation of 168%. While linseed oil-based polyol, with functionality of 6.44, led to PU with tensile strength of 56.3 MPa and breaking elongation of 8%. Due to the larger crosslinking loop originating from the additional carbon introduced during hydroformylation, soybean polyols prepared from Method 2 resulted in PU with less rigidity and 20-30 °C lower T_g than that from Method 1 (*29*). When cured with methylene diphenyl diisocyanate (MDI), triolein-based polyols prepared from Method 1 gave product density of 1.061 g/cm^3 and T_g of 25 °C while that from Method 3 gave 1.156 g/cm^3 and 32 °C (*30*). It can be attributed to the former gives lower crosslink density and dangling chains' plasticizing effect. Triolein-based polyols prepared from Method 4 gave PU T_g of 53 °C, tensile strength of 51 MPa, and breaking elongation of 25% (*36*).

Polyricinoleate diol (Figure 5), which was synthesized via polycondensation of methyl ricinoleate with diethylene glycol as initiator, can be used as soft segment for PU (*37*). When containing 40% soft segment, PU had tensile strength of 26 MPa and breaking elongation of 188%; while containing 70% soft segment, these values decreased to 2.8 MPa and 67%.

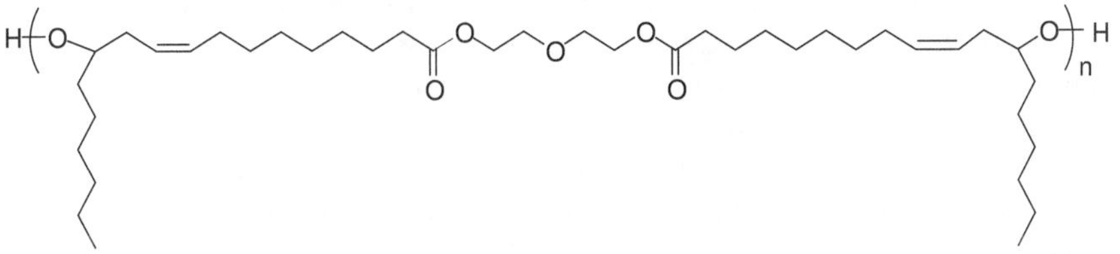

Figure 5. Polyricinoleate diol (37).

Fatty acid derived diisocyanates have also been synthesized. 1,7-Heptamethylene diisocyanate (HPMDI, Figure 6) was synthesized from oleic acid via ozonolysis, oxidation, and Curtius rearrangement (*38*). Fully vegetable oil-based materials were prepared using HPMDI to cure canola oil-based polyol prepared by ozonolysis (Method 4 in Figure 4), and have comparable properties to those prepared from petroleum-derived diisocyanate, 1,6-hexamethylenediisocyanate (HMDI). Cramail group (*39*) synthesized 1-isocyanato-10-[(isocyanatomethyl)thio]decane (DITD, Figure 6) from methyl-10-undecenoate (MUD). DITD cured propanediol PU was amorphous with T_g of -5.2 °C. DITD cured longer chain diols, such as hexanediol and dodecanediol, imparted semi-crystallin PU with specific thermal behaviors. A distinct T_g and several exothermic and endothermic peaks were found in differential scanning calorimetry (DSC) curves. Sulfur atoms in DITD might be responsible for these transitions since sulfur is known to enhance flexibility of backbones. Cramail group (*40*) also developed AB-type monomers composed of acyl-azide, hydroxyl, and methyl urethane functionalities, namely, 10-hydroxy-9-methoxyoctadecanoylazide, 9-hydroxy-10-methoxyoctadecanoylazide (HMODAz), 12-hydroxy-9-cis-octadecenoylazide (HODEAz), and methyl-N-11-hydroxy-9-cis-heptadecenecarbamate (MHHDC)

(Figure 6). HMODAz was prepared from methyl oleate, and the latter two from ricinoleic acid. HMODAz and HODEAz are acyl azide groups of AB-type monomer which can decompose to form in situ isocyanate as an intermediate to be condensed into polyurethane. MHHDC can form PU via the transurethane reaction approach.

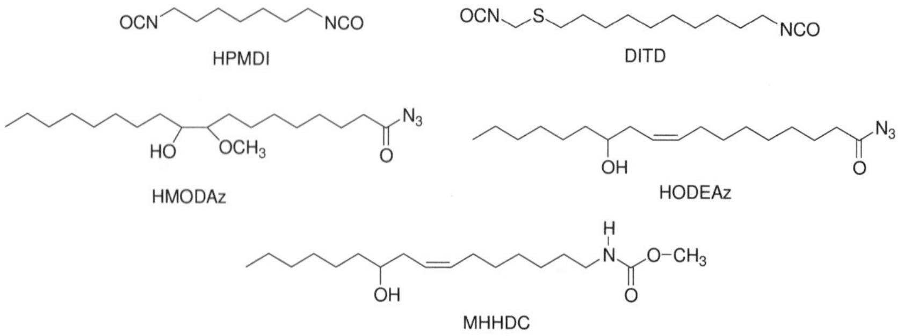

OCN NCO
HPMDI
OCN S NCO
DITD
N_3
HO OCH_3 O
HMODAz
N_3
OH O
HODEAz
H
N O–CH_3
OH O
MHHDC

Figure 6. Fatty acids derivatives for polyurethane synthesis (38–40).

Conventional polyurethane production methods involve isocyanates which are highly reactive and toxic, and commonly produced from an even more dangerous component, phosgene. New kinds of PUs, vegetable oils-based non-isosyanate PU (NIPU), have been developed to overcome the disadvantage of isocyanate chemicals (*41–46*). Javni et al. (*42*) prepared NIPU from carbonated soybean oil (CSBO) with amine. CSBO was prepared by carbonation of epoxidized soybean oil with carbon dioxide. While amines react with cyclic carbonates to form NIPU, they also react with the ester group and produce amides. When the molar ratio of amine to carbonate increased from 1 to 2, 1,6-hexamethylenediamine (HMDA) cured NIPU's T_g decreased from 22.5 to 15.5 °C, tensile strength from 3.80 to 2.67 MPa, while breaking elongation increased from 189 to 207%. It is because amidation decreased the network crosslinking density. Using rigid aromatic or cyclic structures diamines can increase strength and rigidity. NIPUs prepared with CSBO and *m*-xylylene diamine (*m*-XDA), *p*-xylylene diamine (*p*-XDA), and isophorone diamine displayed tensile strength up to 11.1 MPa, and elongation up to 433% (*43*).

Polyesters

Saturated polyesters are the most frequently used synthetic polymers in the manufacture of fibers and films. Unsaturated polyesters are thermosetting polymers, less flexible, and are widely used as casting materials. Three main methods are used to prepare polyesters: Transesterification of hydroxyalkanoates, ring-opening polymerization of lactones (cyclic esters), and polycondensation of dicarboxylic acids with diols or AB-type hydroxyacid. Hydroxylic and carboxylic derivatives are the main monomers for synthesis of polyesters. Since vegetable oils consist of aliphatic acidic esters, they could be good feedstock for hydroxylic and carboxylic derivatives.

Ricinoleic acid, a hydroxylic acid and a main hydrolysis product of castor oil, could be easily polymerized by polycondensation. Poly(ricinoleic acid) (PRA) was a viscous liquid at room temperature with a low T_g of -74.8 °C (*47*). PRA was readily cured using a dicumyl peroxide to produce a crosslinked polyricinoleate with a hardness of 50A using durometer A (*48*). Copolymerization of ricinoleic acid with sebacic or lactic acid increased products' melting points. Different copolymerization methods gave final products with different thermal behaviors (*49–52*). With 40 wt% of RA, copolymer P(LA-RA) synthesized by transesterification followed by condensation repolymerization has a melting temperature at 93 °C, while that by ring-opening polymerization has 140 °C (*49*).

Petrovic et al. (*53*) synthesized 9-hydoxynonanoicacid (HNME) from castor oil via ozonolysis, reduction, and methanolysis. The obtained HNME was used to prepare high molecular weight polyesters (PHNA) via transesterification. Properties of PHNA were strongly dependent on its molecular weight. PHNA with low molecular weight was brittle with low mechanical strength, while that with high molecular weight showed high crystallinity and melting point,and could be spun into fibers of high toughness, strength and elongation. Highly crystalline PHNA had a melting temperature of 70.5 °C, average breaking stress of 9.1 MPa, breaking elongation of 221%, yield stress of 13.9 MPa, yield elongation of 9.7 %, and Young modulus of 172 MPa. Using longer chain length acid, 12-hydroxydodecanoic acid, as starting material increased the melting temperature of polyester to 88 °C (*54*). This hydroxyacid was developed from vernonia oil via oxidative cleavage. Poly(ω-pentadecalactone) (PPDL) was prepared from cyclic 15-carbon lactone (*55*). Increasing the molecular weight (M_w) from 45 to 81 kg/mol, PPDL samples showed a brittle-to-ductile transition. For PPDL with M_w of 81 kg/mol, inter-fibrillar slippage dominated during deformation. PPL with M_w over 189 kg/mol exhibited tough properties with breaking elongation of about 650% and tensile strength at about 60.8 MPa. Strain-hardening phenomenon existed as a result of enhanced entanglement of the molecular network. The mechanical properties of PPDL were comparable to high density linear polyethylene (HDLPE).

A set of renewable monomers (Figure 7) have been developed via thiol-ene additions of methyl 10-undecenoate or 10-undecenol, derivatives of castor oil (*56*). Linear as well as hyperbranched polyesters were obtained via polymerization of the resulted monomers using triazabicyclodecene (TBD) as a catalyst. These polymers also had thio-ether linkages and melting points from 50 to 71 °C. Moreover, they were stable under temperature up to 300 °C. In summary, longer chain length between ester linkages gives polyesters higher crystallinity and higher melting points.

Polyethylene (PE) is the most common polymeric material with melting point in the range of 120-180 °C. To get polyesters with comparable melting point of polyethylene (PE), a series of long-chain α, ω-diene and α, ω-diester monomers (Figure 8) were developed from fatty acids by transesterification and metathesis, respectively (*57–60*). Polyesters with good thermo-mechanical properties were obtained via polymerization of these monomers through acyclic diene metathesis (ADMET) or tansesterification. However, melting points are lower than 100 °C and much lower than that of expected PE. Longer α, ω-dicarboxylic acid and α, ω-diol, consisting of 12-carbon chain (Figure 9), were synthesized via metathesis, hydrogenation, and reduction (*61*). Polycondensation of these monomers resulted in polyesters with melting point of 108 °C.

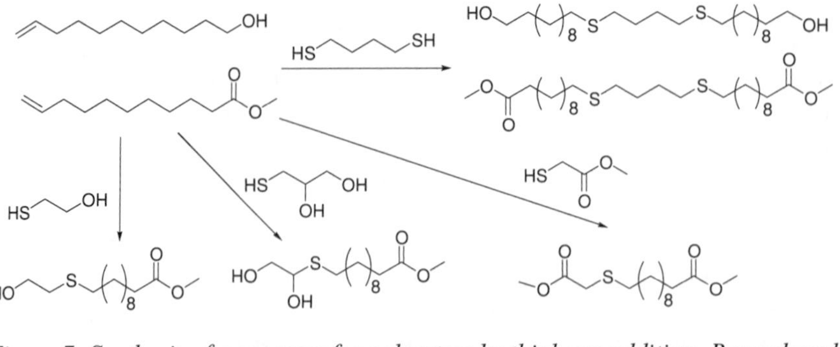

Figure 7. Synthesis of monomers for polyesters by thiol-ene addition. Reproduced with permission from reference (56). Copyright (2010) John Wiley and Sons.

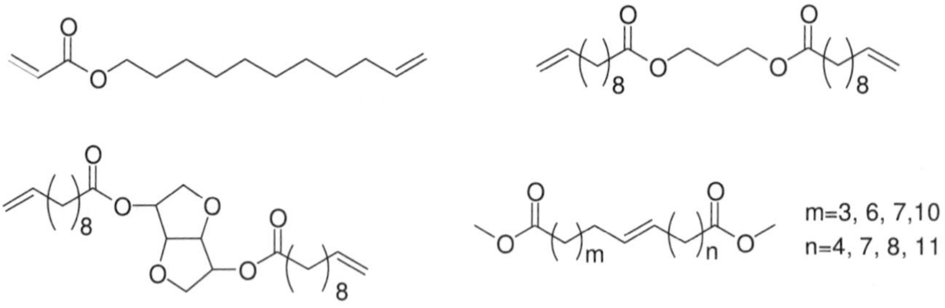

Figure 8. α, ω-Diene and α, ω-diester monomers for polyester synthesis (57–60).

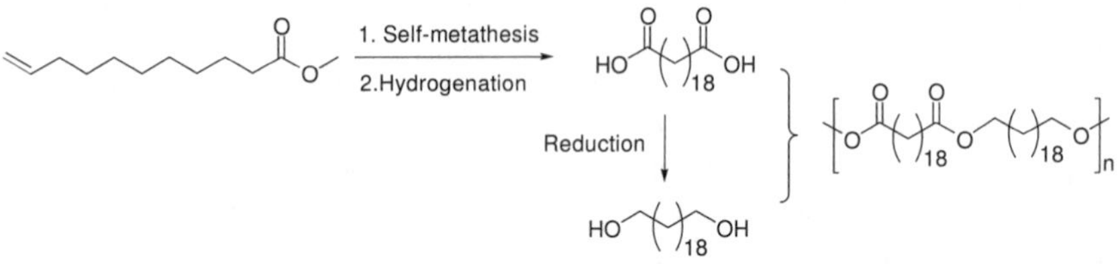

Figure 9. Synthesis of polyester from α, ω-dicarboxylic acid and α, ω-diol. Reproduced with permission from reference (61). Copyright (2011) John Wiley and Sons.

To fine-tune the end-properties of the polymers, Cramail group (*62–64*) designed several α, ω-diols which composed of linear or cyclic central blocks with one or two ester linkages, one or two amide linkages, or both ester and amide linkages (Figure 10). The diols were synthesized via a two-step procedure. First, transesterification/amidation or transamide-esterification occurred on renewable diols (1,3-propanediol or isosorbide), aminoalcohol (1,3-aminopropanol) or diamine (1,4-diaminobutane) to obtain bis-unsaturated diesters, monoester, esteramide or diamide, respectively, with different central blocks. Subsequently, thiol-ene addition was carried out and 2-mercaptoethanol MCET was added on to the terminal double bonds, leading to formation of dihydroxy-telechelic compounds. These diols were copolymerized with a C20 dimethyl ester (shown in Figure 8) by polycondensation. New vegetable oil-based polyesters and poly(ester-amide)s were obtained. M_w of the polymers ranged from 6 to 19 kg/mol. These polymers were stable up to 325°C which was comparable to the thermal stabilities of petroleum-based aliphatic polyesters and poly(ester-amide)s. All polymers were semi-crystalline with melting points varying from 30 to 125°C. When two amides replaced two ester linkages in the monomer, polymers' ultimate strength increased from 2.9 to 10.0 MPa, with quite similar breaking elongations at 4%. The high breaking strength was due to hydrogen bonds formed by amides.

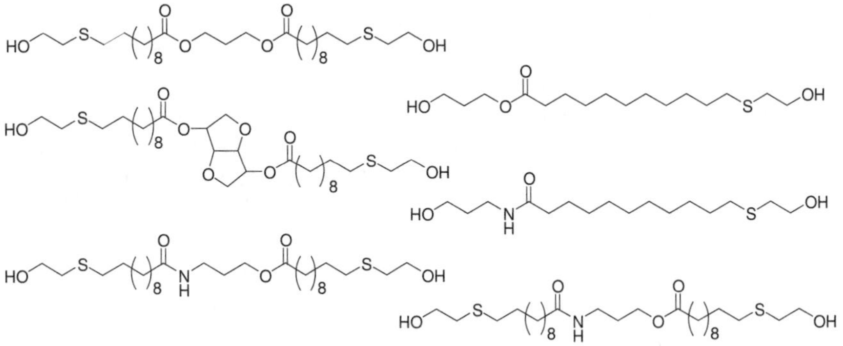

Figure 10. Chemical structure of α, ω-diols for polyester synthesis (62–64).

Polymers Prepared from Other Functionalized Vegetable Oils

As a polyol synthesis method, using a nucleophile with functional groups to open epoxy ring of epoxidized oil or fatty acids can form new functionalized vegetable oils, which can be further homopolymerized or copolymerized to obtain highly cross-linked materials. Acrylated ESBO (Ac-ESBO) and allylated ESBO (Al-ESBO) were obtained via oxirane ring-opening with acrylic acid and allyl alcohol, respectively.

Ac-ESBO was copolymerized with styrene (*65, 66*). With the styrene content increasing from 0 to 40 wt%, the resulting T_g, modulus, and tensile strength of thermosets increased linearly from 40 to 75 °C, 0.44 to1.6 GPa and 6 to 21 MPa, respectively. After hydroxyl groups on Ac-ESBO reacted with maleic anhydride (MA), copolymers of resulted monomer and styrene showed improved thermal and mechanical properties. The T_g ranged from 100 to 115 °C, while storage moduli ranged from 1.9 to 2.2 GPa. It was because the carboxylic acid groups, generated in the reaction between hydroxyl and MA, can react with residual epoxy groups or hydroxy groups on the triglyceride and increase cross-link densities.

Al-ESBO has been copolymerized with maleic anhydride (MA) by free radical polymerization. Polymers had different cross-linking densities depending on the amount of MA and radical initator. T_gs were in the range of 17.9 to 123 °C (*67*). When MA loading was 30 wt%, the resin had high gel content (99%), low water absorption (~ 1%), and low swelling ratio in toluene (~4%), and possessed modulus of 1080 MPa and tensile strength of 29 MPa. When AESBO was transesterified by allyl alcohol, allylated fatty acids (AE-ESBO) were obtained (*68*). At the same MA loading, resins derived from AE-ESBO showed higher mechanical strength than those derived from AESBO because AE-ESBO contains higher double bond content and results in higher cross-linking density. When MA loading was 30 wt%, modulus was of 1452 MPa, and tensile strength was of 37.3 MPa. Resins densities were around 1.1 g/cm^3.

Azidated oils and alkynated oils can be obtained by using NaN_3 and propargyl alcohol to open epoxy rings, respectively. They can undergo thermal [3+2] Huisgen cycloaddition without any solvent and catalyst, a kind of green "click" chemistry (*69–72*). The reaction of azidated oils with alkynated oil produces relatively soft polymer networks. For example, polymer derived from azidated and alkynated soybean oils showed a low tensile strength of 1.34 MPa. However, when short chain diynes were used as cross-linkers, both thermal and mechanical properties improved. Moreover, aromatic diyne (1,4-diethynylbenzene) as across-linker produced polymers with higher T_g and strength than aliphatic diyne (1,7-octadiyne) did. Azidated soybean oil reacted with 1,4-diethynylbenzene gave product tensile strength of 32.1 MPa, 97 °C, while 1,7-octadiyne gave tensile strength of 11.6 MPa and T_g of 27 °C. Densities of these polymers increase slightly with increas of cross-linking densities and are in the range of 1.09-1.17 g/cm^3.

Conclusions

In conclusion, vegetable oils and their derivatives are promising raw materials for lightweight products, owing to their multi-functionality, abundant availability, sustainability and biodegradability. Depending on oil type, stoichiometry, and co-monomers, a wide range of polymers can be developed from vegetable oils and their derivatives, from linear to highly cross-linked, flexible to rigid, thermoplastic to thermosetting, and elastomers to hard plastics. Moreover, these vegetable oil-based materials possess properties comparable to petroleum-based materials.

It needs to be pointed out that even for the same type of vegetable oil, their fatty acids profile varies depending on the local weather, soil, and planting conditions. As a result, products from the same type of oil and the same production process may have slightly different properties. For this reason, it is still a challenge for large-scale production of vegetable oil-based materials in industry. Currently, only epoxidized oils and polyols are produced on an industrial scale. Large-scale production of high-purity fatty esters and acids from vegetable oils is the key step to develop and produce vegetable oil-based lightweight materials.

References

1. Gunstone, F. D.; Norris, F. A. *Lipids in Foods Chemistry, Biochemistry and Technology*; Pergamon Press: Elmsford, NY, 1983; pp 1–14; ISBN: 0-08-025499-3.
2. Maisonneuve, L.; Lebarbe, T.; Grau, E.; Cramail, H. *Polym. Chem.* **2013**, *4*, 5472–5517.
3. Li, F.; Larock, R. C. *Biomacromolecules* **2003**, *4*, 1018–1025.
4. Li, F.; Larock, R. C. *J. Polym. Sci., Part B : Polym. Phys.* **2001**, *39*, 60–77.
5. Li, F.; Larock, R. C. *J. Polym. Environ.* **2002**, *10*, 59–67.
6. Kundu, P. P.; Larock, R. C. *Biomacromolecules* **2005**, *6*, 797–806.
7. Yang, L.; Dai, C.; Ma, L.; Lin, S. *J. Polym Environ.* **2011**, *19*, 189–195.
8. Sacristán, M.; Ronda, J. C.; Galià, M.; Cádiz, V. *Biomacromolecules* **2009**, *10*, 2678–2685.
9. Klass, M. R. G.; Warwel, S. In *Recent Developments in the Synthesis of Fatty Acid Derivatives*; Knothe, G., Derksen, J. T. P., Eds.; AOCS Press: Champaign, IL, 1999; pp 157–181; ISBN: 978-1-893997-00-4.
10. Petrović, Z. S.; Zlatanic, A.; Lava, C. C.; Sinadinović-Fišer, S. *Eur. J. Lipid Sci. Technol.* **2002**, *104*, 293–299.
11. Gerbase, A. E.; Gregorio, J. R.; Martine, M.; Brasil, M. C.; Mendes, A. N. F. *J. Am. Oil Chem. Soc.* **2002**, *79*, 179–181.
12. Vleck T., ; Petrovic, S. Z. *J. Am. Oil Chem. Soc.* **2006**, *83*, 247–252.
13. Biresaw, G.; Liu, Z.; Erhan, S. Z. *J. Appl. Polym. Sci.* **2008**, *108*, 1976–1985.
14. Liu, Z.; Erhan, S. Z. *J. Am. Oil Chem. Soc.* **2010**, *87*, 437–444.
15. Kawakami, Y.; Ogawa, A.; Yamashita, Y. *J. Polym. Sci., Part A: Polym. Chem.* **1979**, *17*, 3785–3792.
16. Lligadas, G.; Ronda, J. C.; Galià, M.; Biermann, U.; Metzger, J. O. *J. Polym. Sci., Part A: Polym. Chem.* **2006**, *44*, 634–645.
17. Rosch, J.; Mulhaupt, R. *Polym. Bull.* **1993**, *31*, 697–685.
18. Gerbase, A. E.; Petzhold, C. L.; Costa, A. P. O. *J. Am. Oil Chem. Soc.* **2002**, *9*, 797–802.
19. Altuna, F. I.; Esposito, L. H.; Ruseckaite, R. A.; Stefani, P. M. *J. Appl. Polym. Sci.* **2011**, *120*, 789–798.
20. Pan, X.; Sengupta, P.; Webster, D. C. *Biomacromolecules* **2011**, *12*, 2416–2428.
21. Desroches, M.; Escouvois, M.; Auvergen, R.; Caillol, S.; Boutevin, B. *Polym. Rev.* **2012**, *52*, 38–79.
22. Mothe, C. G.; Araujo, C. R. *Thermochim. Acta* **2000**, *357-358*, 321–325.
23. Ristić, I. S.; Bjelović, Z. D.; Holló, B.; Szécsényi, K. M.; Budinski-Simendić, J.; Lazić, N.; Kićanović, M. *J. Therm. Anal. Calorim.* **2013**, *111*, 1083–1091.
24. Xie, H. Q.; Guo, J. S. *Eur. Polym. J.* **2002**, *38*, 2271–2277.
25. Petrovic, Z.; Guo, A.; Javni, I. U.S. Patent 6,107,433, 2000.
26. Petrovic, Z.; Javni, I.; Guo, A.; Zhang, W. U.S. Patent 6,433,121, 2002.
27. Petrovic, Z.; Javni, I.; Zlatanic, A.; Guo, A.; U.S. Patent 8,153,746, 2012.
28. Guo, A.; Cho, Y.; Petrovic, Z. *J. Polym Sci., Part A: Polym. Chem.* **2000**, *38*, 3900–3910.

29. Guo, A.; Zhang, W.; Petrovic, Z. *J. Mater. Sci.* **2006**, *41*, 4914–4920.
30. Zlatanic, A.; Petrovic, Z. S.; Dusek, K. *Biomacromolecules* **2002**, *3*, 1048–1056.
31. Cvertkovic, I.; Milic, J.; Ionescu, M.; Petrovic, Z. S. *Hem. Ind.* **2008**, *62*, 319–328.
32. Tran, P.; Graiver, D.; Narayan, R. *J. Am. Oil Chem. Soc.* **2005**, *82*, 653–659.
33. Petrovic, Z.; Guo, A.; Zhang, W. *J. Polym Sci., Part A: Polym. Chem.* **2000**, *38*, 4062–4069.
34. Petrovic, Z. S.; Zhang, W.; Zlatanic, A.; Lava, C. C.; Ilavsky, M. *J. Polym. Environ.* **2002**, *10*, 5–12.
35. Zlatanic, A.; Lava, C.; Zhang, W.; Petrovic, Z. S. *J. Polym Sci., Part B: Polym. Phys.* **2004**, *42*, 809–819.
36. Petrovic, Z. S.; Zhang, W.; Javni, I. *Biomacromolecules* **2005**, *6*, 713–719.
37. Xu, Y.; Petrovic, Z.; Das, S.; Wilkes, G. *Polymer* **2008**, *49*, 4248–4258.
38. Hojabri, L.; Kong, X.; Narine, S. S. *Biomacromolecules* **2009**, *10*, 884–891.
39. More, A. S.; Labarbe, T.; Maisonneuve, L.; Gadenne, B.; Alfos, C.; Cramail, H. *Eur. Polym. J.* **2013**, *49*, 823–833.
40. Palaskar, D. V.; Boyer, A.; Cloutet, E.; Alfos, C.; Cramail, H. *Biomacromolecules* **2010**, *11*, 1202–1211.
41. Tamami, B.; Sohn, S.; Wikes, G. L. *J. Appl. Polym. Sci.* **2004**, *92*, 883–891.
42. Javni, I.; Hong, D. P.; Petrovic, Z. S. *J. Appl. Polym. Sci.* **2008**, *108*, 3867–3875.
43. Javni, I.; Hong, D. P.; Petrovic, Z. S. *J. Appl. Polym. Sci.* **2013**, *128*, 566–571.
44. Bähr, M.; Mülhaupt, R. *Green Chem.* **2012**, *14*, 483–489.
45. Boyer, A.; Cloutet, E.; Tassaing, T.; Gadenne, B.; Alfos, C.; Cramail, H. *J. Green Chem.* **2010**, *12*, 2205–2213.
46. Maisonneuve, L.; More, A. S.; Foltran, S.; Alfos, C.; Robert, F.; Landais, Y.; Tassaing, T.; Grau, E.; Cramail, H. *RSC Adv.* **2014**, *4*, 25795–25803.
47. Kelly, A. R.; Hayes, D. G. *J. Appl. Polym. Sci.* **2006**, *101*, 1646–1656.
48. Ebata, H.; Toshima, K.; Matsumura, S. *Macromol. Biosci.* **2007**, *7*, 798–803..
49. Slivniak, R.; Domb, A. J. *Macromolecules* **2005**, *38*, 5545–5553.
50. Slivniak, R.; Domb, A. J. *Macromolecules* **2005**, *6*, 1679–1688.
51. Slivniak, R.; Domb, A. J. *Biomacromolecules* **2006**, *7*, 288–296.
52. Lebarbe, T.; Ibarboure, E.; Gadenne, B.; Alfos, C.; Cramail, H. *Polym. Chem.* **2013**, *4*, 3357–3369.
53. Petrovic, Z.; Milic, J.; Xu, Y.; Cvertkovic, I. *Macromolecules* **2010**, *43*, 4120–4125.
54. Ebata, H.; Toshima, K.; Matsumura, S. *Macromol. Biosci.* **2008**, *8*, 38–45.
55. Cai, J.; Liu, C.; Cai, M.; Zhu, J.; Zuo, F.; Hsiao, B. S.; Gross, R. A. *Polymer* **2010**, *51*, 1088–1099.
56. Turunc, O.; Meier, M. A. R. *Macrom. Rapid Commun.* **2010**, *31*, 1822–1826.
57. Warwel, S.; Tillack, J.; Demes, C.; Kunz, M. *Macromol. Chem. Phys.* **2001**, *202*, 1114–1121.
58. Montero de Espinosa, L.; Meier, M. A. R. *Chem. Commun* **2011**, *47*, 1908–1910.

59. Fokou, P. A.; Meier, M. A. R. *J. Am. Chem. Soc.* **2009**, *131*, 1664–1665.
60. Montero de Espinosa, L.; Meier, M. A. R. *Eur. Polym. J* **2011**, *47*, 837–852.
61. Trzaskowski, J.; Quinzler, D.; Bährle, C.; Mecking, S. *Macromol. Rapid Commun.* **2011**, *32*, 1352–1356.
62. Palaskar, D. V.; Boyer, A.; Cloutet, E.; Meins, J. L.; Gadenne, B.; Alfos, C.; Farcet, C.; Cramail, H. *J. Polym Sci., Part A: Polym. Chem.* **2012**, *50*, 1766–1782.
63. Maisonneuve, L.; Lebarbe, T.; Nguyen, T. H.; Cloutet, E.; Gadenne, B.; Alfos, C.; Cramail, H. *Polym. Chem.* **2012**, *3*, 2583–2595.
64. Lebarbe, T.; Maisonneuve, L.; Nguyen, T. H.; Gadenne, B.; Alfos, C.; Cramail, H. *Polym. Chem.* **2012**, *3*, 2842–2851.
65. Lu, J.; Khot, S.; Wool, R. P. *Polymer* **2005**, *46*, 71–80.
66. Wool, R.; Kusefoglu, S.; Palmese, G.; Khot, S. Zhao, R. U.S. Patent 6,121,398, 2000.
67. Luo, Q.; Liu, M.; Xu, Y.; Ionescu, M.; Petrovic, Z. S. *Macromolecules* **2011**, *44*, 7149–7157.
68. Luo, Q.; Liu, M.; Xu, Y.; Ionescu, M.; Petrovic, Z. S. *J. Appl. Polym. Sci.* **2013**, *127*, 432–438.
69. Hong, J.; Luo, Q.; Shah, B. K. *Biomacromolecules* **2010**, *11*, 2960–2965.
70. Hong, J.; Luo, Q.; Shah, B. K. *Biomacromolecules* **2010**, *11*, 2960–2965.
71. Hong, J.; Luo, Q.; Wan, X.; Petrovic, Z. S.; Shah, B. K. *Biomacromolecules* **2012**, *13*, 261–266.
72. Hong, J.; Shah, B. K.; Petrovic, Z. S. *Eur. J. Lipid Sci. Technol.* **2013**, *115*, 55–60.

Chapter 5

Environmentally Benign Pretreatments for Producing Microfibrillated Cellulose Fibers from Hemp

Renuka Dhandapani[1] and Suraj Sharma[*,1]

[1]Department of Textiles, Merchandising and Interiors, University of Georgia, Athens, Georgia 30602, United States
[*]Tel.: +1 706 542 7353. Fax: +1 706 542 4890.
E-mail: ssharma@uga.edu.

Nanocellulose has been receiving significant attention because of the emerging interest in nanomaterials and nanotechnology for various applications, such as biomedical and biological uses. Nanocellulose fibrils can be isolated from the most abundant biopolymer using a combination of chemical, enzymatic, and mechanical treatments. The harsh chemical treatments to isolate the fibers negatively impacted both fiber properties and environment. The main purpose of this study, therefore, was to use an environmentally friendly approach to produce nano-scale cellulosic fibers. We processed hemp fibers (~78% cellulose) using enzymes combined with a mechanical treatment to produce microfibrillated cellulose (MFC) with an average diameter of 29.5 nm. We selected relatively environmentally benign conditions to obtain nanocellulose. We used enzymes with specificity to pectin, hemicellulose and cellulose during pretreatment and then break down the pretreated fibers to nanocellulose via homogenization. After each treatment, we analyzed the samples using scanning electron microscopy (SEM), differential scanning calorimetry (DSC), and thermogravimetric analysis (TGA). Effective combinations of enzyme and mechanical treatments required to obtain nanocellulose are described in this paper.

Introduction

Cellulose, a polysaccharide, is composed of β-D-glucopyranose linked via β-1-4 linkages. It is abundantly available from various sources, including wood, cotton, plant-based materials, marine animals, bacteria, and industrial waste. Due to its good mechanical properties, cellulose is used as reinforcing material in biocomposites. By adding nanoscale cellulosic fibers as reinforcing material, researchers have obtained an increase in the strength and stiffness of the material and a decrease in the final product weight; the obtained composites were biodegradable and renewable in nature. The cellulose obtained in its native form is referred to as cellulose I, whereas that obtained in regenerated form is cellulose II. In native cellulose, two different crystalline forms of cellulose exist, cellulose I_{α}, obtained primarily from algae and bacterial cellulose, and cellulose I_{β}, obtained mainly from plants (*1*).

Fibers from plant-based sources are composed of cellulose, hemicellulose, lignin, pectin, pigments, and extractives. Cellulose presents as rigid microfibrils enclosed in a protective amorphous layer of lignin-hemicellulose matrix. Hemp fibers mainly consist of cellulose (53%–72%), hemicellulose (7%–19%), pectin (4%–8%), lignin (2%–5%), ash (4%), and other water-soluble materials (2%) (*2*). Microfibrils comprise both amorphous and crystalline regions arranged in an intermittent fashion. Nanocellulose, or microfibrillated cellulose (MFC), is produced by breaking down this structural group of cellulose into individualized bundles of nanofibrils, as shown in the schematics in Figure 1. Biocomposites produced using nanocellulose as filler have potential application in areas such as electronic and electrical, paper, biomedicine, cosmetics, packaging, textile, construction, and building material industries (*3*).

Nanoscale cellulose fibers are classified into two groups: cellulose nanocrystals (CNXL) and microfibrillated cellulose (MFC). Microfibrillated, or microcrystalline cellulose (MCC), fibers are obtained by mechanical shearing of cellulose fibers. They do not display a regular surface due to the presence of a non-crystalline, amorphous structure in a random, spaghetti-like arrangement, along with a crystalline domain. Cellulose nanocrystals, or "whiskers," are pure crystalline cellulose domains extracted by acid digestion of the amorphous regions from the microfibrils. The microfibrils have lower density than nanocrystals due to presence of amorphous regions which is susceptible to chemical modifications, swelling and adsorption phenomenon (*4*).

When cellulose nanocrystals are extracted with acid hydrolysis, amorphous regions of the fibrils are cleaved perpendicular to the cellulose chain. Cellulose whiskers obtained this way, have typical diameters ranging from 8 to 20 nm, and varied length depends on the starting material. When extracted from wood, length of cellulose whiskers is in the range of 180–200 nm; from cotton it is in the range of 100–120 nm; and from tunicates, it is more than 1000 nm. Strong hydrogen bonding occurs easily between individual cellulose nanocrystals, resulting in aggregation of nanocrystals, and thus increasing the width of the nanocrystals by 1–100 nm (*1*, *5*, *6*). This bonding process is called hornification, whereby tight cohesion occurs among nanocrystals, resulting in saturation of hydroxyl groups. Hornification causes irreversible reduction in size and volume of the pores present on the surface. When in suspension, the free hydroxyl groups are bound by water molecules, but removal of water aids hydrogen bond formation between hydroxyl groups, resulting in a tight packing of the structure (*7–9*). Cellulose nanocrystals are separated from the starting raw material using the chemical, mechanical, or enzymatic approach.

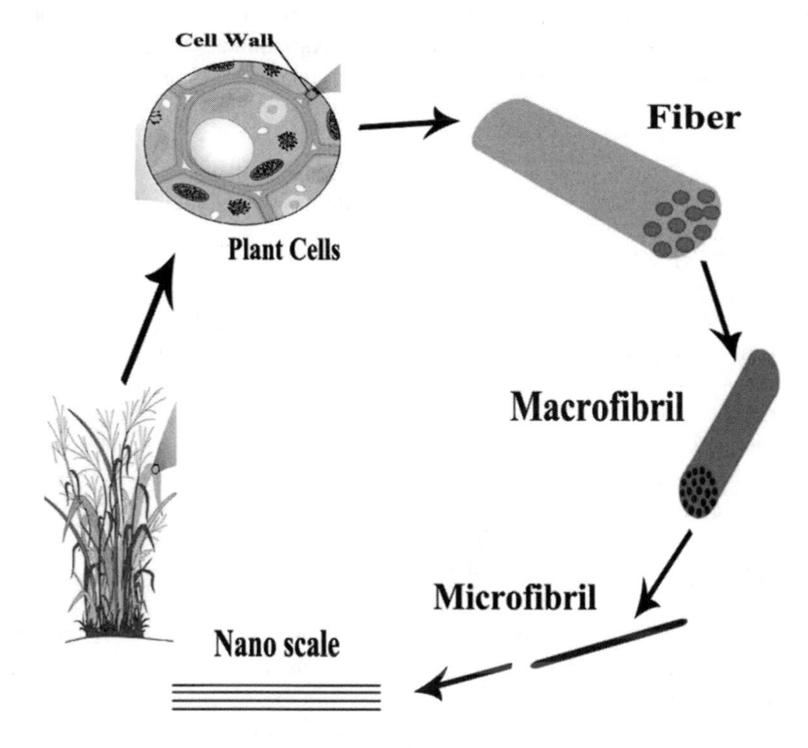

Figure 1. Schematic diagram of nanoscale fibrils from starting material plant to its final elementary state.

Chemical Approach

Acid hydrolysis using mineral acids, such as sulfuric acid, hydrochloric acid, and phosphoric acid, is employed to remove amorphous regions, leaving behind highly crystalline whiskers. Treatment using sulfuric acid produces a negatively charged surface on the cellulose whiskers, resulting in a more stable whisker than those produced by the other acids. Important factors, such as treatment time, temperature, and acid concentration, affect the morphology of nanocellulose. In other studies to isolate cellulose whiskers, researchers treated raw materials such as microcrystalline cellulose (MCC) and juvenile poplar with sulfuric acid in concentrations of 60% (v/v) and 63.5% (v/v), respectively, for 2 h, followed by neutralization and centrifugation (*3*, *10*).

Alkaline treatment, or mercerization, is a process of soaking cellulose fibers in highly concentrated (12%–17.5%) sodium hydroxide solution at room temperature for four minutes. This treatment helps to increase the surface area of lignocellulosic material and makes polysaccharides more susceptible to hydrolysis as the chemical molecules are able to penetrate through the protective lignin-hemicellulose matrix. This increase in access occurs because some of the lignin, most of the hemicellulose, and most of the pectin in the lignocellulosic material is removed by the alkaline process (*11–15*). After alkaline treatment, the remaining lignin in the lignocellulosic material is removed through a bleaching process, using chlorine- or non-chlorine-based compounds. In this process, the chlorine rapidly oxidizes lignin, leading to lignin degradation and solubilization (*16*).

Mechanical Approach

Mechanical treatments apply shear force to separate the bundle of microfibrils present in the cell walls (*15*, *17*). In cryocrushing, the fibers are frozen in liquid nitrogen and manually crushed using a mortar and pestle. By using liquid nitrogen, the water molecules present within the fibers are frozen to form ice crystals surrounding the microfibrils. These ice crystals are then ruptured by the crushing action of the pestle that breaks apart the fibers, releasing the microfibers from the enclosed cell wall. The cryocrushed fibers are then blended in a disintegrator or blender for the selected period to obtain uniformly distributed fibers suspended in water (*3*, *11*, *16*, *17*). The cryocrushed fibers, in concentrations of 1%–2% in suspension, are then passed through a homogenizer that allows the fibers to pass through a small opening under very high pressure. The combined action of high pressure, high velocity, high energy due to particle collision, and heat produced within the suspension generates shearing force that reduces the fiber material to nanoscale. The microfibrils have to pass through the homogenizer 10–20 times before nano-dimensional fibers can be obtained. As the fibers cycle through the homogenizer, the suspension viscosity increases; at the same time, the temperature in the homogenizer increases from room temperature to 70°–80 °C. After homogenization, the fibers in suspension remain stable. The homogenization step can, therefore, be designated as a main requisite in nanoscale fiber production (*3*, *16–18*).

Enzymatic Approach

The starting raw material for nanoscale fibers consists of lignin, hemicellulose, and pectin, as well as cellulose. To obtain pure cellulosic material, enzymatic treatment would be an ideal approach as the enzymes attack specific components in the substrate through selective hydrolysis (*19–21*). Researchers have observed an increase in crystallinity, thermal stability, and the amount of OH groups exposed on the fiber surface due to the removal of cellulosic and non-cellulosic components by various enzymatic treatments (*22*). These can be degraded in the presence of microorganisms, such as fungi, bacteria, and genetically modified enzymes. Enzymes used in hydrolysis include xylanase, pectinase, cellulase, and laccase. Enzymes have various advantages compared to chemicals used for the same purpose under similar conditions, because of the low concentrations of enzyme, low power consumption, and low amount of pollutants released after a treatment (*23*, *24*). However, an enzyme is only fully effective in optimum pH and temperature conditions; any deviation from these conditions instantly reduces the enzyme's activity and further deactivates the enzyme. Endoglucanase cellulase enzymes specifically attack the amorphous regions in cellulose substrate. Researchers have found that, using cellulase enzymes, they can easily separate microfibrils from the fiber substrate and, when combined with mechanical shearing, the treatment effectively reduces the size of cellulose fibril (*6*, *25–27*). Moreover, another advantage of combining treatments is that the number of passes through the homogenizer is considerably reduced, resulting in lower energy consumption (*18*, *26*).

Nanocellulose are finding limited applications despite their vast application possibilities due to the high energy demands of the complex process of converting plant fibers into corresponding small elementary entities (*1*) and environmentally unfriendly pretreatments. The objective of this study was to use environmentally benign pretreatment processes in combination with mechanical delamination to produce microfibrillated cellulose.

Materials and Experimental

Materials

Raw hemp fiber used in this study was obtained from the University of Mississippi, and the cross-section of the fiber is given in Figure 2. The cross sectional structure seen in the Figure is characteristic feature of a hemp fiber. Filter paper (Whatman #1) was purchased from Sigma Aldrich to represent pure cellulose. Cellulose powder (310697) from cotton linter was purchased from Sigma Aldrich to represent microcrystalline structure. The different enzymes used in this study included cellulase (C2730) purchased from Sigma Aldrich, and xylanase (Pulpzyme® HC) and pectinase (Scourzyme™ L) obtained from Novozymes.

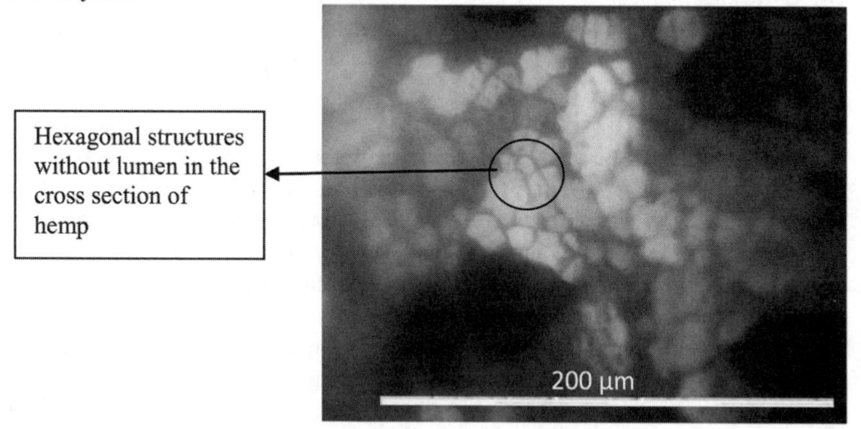

Figure 2. Cross sectional view of hemp fiber.

Preparation of Microfibrillated Cellulose

A Wiley Mill was used to initially cut the long hemp fibers to ensure they could easily pass through a mesh of size 10 (less than 2 mm). Ten grams of fibers were impregnated in 4% sodium hydroxide solution under constant stirring for 2 h at room temperature. At the end of the impregnation period, the fibers were irradiated using microwave for 5 min at 50% power. Microwave irradiation has been widely used as an energy source to speed up chemical reactions (*19*, *28*). The sodium hydroxide-treated fibers were then separated from the supernatant via vacuum suction, followed by multiple washings. The next step in the treatment was pressure cooking (a Cuisinart electric pressure cooker; EPC 1200 pc) using a material to liquor ratio of 1:40 at a pressure of 10 psi for 10 min followed by instant release of pressure. The purpose of using a pressure cooker was to solubilize the hemicellulose and modify or dissociate lignin to increase the accessibility to cellulose in later processing treatments, such as the enzymatic treatments. After pressure-cooking, we blended the fibers in a Waring blender for 20 min to obtain uniform pulp, followed by vacuum suction to separate supernatant from the fibers. The rationale of blending was to further break down (or open) the complex cell structure, since we used a lower concentration of sodium hydroxide. The blended fibers were then enzymatically treated with different enzymes (material/liquor 1:40): pectinase, xylanase or hemicellulase, and cellulase. The enzymatic treatments were carried out in a shaking water bath at 180 rpm, using three enzymes: pectinase enzyme (Scourzyme™ L), xylanase enzyme (Pulpzyme® HC) and cellulase enzyme at material-to-liquor of 1:40.

Here, we describe different parameters in the enzymatic treatments. We used a boric acid–borax buffer at pH 8 in the pectinase and xylanase enzyme treatments along with 10 mM ethylenediamine-tetra-acetic acid (EDTA) and 0.05% nonionic surfactant (Triton-X-100) in the solution. We used a low concentration, 0.6% ows (based on weight of substrate), of pectinase enzyme for 30 min and 6 h treatments at 55 °C, and a low concentration, 0.06% owl (based on weight of liquor), of xylanase enzyme for the 1-h treatment at 60°C. For cellulase treatment, Britton-Robinson buffer (pH-5) was used along with 0.01% surfactant and 0.2% owl enzyme at 55 °C for 2 h. The treatments were investigated in terms of thermal stability in the following combinations: only pectinase enzyme for 30 min and 6 h; only xylanase enzyme for 1 h; pectinase enzyme (6 h) followed by xylanase enzyme (1 h); xylanase enzyme (1 h) followed by cellulase enzyme (2 h); pectinase enzyme (6 h) followed by xylanase enzyme (1 h), followed by cellulase enzyme (2 h). After the initial treatments, we selected the duration for each enzymatic treatment as described above. The optimization of the treatments was not of paramount importance at this stage. Toward the end of each treatment, the fibers were separated from the supernatant solution by vacuum suction and then washed in cold water. The optimum temperature range for the enzymes used in this study is between 40 and 60 °C. The cold water temperature was between 5 and 15 °C. Finally, after the enzymatic combinations were applied to the hemp fibers, the filtered fibers were blended for 5 min in a Waring blender to obtain a uniform fiber pulp. In the next step, cryocrushing, we subjected the hemp pulp to crushing action in a cryo-cup grinder (Cryocrushing cup, Biospec). The samples were frozen under liquid nitrogen and ground for 2 min using a pestle. These cooling conditions affect the 3D structure of any residual enzymes on fibers, leading to inactivation.

In the final step, the cryocrushed fibers were dispersed in water to a 2% suspension and homogenized using an APV 1000 homogenizer for 20 cycles with pressure starting at 500 bar and reaching 750 bar toward the end of the cycle. The temperature of the homogenizer reached a maximum of 40°C toward the end of 20 cycles. After each treatment step, a small sample was saved, diluted (0.1 % suspension), and centrifuged twice before the samples were oven dried at 80°C overnight. These samples were then conditioned in a Dry-Keeper desiccator chamber for 24 h at room temperature, where the moisture level in the chamber was controlled at 20% using anhydrous calcium sulfate (Drierite). After conditioning, the samples were characterized and analyzed using scanning electron microscopy (SEM), differential scanning calorimetry (DSC), and thermogravimetric analysis (TGA). In DSC, the samples were analyzed in temperatures ranging from -50–250 °C at a constant heating rate of 10°C/min under air. In TGA, the samples were analyzed in a temperature range from 25°C to 800°C at the heating rate of 10°C/min under a nitrogen-only environment.

Further analysis to determine the effect of homogenizing on the formation of microfibrillated cellulose was carried out by determining the particle size of the fibers and by observing with atomic force microscopy (AFM) and scanning electron microscopy (SEM), and by using an X-ray diffractometer (XRD). The particle size of cryocrushed fiber and homogenized fiber of the hemp fibers treated with combined xylanase and cellulase enzymes were analyzed using Nicomp 380, a nanoparticle size analyzer (by Particle Sizing Systems). This analyzer uses a dynamic light scattering technique and measures changes in the amplitude of scattered laser light as a function of time.

Results and Discussion

Characterization of MFC from Hemp Using SEM

We obtained MFC from hemp using both enzymatic and mechanical treatments. The fibers thus obtained after each treatment were dried and subjected to SEM.

Figure 3 shows the morphologies of hemp fibers from microfibers into nanosized fibers at various stages of the treatments. Figure 3a represents an untreated hemp fiber of 70 μm in cross-section, where the bundles of fibers are seen to be held closely together by a waxy layer. In Figure 3b, following a preliminary treatment with 4% sodium hydroxide and 5 min microwave irradiation, the fibers exhibits a more open structure. In Figure 3c, further separation or delamination of the fibers reduced the fiber size to 10–25 μm. Following blending action, the individual fibers further reduced in size to 8–15 μm (Figure 3d). Enzymatic treatments (Figure 3e), such as consecutive treatments with pectinase and hemicellulase enzymes, caused further reduction in the fiber size to 6–10 μm. After cryocrushing (Figure 3f), the fibers appeared to be of the same thickness but flatter in cross-section. For the final step, i.e., homogenizing the sample resulted in nanocellulose with a size of less than 1 μm as seen in Figure 3g. The fibers formed a gelatinous liquid after final homogenization step. The AFM images in Figure 3(h) also confirm the isolation of the nanocellulose by the Gaulin homogenizer.

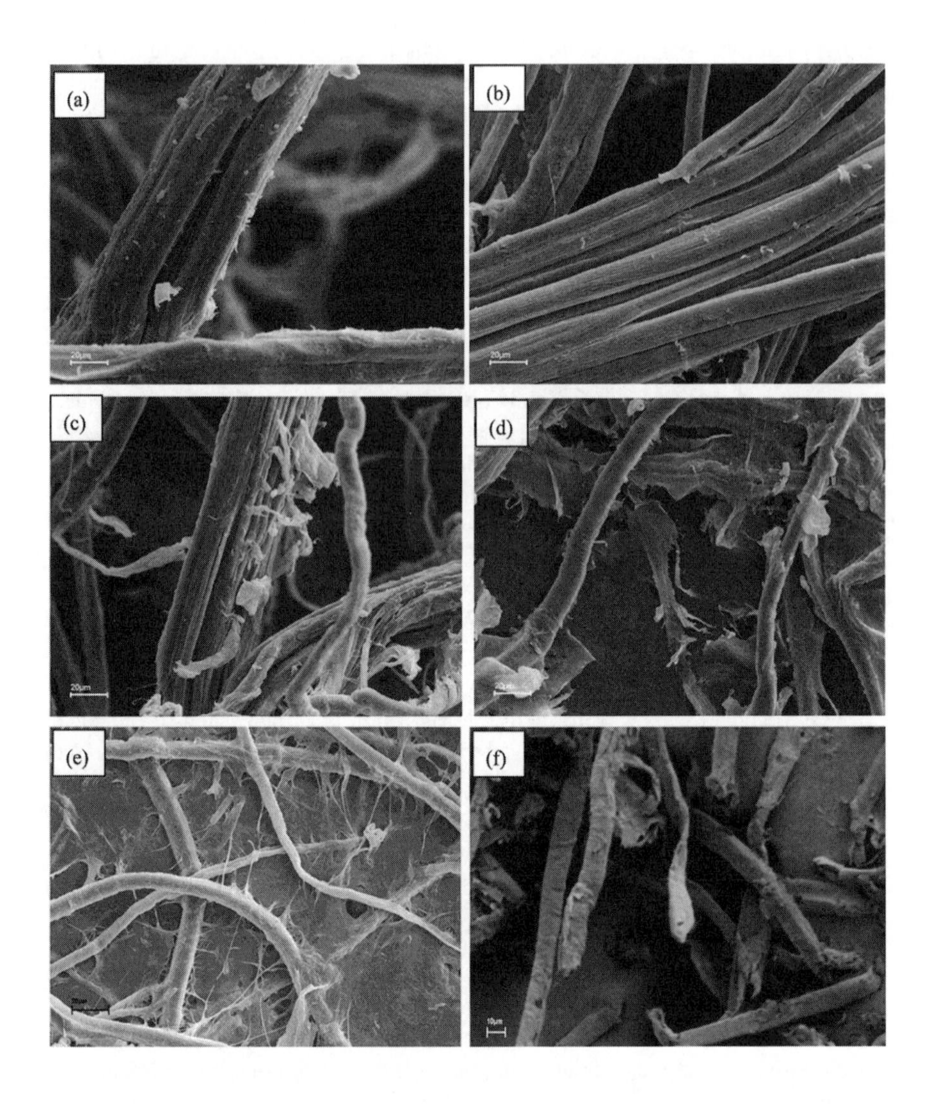

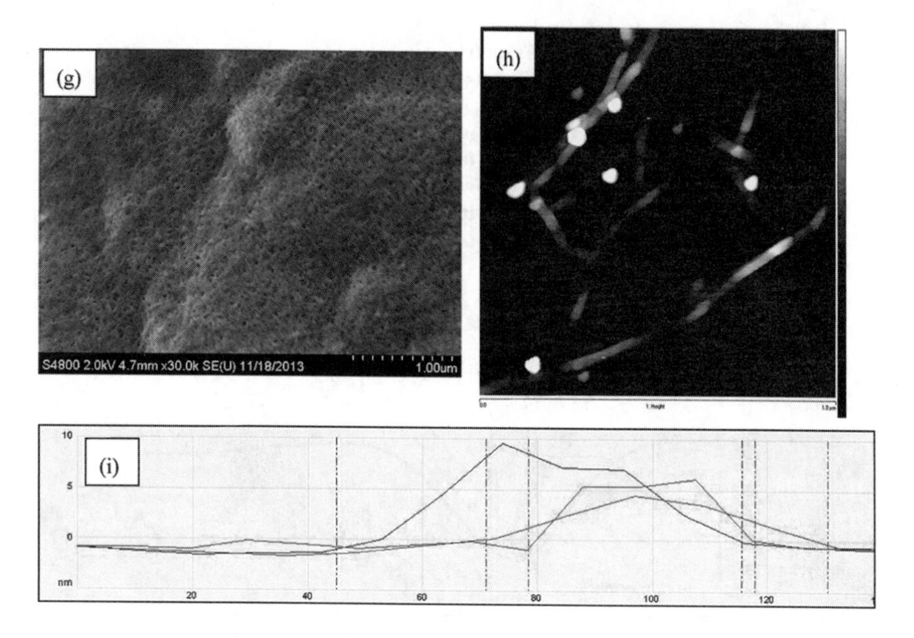

Figure 3. SEM (500X) of hemp fibers at different processing steps: (a) untreated; (b) 4% NaOH and microwave treated; (c) 10 min pressure cooked; (d) 20 min blended; (e) enzymatic treated; (f) cryocrushed; (g) SEM-FE (80KX) of Microfibrillated cellulose after homogenization; (h, i) Atomic force microscopy (AFM) of microfibrillated cellulose after homogenization.

Enthalpy Values of Hemp Fibers before Homogenization

Non-freezing water is the fraction that is most closely associated with cellulose matrix, with the molecules directly attached to the hydroxyl group in the amorphous region. Figure 4 shows the evaporation of non-freezing water as an endothermic peak between 0°C and 150°C (*29*, *30*) for various pretreatments steps as well as for various enzyme combinations. The water molecules are bound mainly with the free hydroxyl groups present in the amorphous region. Therefore, the area under the endothermic peak represents the bound-water enthalpy of evaporation, correlating to the amount of amorphous content in the cellulose structure (*31*). The non-freezing bound water, being closer to the surface, is trapped in nanosized pores on the cellulose, and hence requires higher energy to remove the water molecules from the structure (*13*, *32*).

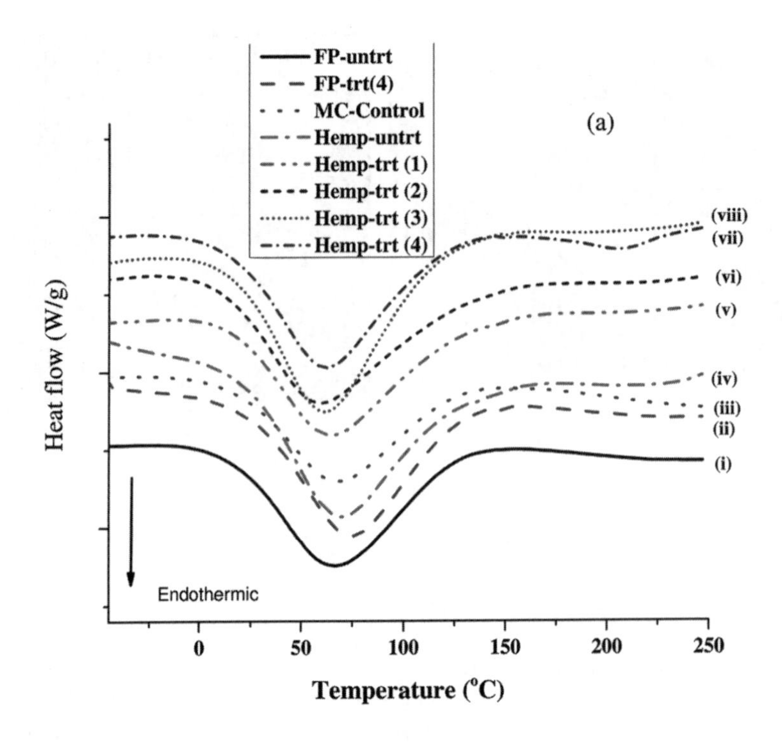

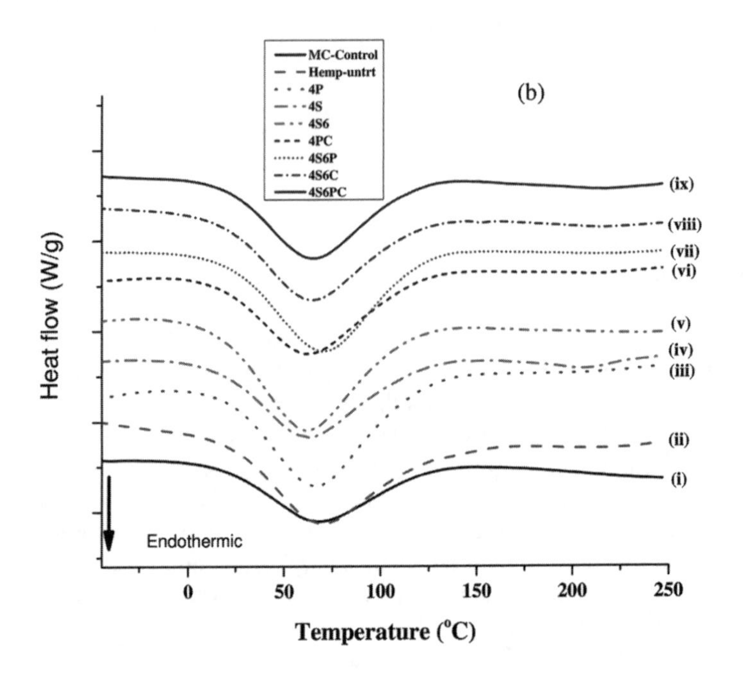

Figure 4. (a) DSC curves obtained for untreated and treated hemp fibers processed through the initial stages of the pretreatments. Note: Stacked lines by Y

offset; Graph legends: (i) Filter paper untreated; (ii) Enzymatically treated filter paper; (iii) Pure cellulose from cotton linters; (iv) Hemp untreated; (v) NaOH and microwave treated hemp; (vi) 10 min pressure-cooked hemp; (vii) 20 min blended hemp; (viii) Enzymatically treated hemp. (b) DSC curves for untreated and treated hemp fibers processed through various enzymatic treatments. Note: Stacked lines by Y offset; Graph legends: (i) Pure cellulose from cotton linters; (ii) Untreated hemp; (iii) 1h xylanase; (iv) 30min pectinase; (v) 6h pectinase; (vi) 1h xylanase + 2 h cellulase; (vii) 6h pectinase + 2h cellulase; (viii) 6h pectinase + 1h xylanase; (ix) 6h pectinase+ 1h xylanase + 2h cellulase.

From the enthalpy values measured after each treatment, as given in Table 1, a typical trend was observed in the enthalpy values. Initially, after treating the fibers using sodium hydroxide, pressure cooking and blending, the enthalpy values increased. One of the reasons for this increase could be that these treatments performed on the hemp fiber caused the fibers to swell, unfolding its basic structure and exposing more hydroxyl groups in the amorphous regions. Therefore, more energy was required to evaporate the bound water molecules. This was found to be true even with the filter paper where, after treatment, the enthalpy value increased. After enzymatic treatments, particularly with xylanase and cellulase enzymes, the enthalpy value dropped. A plausible reason for this could be that, by using enzymes, the hemicellulose and the amorphous cellulose present in the hemp fibers were broken down. This, in turn, reduced the number of binding sites for the water molecules, leading to increased ordered structure (crystalline) and less ordered structure (amorphous).

Thermal degradation characteristics of hemp fibers were analyzed using TGA at various steps of pretreatments, as shown in Figure 5. The initial weight loss, which occurred below 100°C, was primarily due to the moisture evaporation and release of water from the fibers. Major degradation occurred in a temperature region between 220°C to 450°C and mainly attributes to the thermal degradation of hemicelluloses, a small fraction of lignin components, and cellulose (*29*). Table 2 shows the $T_{10\%}$ and $T_{50\%}$ weight losses, and % ash content for various enzymatic treatments to determine which was the most efficient. As given in Table 2, the $T_{10\%}$ and $T_{50\%}$ for untreated hemp were 275°C and 341°C, respectively, whereas, for the cryocrushed samples (before the homogenization step), the temperatures shifted to 309°C and 370°C, respectively. This increase in degradation temperatures indicate the presence of a stable cellulose structure resulting from the rediction of fractions of hemicellulose and amorphous regions (*33*). Figure 5 also shows the shift in temperature to the right with respect to the various enzymatic treatments. Toward the end of the treatments, when compared to the MC-control, higher amounts of residues were obtained for the hemp samples. The residue could contain hemicellulose and lignin present in hemp fibers. In Figure 5, there is a trend of decreasing residual weight between the untreated and the treated hemp samples. This decrease highlights the removal of hemicellulose and lignin from the samples (*34*). Table 2 shows the thermal behavior of untreated and treated hemp fibers.

Table 1. Thermal parameters obtained for untreated and treated samples using DSC

Treatment	*Avg. Enthalpy (ΔH) (J/g)*
Filter paper untrt	-146.44
Filter paper trt-(4)	-160.65
Pure cellulose	-122.07
Hemp untrt	-144.53
Hemp trt-(1)	-134.89
Hemp trt-(2)	-162.51
Hemp trt-(3)	-205.22
Hemp trt-(4S)	-199.60
Hemp trt-(4P)	-202.77
Hemp trt-(4C)	-160.07
Hemp trt-(4SP)	-199.94
Hemp trt-(4SC)	-172.48
Hemp trt-(4PC)	**-160.70**
Hemp trt-(4SPC)	**-155.59**

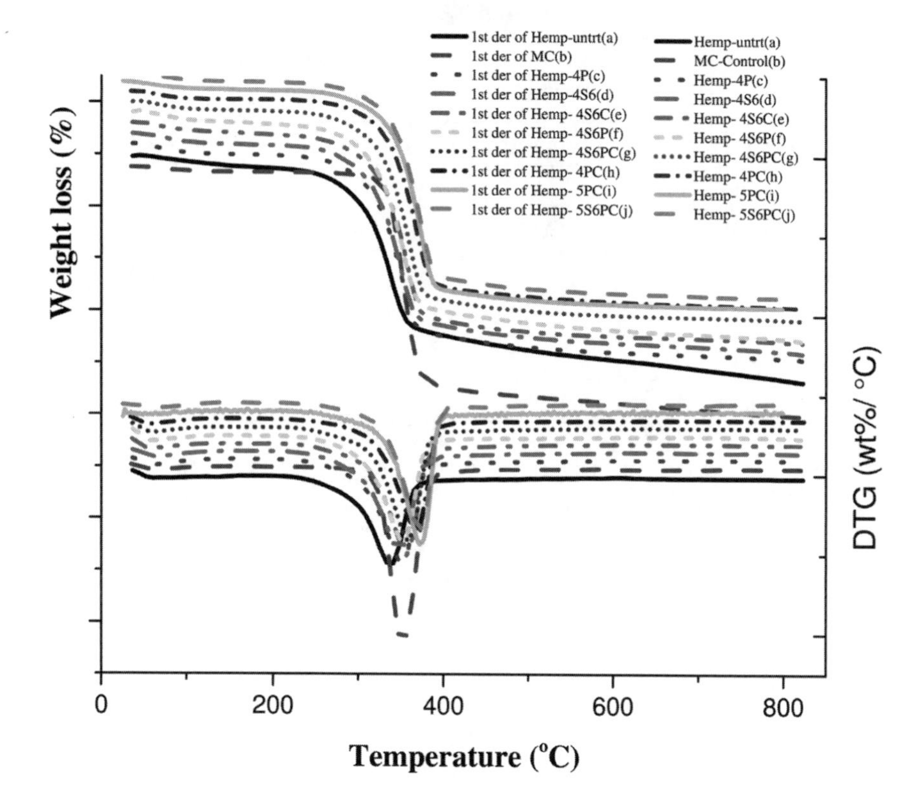

Figure 5. TGA and DTG curves for untreated and treated hemp fibers. Note: Stacked lines by Y offset; Graph legends: Hemp-untrt (a): untreated hemp fibers; MC-Control (b): Pure cellulose from cotton linters; Hemp-4P (c): 1 h pulpzyme; Hemp 4S6 (d): 6 h pectinase; Hemp 4S6C (e): 6 h pectinase + 2 h cellulase; Hemp 4S6P (f): 6 h pectinase + 1 h xylanase; Hemp 4S6PC (g): 6 h pectinase + 1 h xylanase + 2 h cellulase; Hemp 4PC (h): 1 h pulpzyme + 2 h cellulose; Hemp 5PC (i): Cryocrushed 1 h xylanase +2 h cellulase; Hemp 5S6PC (j): Cryocrushed 1 h xylanase +1 h xylanase +2 h cellulose.

Table 2. Thermal behavior of untreated and treated hemp fibers at $T_{10\%}$ and $T_{50\%}$

Treatments	*(%) Ash content at 800°C*	*Temperature (°C)*	
		$T_{10\%}$	$T_{50\%}$
Hemp untrt	14.95	273.74	341.11
MC-control	5.18	335.84	352.80
Hemp-4P	17.85	287.03	349.13
Hemp-4PC	17.51	309.49	366.32
Hemp-4S6	16.36	284.74	349.36
Hemp-4S6C	15.98	287.72	353.25
Hemp-4S6P	12.95	284.28	351.65
Hemp-4S6PC	16.58	293.68	356.92
Hemp-5PC	13.14	308.75	370.36
Hemp-5S6PC	12.95	306.14	369.05

X-ray Diffraction (XRD) Analysis

XRD patterns for the isolated nanocellulose and other intermediates are shown in Figure 6. The graph suggests the samples (i.e., untreated; NaOH treated; NaOH treated plus microwaved; and nanocellulose) follow a similar pattern to that of cellulose I_{β} (*35, 36*), indicating that a conversion to cellulose II did not occur during the microwave / NaOH treatments.

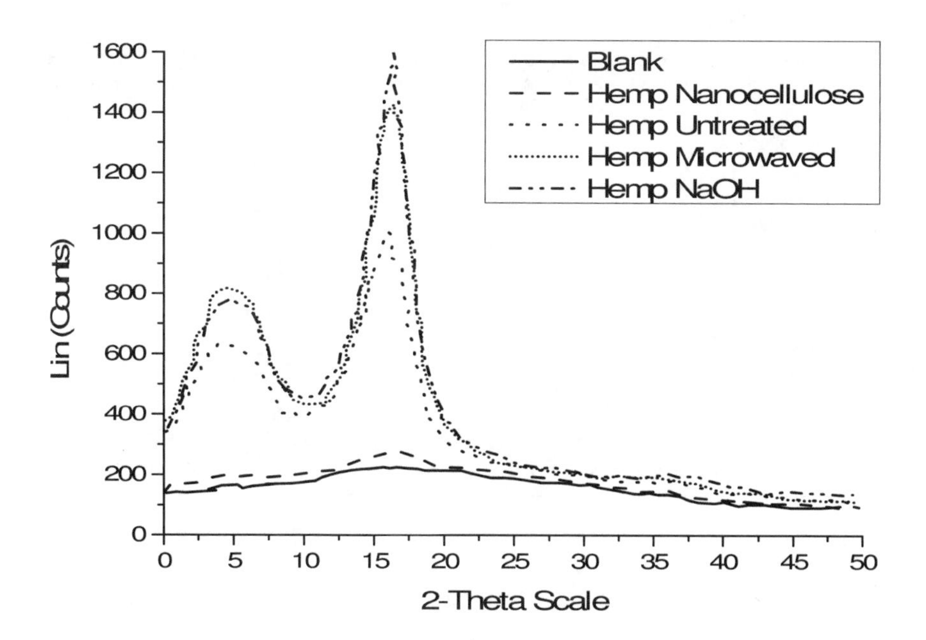

Figure 6. XRD patterns of microfibrillated cellulose and other intermediates.

Particle Size Analysis

The size analysis results given in Figure 7 are for xylanase-cellulase treated fibers after the cryocrushing stage (a) and final homogenizing stage (b). For cryocrushed, xylanase -cellulase (PC) treated fibers, the particle size diameter (PSD) varied from 32.9 nm (2.9%) to 127.7 nm (97.1%). Since the bulk size of the fibers are 130 nm, we can conclude that these were obtained after cryocrushing the fibers. For the homogenized, cryocrushed, xylanase-cellulase (PC) treated fibers three particle sizes—29.5 nm (58.0%), 362.9 nm (15.4%), and 1862.0 nm (26.6%)—were obtained. Since the bulk of the fibers have average size 30 nm, this is the particle size obtained after homogenizing the fibers. The particle size greater than 1000 nm could be obtained because of hornifaction occurring in the sample during storage.

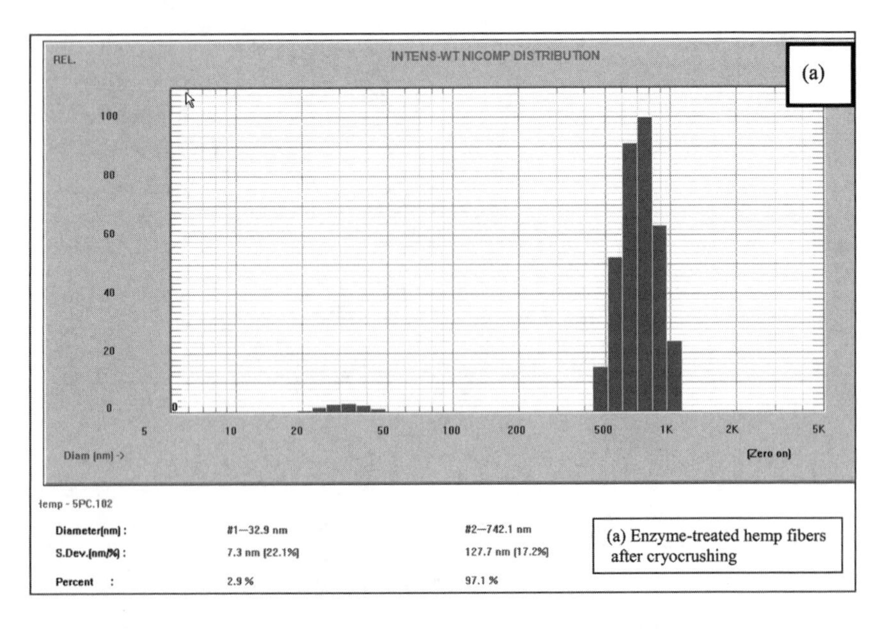

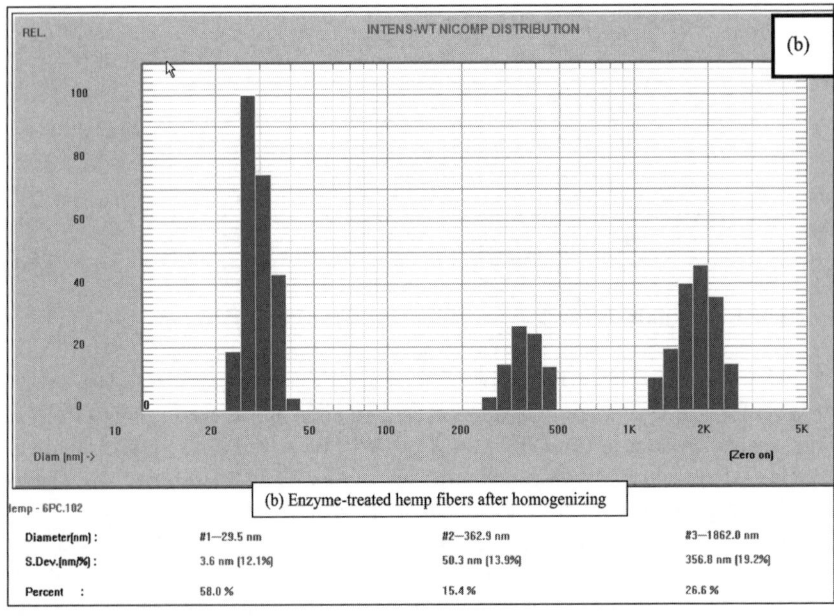

Figure 7. Particle size analysis results obtained for hemp fibers after cryocrushing (a) and homogenizing (b) the samples.

Conclusions

In this paper, the research focused on developing environmentally friendly pretreatments for producing nanocellulose or microfibrillated cellulose. We substituted eco-friendly enzymes for the reduction of caustic soda, typically employed in the traditional microfibrillating process. Of the different enzyme combinations tested, 1 h xylanase enzyme followed by 2 h cellulase enzyme combination (PC) appeared to give the best results, as confirmed using a particle size analyzer. Since this was a proof-of-concept study to plan an environmentally-friendly approach to produce nanoscaled fibers, the conditions in the enzymatic treatments described in this paper were not completely optimized. Thus, a necessary future step would be to optimize the enzymatic parameters to be applied to the plant fibers, and to select other plant sources as raw material. Moreover, economic and environmental benefits are required to be studied.

References

1. Siró, I.; Plackett, D. Microfibrillated cellulose and new nanocomposite materials: A review. *Cellulose* **2010**, *17*, 459–494.
2. Thomsen, A. B.; Rasmussen, S.; Bohn, V.; Nielsen, K. V.; Thygesen, A. *Hemp Raw Materials: The Effect of Cultivar, Growth Conditions and Pretreatment on the Chemical Composition of the Fibers*; Riso National Laboratory: Roskilde, Denmark, 2005.
3. Frone, A. N.; Panaitescu, D. M.; Donescu, D. Some aspects concerning the isolation of cellulose micro- and nano-fibers. *U.P.B. Sci. Bull.* **2011**, *73*, 133–152.
4. de Souza Lima, M. M.; Borsali, R. Rodlike cellulose microcrystals: structure, properties, and applications. *Macromol. Rapid Commun.* **2004**, *25*, 771–787.
5. Favier, V.; Chanzy, H.; Cavaillé, J. Y. Polymer nanocomposites reinforced by cellulose whiskers. *Macromolecules* **1995**, *28*, 6365–6367.
6. Hubbe, M. A.; Rojas, O. J.; Lucia, L. A.; Sain, M. Cellulosic nanocomposites, review. *BioResources* **2008**, *3*, 929–980.
7. Chakraborty, A.; Sain, M.; Kortschot, M. Cellulose microfibrils: A novel method of preparation using high shear refining and cryocrushing. *Holzforschung* **2005**, *59*, 102–107.
8. Diniz, J. M. B. F.; Gil, M. H.; Castro, J. A. A. M. Hornification—Its origin and interpretation in wood pulps. *Wood Sci. Technol.* **2004**, *37*, 489–494.
9. Röder, T.; Sixta, H. Thermal treatment of cellulose pulps and its influence to cellulose reactivity. *Lenzinger Ber.* **2004**, *83*, 79–83.
10. Cheng, Q.; Wang, J. In *Green Nanocomposites Reinforced with Cellulosic Crystals Isolated from Juvenile Poplar*; International Convention of Society of Wood Science and Technology, United Nations Economic Commission for Europe – Timber Committee: Geneva, Switzerland, 2010.
11. Wang, B.; Sain, M. Dispersion of soybean stock-based nanofiber in a plastic matrix. *Polym. Int.* **2007**, *56*, 538–546.

12. Souza, S. F.; Leão, A. L.; Cai, J. H.; Wu, C.; Sain, M.; Cherian, B. M. Nanocellulose from Curava fibers and their nanocomposites. *Mol. Cryst. Liq. Cryst.* **2010**, *522*, 342–352.
13. Alemdar, A.; Sain, M. Biocomposites from wheat straw nanofibers: morphology, thermal and mechanical properties. *Compos. Sci. Technol.* **2007**, *68*, 557–565.
14. Alemdar, A.; Sain, M. Isolation and characterization of nanofibers from agricultural residues – Wheat straw and soy hulls. *Bioresour. Technol.* **2008**, *99*, 1664–1671.
15. Bhatnagar, A.; Sain, M. Processing of cellulose nanofiber-reinforced composites. *J. Reinf. Plast. Compos.* **2005**, *24*, 1259–1268.
16. Wang, B.; Sain, M. Isolation of nanofibers from soybean source and their reinforcing capacity on synthetic polymers. *Compos. Sci. Technol.* **2007**, *67*, 2521–2527.
17. Dufresne, A.; Cavaillé, J. Y.; Vignon, M. R. Mechanical behavior of sheets prepared from sugar beet cellulose microfibrils. *J. Appl. Polym. Sci.* **1997**, *64*, 1185–1194.
18. Henriksson, M.; Henriksson, G.; Brerglund, L. A.; Lindström, T. An environmentally friendly method for enzyme-assisted preparation of microfibrillated cellulose (MFC) nanofibers. *Eur. Polym. J.* **2007**, *43*, 3434–3441.
19. Zhu, S.; Wu, Y.; Yu, Z.; Liao, J.; Zhang, Y. Pretreatment by microwave/alkali of rice straw and its enzymatic hydrolysis. *Process Biochem.* **2005**, *40*, 3082–3086.
20. Satyamurthy, P.; Vigneshwaran, N. A novel process for synthesis of spherical nanocellulose by controlled hydrolysis of microcrystalline cellulose using anaerobic microbial consortium. *Enzyme Microb. Technol.* **2013**, *52*, 20–25.
21. Li, Q.; Liu, Y. X.; Yu, H. P.; Chen, W. S. The mechanical and it combined with enzymatic preparation methods of microfibrillated cellulose. *Adv. Mater. Res.* **2010**, *87*, 393–397.
22. Li, Y.; Pickering, K. L. Hemp fiber reinforced compositiess using chelator and enzyme treatments. *Compos. Sci. Technol.* **2008**, *68*, 3293–3298.
23. Lee, S. H.; Inoue, S.; Teramoto, Y.; Endo, T. Enzymatic saccharification of woody biomass micro/nanofibrillated by continuous extrusion process II: Effect of hot-compressed water treatment. *Bioresour. Technol.* **2010**, *101*, 9645–9649.
24. Janardhnan, S.; Sain, M. Isolation of cellulose nanofibers: Effect of biotreatment on hydrogen bonding network in wood fibers. *Int. J. Polym. Sci.* **2011**, DOI: 10.1155/2011/279610.
25. Janardhan, S.; Sain, M. Isolation of cellulose microfibrils – An enzymatic approach. *BioResources* **2006**, *1*, 176–188.
26. Pääkkö, M.; Ankerfors, M.; Nykänen, H.; Ahola, S.; Österberg, M.; Ruokolainen, J.; Laine, J.; Ikkala, O.; Lindstrom, T. Enzymatic hydrolysis combined with mechanical shearing and high-pressure homogenization for nanoscale cellulose fibrils and strong gels. *Biomacromolecules* **2007**, *8*, 1934–1941.

27. Hassan, M. L.; Mathew, A. P.; Hassan, E. A.; Oksman, K. Effect of pretreatment of bagasse pulp on properties of isolated nanofibers and nanopaper sheets. *Wood Fiber Sci.* **2010**, *42*, 362–376.
28. Keshwani, D. R.; Cheng, J. J. Modeling changes in biomass composition during microwave-based alkali pretreatment of switchgrass. *Biotechnol. Bioeng.* **2010**, *105*, 88–97.
29. Deepa, B.; Abraham, E.; Cherian, B. M.; Bismarck, A.; Blaker, J.; Pothan, L. A.; Leao, A. L.; Sde Souza, S. F.; Kottaiasamy, M. Structure, morphology and thermal characteristics of banana nano fibers obtained by steam explosion. *Bioresour. Technol.* **2011**, *102*, 1988–1997.
30. Hatakeyama, T.; Nakamura, K.; Hatakeyama, H. Vaporisation of bound water associated with cellulose fibers. *Thermochim. Acta* **2000**, *352-353*, 233–239.
31. Ciolacu, D.; Ciolacu, F.; Popa, V. I. Amorphous cellulose – Structure and characterization. *Cellul. Chem. Technol.* **2011**, *45*, 13–21.
32. Park, S.; Venditti, R. A.; Jameel, H.; Pawlak, J. J. Studies of the heat of vaporization of water associated with cellulose fibers characterized by thermal analysis. *Cellulose* **2007**, *14*, 195–204.
33. Ashori, A.; Harun, J.; Raverty, W.; Yusoff, M. Chemical and morphological characteristics of Malaysian cultivated kenaf (*Hibiscus cannabinus*) fiber. *Polym.-Plast. Technol. Eng.* **2006**, *45*, 131–134.
34. Moran, J. I.; Alvarez, V. A.; Cyras, V. P.; Vazquez, A. Extraction of cellulose and preparation of nanocellulose from sisal fibers. *Cellulose* **2008**, *15*, 149–159.
35. French, A. D. Idealized powder diffraction patterns for cellulose polymorphs. *Cellulose* **2014**, *21*, 885–896.
36. French, A. D.; Cintrón, M. S. Cellulose polymorphy, crystallite size, and the Segal Crystallinity Index. *Cellulose* **2013**, *20*, 583–588.

Chapter 6

Biothermoplastics from Coproducts of Biofuel Production

Narendra Reddy*

**Center for Emerging Technologies, Jain University,
Jain Global Campus, Jakkasandra Post,
Bangalore-562112, India
*E-mail: nreddy3@outlook.com**

Utilizing coproducts of biofuel production is essential to reduce cost and make biofuels competitive to petroleum based fuels. About 30-50% of grains processed for ethanol or biodiesel end up as coproducts, generally called meals. These meals contain 25-50% proteins, carbohydrates and other valuable components. In addition, the coproducts are relatively cheap and derived from renewable and sustainable sources. So far, biofuel coproducts have mainly been used as animal feed, a limited market and low value application. Limited attempts have been made to use the coproducts for industrial applications. However, researchers have shown that the components in the coproducts can be extracted with properties suitable for high value end uses. Similarly, the coproducts have been used as reinforcement for composites or chemically modified to make thermoplastics. This paper presents a brief review on the attempts made to develop thermoplastic products from the coproducts of biofuel production.

Introduction

Production of biofuels from cereal grains generates considerable amounts (upto 35% of the feedstock)ofcoproductsthat currently have limited applications. Distillers dried grains obtained as the coproducts of ethanol production, soymeal obtained after processing soybeans for food or fuel are some examples of coproducts that are currently available on the market and mainly used as animal feed. About 40 million tons of corn DDGS and35 tons of soymeal are generated in the United States every year. Current selling prices of DDGS are between \$85 and \$130 and soymeal are between \$440 and \$470, making them two of the most inexpensive resources. Composition of some of the oil meals produced from common cereal grains are given in Table 1. As seen from the table, major components in the meals are proteins and carbohydrates which are useful for various industrial applications. Attempts have been made to develop non-food industrial applications with a view to add value to the meals, develop biodegradable and environmentally friendly bioproducts. This chapter provides an overview of approaches used to modify the coproducts of biofuel production, develop products and properties of the products developed from the modified coproducts.

Table 1. Availability, cost and composition of coproducts obtained after processing common cereal grains

Type of Coproduct	*Annual Availability, MT*		*Price per Ton, US \$*	*Major Composition, %*		
	World	*US*		*Proteins*	*Carbohydrates*	*Oil*
Soybean	268	36	440-470	40-50	25-30	10-15
Canola	70	8.3	330-360	35	20-30	3.5-5
Cotton seed	44	0.8	345-375	45	40	2
Sunflower seed	43	-	250-280	-	-	-
Corn DDGS	-	32.5	85-130	25-30	35-50	10-12

Bioproducts from Corn Distillers Dried Grains with Solubles (DDGS)

Blending of DDGS with Other Polymers

DDGSis the coproduct obtained during the production of ethanol from corn. United States being the largest producer of ethanol, about 32 million tons of DDGS was produced in 2011 as seen from Table 1. Currently, DDG is mainly used as animal feed at about \$85-130 per ton. Typically, about 30% of the corn used is obtained as DDGS that is roughly composed of 25-30% proteins, 35-50% cellulose and hemicelluloses and 10-12% oil (*1*, *2*). Although DDGS contain valuable proteins and carbohydrates, DDGS is non-thermoplastic and therefore difficult to be made into bioproducts. The components in DDGS were reported to thermally degrade in three different stages as seen from Figure 1. The first stage was observed between 26-130 °C due to the moisture loss and volatilization of light molecules (*3*). Second stage of degradation was between 130-530 °C and was attributed due to the degradation of polysaccharides and proteins. A final stage of degradation between 530-775 °C was attributed to lignin and presence of some metals (*3*).

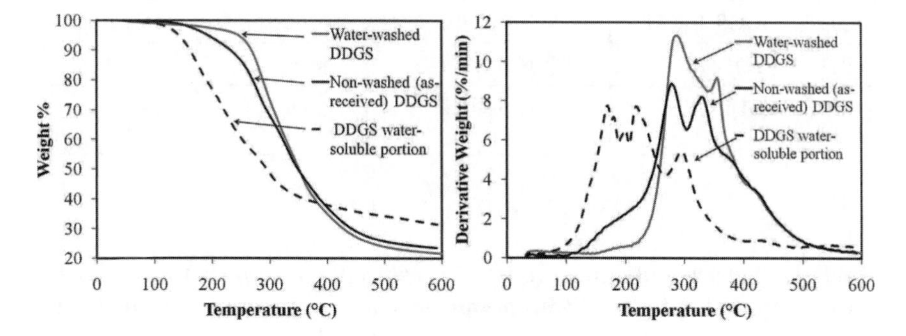

Figure 1. Thermal behavior (TGA and derivative TGA curves) of native and treated DDGS. Reproduced with permission from reference (3).Copyright (2011) Wiley.

Due to its large availability and relatively low cost, DDGS has been more commonly used to develop bioproducts than the other coproducts of biofuel production. In a simple approach, DDGS (25%) was mixed with high density polyethylene (HDPE) and maleated polyethylene (MAPE) and extruded in a twin screw extruder (*4*). In addition, DDGS was treated with hexane to remove oil and later with dichloromethane to improve the compounding. The compounded product was injection molded into ASTM D638 type tensile bars and ASTM D790 type of flexural bars. DDGS particles were heterogeneously distributed and even seen in clumps in the composites developed. It was observed that addition of MAPE facilitated better adhesion between the reinforcement and matrix. The mechanical properties of the DDGS composites developed are given in Table 2. As seen from the Table, addition of DDGS into HDPE decreased the strength of the composites considerably and the elongation by nearly 85% suggesting that the DDGS acted as a filler rather than reinforcement. Although no major improvement was seen in the tensile properties with the addition of MAPE, solvent treated DDGS combined with HDPE-MAPE showed tensile strength similar to that of pure HDPE. Solvent treatment probably removed some of the hydrophobic materials and facilitated better adhesion between the matrix and reinforcement.

Similar results were also observed for the flexural properties as seen from Table 3. Composites reinforced with solvent treated DDGS and MAPE showed higher strength than neat HDPE, about 30% higher modulus of elasticity and 30% lower impact energy.

Table 2. Tensile properties of composites reinforced with DDGS. Reproduced with permission from Reference (*4*). Copyright (2013) North Carolina State University.

Composition	*Tensile Strength, MPa*	*Modulus, MPa*	*Elongation, %*
HDPE	21.5 ± 0.1	339 ± 10	105 ± 1
HDPE + MAPE	20.6 ± 0.2	333 ± 15	103 ± 13
HDPE + DDGS	14.7 ± 0.0	356 ± 9	17.9 ± 0.5
HDPE + DDGS + MAPE	20.3 ± 0.3	366 ± 6	16.2 ± 0.7
HDPE + STDDGS	16.6 ± 0.1	478 ± 68	15.1 ± 0.3
HDPE + STDDGS + MAPE	23.8 ± 0.2	446 ± 4	14.5 ± 0.4

Table 3. Flexural and impact resistance properties of composites reinforced with DDGS. Reproduced with permission from reference (*4*). Copyright (2013) North Carolina State University.

Composition	*Flexural strength, MPa*	*Modulus of Elasticity, MPa*	*Impact Energy, J/m*
HDPE	27.9 ± 0.1	894 ± 15	38.7 ± 0.1
HDPE + MAPE	26.1 ± 0.1	804 ± 8	38.5 ± 0.5
HDPE + DDGS	24.1 ± 0.1	954 ± 6	31.7 ± 1.2
HDPE + DDGS + MAPE	28.6 ± 0.1	937 ± 3	28.7 ± 1.3
HDPE + STDDGS	27.8 ± 0.1	1280 ± 6	30.4 ± 2.0
HDPE + STDDGS + MAPE	34.4 ± 0.2	1231 ± 14	30.0 ± 1.7

In addition to the mechanical properties, water absorption of the composites at room temperature was also reported. Water absorption of the composites after 672 hours soaking in water was between 2.2-4% depending on the condition of the DDGS. Removal of oil resulted in composites that were more resistant to water. Although DDGS is considerably hydrophilic, the composites showed good strength and modulus retention even after being in water for 672 hours. Presence of the synthetic polymers on the surface of the composite should prevent water penetration and therefore provide better retention.

In a similar study, biodegradable green composites were developed from DDGS and polyhydroxy(butyrate-co-valerate) (PHBV) and poly(butylenes succinate)(PBS) (*3*). DDGS (30%) without any treatment and that was also washed in water and blended with PHBV in a twin screw extruder. A compatibilizer (polymeric methylene diphenyldiisocyanate, PMDI) was also added to the water washed DDGS to improve compatibility. Addition of DDGS resulted in composites that had tensile, flexural and impact resistance properties between that of neat PHBV and PBS. The water washed DDGS was found to provide considerably better properties to the composites compared to unwashed DDGS. Similarly, addition of a compatibilizer was also observed to increase the properties of the composites due to improved interfacial adhesion. SEM images of the fracture surface of composites containing the plasticizer and water washed DDGS did not show any pull out of DDGS and no gaps between the reinforcement and matrix was observed (*3*).

DDGS powdered to a particle size of about 0.34 mm was utilized as inexpensive filler with conventional wood glue and phenolic resin as the matrix (*2*).The filler and glue were mixed by hand and then compression molded at a pressure of 25 kPa. Samples formed were later heat treated at 75 °C for 5 hours in an oven. The type of glue and the compression temperature played a major role on the morphology and composites. Composites formed using phenolic resin glue had tensile strength between 150-380 kPa compared to 6-35 kPa for the glue composites. Decreasing the size of DDGS and increasing compression time resulted in improved composite properties. Similarly, increasing the content of DDGS increased flexural strength but decreased modulus (*2*).

Low density polyethylene (LDPE) was also reinforced with alkali treated DDGS (*5*).Treatment with NaOH was found necessary to obtain composite properties when LDPE was reinforced with 40% DDGS. As seen from Table 4, the tensile moduli of the DDGS reinforced composites were nearly 3 times higher than that of the neat LDPE whereas the strength and elongation decreased considerably. Higher mass loss was also observed for the composites containing DDGS since the proteins and carbohydrates in DDGS were more susceptible to heat compared to neat LDPE.

Completely biodegradable DDGS composites were developed using DDGS as the reinforcement and poly(lactic acid)as the matrix (*6*). PLA and DDG were ground in a Wiley mill and mixed in ratios of 20, 30, 40 and 50% at 180 °C. To improve compatibility between PLA and DDGS, methylene diphenyldiisocyanate (MDI) was used as a coupling agent. The blends were compression molded into ASTM D 638 standard dog bone shaped specimens. FTIR studies indicated that the compatibilizer was well reacted with the polymers. Tensile properties of the PLA/DDGS composites without any compatibilizer are given in Table 5. Similarly, the tensile properties of composites containing 20% DDGS and various amounts of compatibilizer are given in Table 6.

As seen from Table 5, addition of DDGS drastically reduces the tensile strength but more than doubles the elongation at low concentrations of DDGS. Modulus of the composites is considerably low and decreases with increase in the amount of DDGS. However, addition of the compatibilizer even at a low level of 0.25% increases the strength more than twice and also causes a considerable decrease in elongation as seen in Table 6. Increasing the concentration of the compatibilizer increases the elongation but does not show any particular trend with increase in strength. Authors concluded that the poor interfacial adhesion between DDGS and PLA was responsible for the inferior properties of the composites without compatibilizer. SEM images (Figure 2) showed debonding and void formation suggesting that the matrix and reinforcement were incompatible. Formation of covalent linkages between hydroxyl and carboxyl groups in the polymers and compatibilizers was suggested to the reason for the improvement in tensile properties.

Table 4. Tensile and impact resistance of DDGS reinforced LDPE composites. Reproduced with permission from Reference (*5*). Copyright (2013) Elsevier.

Composition	*Tensile strength, MPa*	*Elongation, %*	*Modulus, MPa*	*Impact strength, kJ/m²*
LDPE	14.5 ± 0.5	109 ± 5.1	111 ± 4.9	60 ± 2.9
LDPE + 10% DDGS	11.1 ± 0.5	90 ± 5.3	147 ± 6.9	58 ± 2.5
LDPE + 20% DDGS	10.6 ± 0.3	54 ± 3.7	196 ± 8.1	39 ± 2.1
LDPE + 30% DDGS	9.9 ± 0.4	32 ± 3	243 ± 9.7	28 ± 1.7
LDPE + 40% DDGS	9.2 ± 0.4	21 ± 2.3	294 ± 10.3	27 ± 1.9

Table 5. Tensile properties of PLA composites reinforced with various ratios of DDGS and without any compatibilizer. Reproduced with permission from reference (*4*). Copyright (2011) Wiley.

Property	*% of DDGS in PLA*				
	0	*20*	*30*	*40*	*50*
Tensile Strength, MPa	77 ± 0.8	27 ± 1.0	20 ± 1	14 ± 1	10 ± 1
Breaking elongation, %	6.2 ± 0.4	15.7 ± 2.4	11.6 ± 2.1	11.0 ± 3.6	7.8 ± 1.9
Modulus, MPa	2.0 ± 0.02	1.8 ± 0.04	1.5 ± 0.05	1.5 ± 0.14	0.8 ± 0.05

Table 6. Tensile properties of PLA composites reinforced with 20% DDGS and various amounts of compatibilizer. Reproduced with permission from reference (*6*). Copyright (2011) Wiley.

Property	*% of compatibilizer*			
	0.25	*0.5*	*1*	*2*
Tensile Strength, MPa	54 ± 1	68 ± 1	77 ± 2	69 ± 1
Breaking elongation, %	3.9 ± 0.1	4.6 ± 0.2	5.0 ± 0.2	4.8 ± 0.4
Modulus, MPa	2.4 ± 0.09	2.4 ± 0.05	2.5 ± 1.26	2.4 ± 0.04

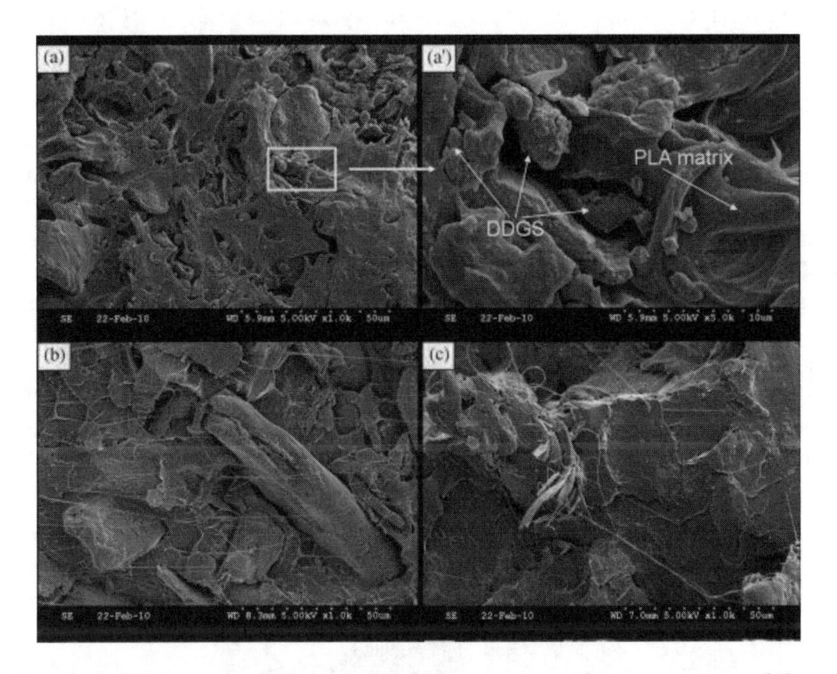

Figure 2. SEM image of PLA (a), DDGS composite without any compatibilizer (a') and with 0.25% (b) and 0.5% (c) compatibilizer. Reproduced with permission from Reference (6). Copyright (2011) Wiley Interscience.

Plastic fiber composites were prepared from polypropylene (PP) or polyethylene (PE) combined with various biomasses including DDGS (*7*). Composites containing 20 and 30% of DDGS ground into 30 mesh size had properties comparable to that of neat PP composites. Tensile strength of the PE composites was between 25-28 MPa and modulus was between 3.9-4.3 MPa depending on the % of DDGS in the composites. For PP composites reinforced with DDGS, the flexural strength was between 42-43 MPa. DDGS reinforced PP composites had considerably higher impact strength than the other reinforcements studied (*7*).

Chemical Modifications of DDGS

Unlike the above discussed reports that used DDGS as reinforcement for composites, our research group has demonstrated that DDGS can be chemically modified and made into thermoplastic films. Green thermoplastics were developed from DDGS after simultaneous acetylation of the proteins and carbohydrates under alkaline conditions using acetic acid and acetic anhydride (*8*, *9*). Acetylation of DDGS resulted in formation of products that were soluble and insoluble in acetic acid. The soluble products obtained, up to 63% of the DDGS used, had an high acetyl content of 45% indicating that it was thermoplastic. A combination of the insoluble and soluble product was termed as total product which had a lower melting enthalpy of 2.7 J/g considerably lower than that for the soluble product. Compression molding the acetylated DDGS at 138 °C resulted in the formation of thermoplastic films shown in Figure 2. The soluble product that had better thermoplasticity (higher enthalpy and degree of substitution) formed transparent films compared to the total product. Another important aspect of the acetylation of DDGS was the use of low ratios of acetic anhydride compared to the acetylation of starch and other biopolymers suggesting that the acetylated DDGS would be inexpensive. Further studies showed that acetylation under acidic conditions provided higher degree of substitution and better thermoplasticity to DDGS even at low ratios of acetic anhydride compared to acetylation under alkaline conditions (*8*, *9*).An acetyl content of 36% (equivalent to a degree of substitution of 2.1) was obtained under acidic conditions. Thermal behavior studies showed that acid catalyzed DDGS acetates had a melting peak at 125 ° compared to 147 ° for the alkaline acetates. As seen from Figure 3, DDGS acetates produced using acidic catalysts formed better thermoplastics compared to the alkaline DDGS acetates.

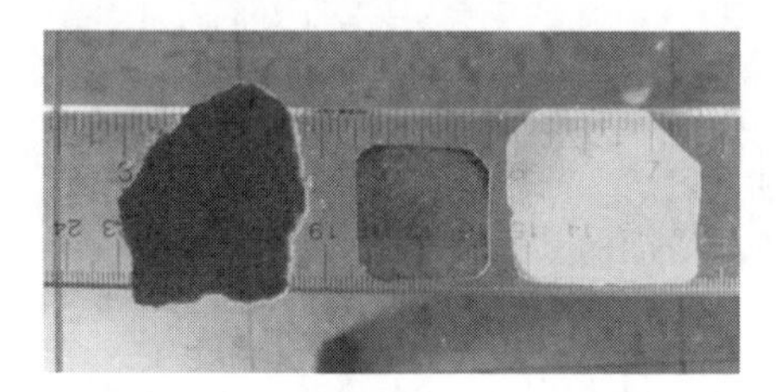

Figure 3. Digital image showing unmodified DDGS (left), DDGS acetylated under acidic conditions (center) and DDGS acetylated under alkaline conditions (right).

The potential of converting DDGS into thermoplastics by etherification using acrylonitrile was studied by Hu et al (*10*).Etherification was considered to be a milder chemical process compared to acetylation and it was expected that the properties of the proteins and carbohydrates could be preserved. Etherification of DDGS resulted in a weight gain of 42% and the modified DDGS had a melting peak at 140 °C. The modified DDGS was compression molded into highly transparent films (Figure 4) with excellent elongation. In an ongoing study, we have developed thermoplastic DDGS by grafting various methacrylates. Films compression molded from grafted DDGS have tensile strengths between 2.8-3.9

MPa and elongation between 1.5 to 3.4 % in the absence of any homopolymer. Unlike the DDGS thermoplastics developed by acetylation and etherification, the films developed from grafted DDGS were found to stable in water. Similar to grafting, various chemical modifications such as carboxymethylation, glutaration, maleiation, phthallation or succination were done on DDGS to enable DDGS to be blended with soyprotein isolates (*11*).The modified DDGS was blended with soyprotein dispersed in water and the mixture was later compression molded at 1.2 GPa for 5 minutes. Mean tensile strength of the pellets obtained after compression molding ranged from 0.2 to 1.7 MPa with glutarated DDGS producing samples with the highest strength.

Figure 4. Digital image of a transparent thermoplastic film obtained after DDGS was cyanoethylated using acrylonitrile. Reproduced with permission from Reference (8). Copyright (2011) American Chemical Society.

DDGS from Other Cereal Grains

Although corn has been the most popular cereal used for ethanol production, reports are available on the use of other cereals such as wheat as a source for ethanol production. Sorghum is a crop that is considered to be an excellent alternative to corn for ethanol production because sorghum is easier to grow and is also less nutritious than corn. A few ethanol plants in the United States use sorghum for ethanol on a commercial scale. Similar to corn DDGS, production of ethanol from sorghum also generates distillers grains. Sorghum DDGS contains highly indigestible proteins and is not preferred even as animal feed. It would therefore be appropriate that sorghum DDGS be used to develop bioproducts. We have recently used sorghum DDGS to develop thermoplastics by grafting various vinyl monomers (*12*). Methyl methacrylate (MMA), ethyl methacrylate (EMA) and butyl methacrylate (BMA) were grafted onto sorghum DDGS and the grafted product was compression molded into thermoplastics films. The influences of various levels of grafting BMA on the wet and dry tensile properties of the films were investigated. As seen from Table 7, increasing grafting ratio up to 40% increased the strength and elongation by making the DDGS more thermoplastic.

The wet strength also increases with increasing grafting ratio since the films become more hydrophobic with the addition of the grafted polymers.

Table 7. Influence of the level of BMA grafting on the tensile properties of grafted sorghum DDGS films. Reproduced with permission from Reference (*12*). Copyright (2014) American Chemical Society.

Grafting ratio, % (mmol/g)	*Tensile strength, MPa*		*Breaking Elongation, %*		*Young's Modulus, MPa*	
	Dry	*Wet*	*Dry*	*Wet*	*Dry*	*Wet*
23 (2.01)	2.0 ± 0.4	0.3 ± 0.09	1.1 ± 0.3	2.5 ± 0.5	700 ± 80	46 ± 15
30 (2.63)	2.8 ± 0.9	0.84 ± 0.2	1.6 ± 0.4	4.4 ± 0.9	427 ± 79	60 ± 20
40 (3.50)	4.3 ± 0.8	1.84 ± 0.4	2.3 ± 0.5	4.0 ± 0.9	550 ± 50	155 ± 57
53 (4.64	3.1 ± 0.9	1.9 ± 0.8	2.3 ± 0.6	4.3 ± 1.3	580 ± 10	170 ± 91

Thermoplastics from Coproducts of Biodiesel Production

Processing of oil seeds such as soybeans, peanuts and camelina for food or biodiesel also generates coproducts in substantial quantities. The oil meals contain up to 50% proteins and are commonly used as animal feeds. Efforts have been made to convert the oil meals into various types of high value products. Qin et al. attempted to develop adhesives from soymeal by treating with enzymes and also by grafting glycidyl methacrylate (*13*).The aim of the study was to reduce the viscosity of the protein, increase the shear wet strength and make soymeal adhesive useable in the wood industry. Wet strength of the soymeal was about 0.45 MPa and enzyme treated soymeal adhesive had strength of about 0.25 MPa. Modifying the enzyme treated soymeal with various levels of GMA increased the adhesive strength up to 1.05 MPa. It was suggested that addition of GMA increased crosslinking density and decreased water sorption and therefore increased the wet strength. Chemical modifications of the meal were also found to improve the thermal stability of the adhesive.

Soymeal based biodegradable plastics were developed by melt extrusion and the effect of various processing conditions on the properties of the composites developed were studied. Thermoplastic soymeal was prepared by blending a plasticizer (glycerol), urea, sodium sulfite and pH 10 water and extruding at 100 °C. Prepared thermoplastic soymeal was subsequently blended with synthetic polymers such as poly(butylene succinate), poly(caprolactone) and poly(butylenesadipateterepthlate (PBAT) and injection molded at 135 °C and 6 MPa into test samples (*14*). Tensile tests showed that sodium sulfite did not have any effect on soymeal whereas glycerol and urea improved the strength. The type of biodegradable polyester used also had a significant influence on tensile properties. Under the optimum conditions (treating soymeal with 12.5% urea and 30% glycerol), the composites developed had a strength of 30 MPa and elongation of about 1100% (*14*). In a similar study, binary and tertiary blends of soymeal treated with urea and polycaprolactone and poly(butylenes succinate) were prepared (*15*).Tensile strength of the samples varied between 22-30 MPa and elongation was from 100-600%. Impact strength of the samples was in the range of 70-160 J/M with soyprotein and polycaprolactone blends providing the highest tensile and flexural strength and also high elongations. This was suggested to be due to the formation of fibril morphology compared to the droplet or particle morphology in the other blends.

In a unique approach, thermoplasticized soymeal was blended with natural rubber and vulcanized rubber in various ratios and injection molded into thermoplastics. Calcium sulfate dihydrate was added as a compatibilizer to improve mechanical properties (*16*). Table 8 shows the tensile properties of the soymeal rubber composites under dry and wet conditions. It is remarkable to note that the composites have high strength retention even when wet and after absorbing up to 40% water. An increase in the glass transition temperature of the rubber component in the blend was observed based on dynamic mechanical analysis and SEM pictures.

Table 8. Dry and wet tensile properties of thermoplastic soymeal blended with natural rubber and a compatibilizer. Reproduced with permission from reference (*16*). Copyright (2007) Wiley.

Type of sample	*Dry tensile properties*		*Wet tensile properties*	
	Strength, MPa	*Elongation, %*	*Strength, MPa*	*Elongation, %*
Thermoplastic soymeal (TPS)	2.0 ± 1.0	20 ± 0.4	-	-
TPS (48.7%) + Natural rubber	2.65 ± 0.09	430 ± 21	2.1 ± 0.1	415 ± 54
TPS (54.5%) + Natural rubber	2.33 ± 0.03	371 ± 8	1.54 ± 0.13	273 ± 20
TPS (54.5%) + Natural rubber-Vulcanized	1.81 ± 0.12	345 ± 6	1.04 ± 0.08	288 ± 15
TPS (47.2%) + Natural rubber+ compatibilizer	3.29 ± 0.06	356 ± 7	3.10 ± 0.2	472 ± 24

Acceptance and the projected increase in the demand of biodiesel as an alternative fuel have led to the consideration of using other feedstocks as source of biodiesel. Camelina is considered to be a better choice for biodiesel production than soybeans since camelina requires fewer resources and yields more oil. Pilot scale production of biodiesel from camelina oil has been reported in the United States and Canada. Coproducts obtained after processing camelina for oil are similar to soymeal and could be used for industrial applications. Camelina meal was compression molded into films with the addition of glycerol but without any chemical modifications (*17*). Table 9 lists the properties of films obtained from camelina meal. Although films could be compression molded from the meal, the films were unstable and disintegrated in water. Increasing glycerol ratio made the meal more thermoplastic and increased the elongation. However, substantial decrease in strength was observed due to the plasticizing phenomenon of glycerol that has been commonly observed. Grafting of BMA onto the meal and addition of various levels of homopolymers resulted in thermoplastic films with good tensile strength and elongation as seen from Table 10. Grafting and addition of homopolymer was necessary to obtain films with high tensile strength and elongation as seen from the table.

Table 9. Properties of thermoplastic films developed from camelina meal with different levels of glycerol. The meal was compression molded at 160 °C for 2 minutes to form thermoplastics. Reproduced with permission from reference (*17*). Copyright (2012) American Chemical Society.

% Glycerol	*Peak Stress, MPa*	*Breaking Elongation, %*	*Modulus, MPa*
0	3.0 ± 1.0	0.7 ± 0.3	1241± 400
5	2.8 ± 1.1	0.8 ± 0.4	1063 ± 194
10	2.3 ± 0.6	1.3 ± 0.4	566 ± 70
20	0.4 ± 0.1	1.0 ± 0.4	93 ± 26

Table 10. Properties of thermoplastic camelina films compression molded before and after grafting and containing various levels of the homopolymer. Reproduced with permission from reference (*17*). Copyright (2012) American Chemical Society.

% of Homopolymer	*Homopolymer + grafted camelina*		*Homopolymer + ungraftedcamelina*	
	Peak Stress, MPa	*Elongation, %*	*Peak Stress, MPa*	*Elongation, %*
0	7.0 ± 1.4	2.5 ± 0.7	7.0 ± 1.4	2.5 ± 0.7
25	48.4 ± 3.7	3.0 ± 0.6	1.7 ± 0.3	1.7 ± 0.2
50	53.7 ± 4.4	3.1 ± 0.7	1.7 ± 0.3	1.8 ± 0.3
75	1.7 ± 0.1	42.7 ± 9.7	1.3 ± 0.3	2.1 ± 0.6
100	1.0 ± 0.2	257 ± 44	1.0 ± 0.2	257 ± 44

Conclusions

Projected increase in the demand for petroleum, especially in the developing countries, combined with the mandatory use of biofuels in the near future will inevitably create a surplus of the biofuel coproducts in the market. Since the traditional means of using the coproducts as animal feed is a limited market and low value application, it is imperative that the coproducts be used for industrial applications. The coproducts contain valuable components and combined with the low cost and large availability are ideal sources to develop bioproducts. However, the complexity of the components in the coproducts also makes it difficult to modify and develop products. Research so far has demonstrated the possibility of developing thermoplastics on a laboratory scale. Further research is necessary to demonstrate that the products developed have the properties for particular end use applications. Under the current circumstances, it would be reasonable to assume that biodiesel will not be competitive, cost-wise, to petroleum based diesel unless the coproducts are used for high value applications.

References

1. Xu, W.; Reddy, N.; Yang, Y. *J. Agric. Food Chem.* **2007**, *55*, 6279–6284.
2. Cheesbrough, V.; Rosentrater, K. A.; Visser, J. *J. Polym. Environ.* **2008**, *16*, 40–50.
3. Zarrinbakhsh, N.; Misra, M.; Mohanty, A. K. *Macromol. Mater. Eng.* **2011**, *296*, 1035–1045.
4. Tisserat, B.; Reifschneider, L.; O'Kuru, R. H.; Finkenstadt, V. L. *Bioresources* **2013**, *8*, 59–75.
5. Luo, X.; Li, J.; Feng, J.; Xie, S.; Lin, X. *Comp. Sci. Technol.* **2013**, *89*, 175–179.
6. Li, Y.; Sun, X. S. *J. Appl. Polym. Sci.* **2011**, *121*, 589–597.

7. Julson, J. L.; Subbaran, G.; Stokke, D. D.; Gieselman, H. H.; Muthukumarappan, K. *J. Appl. Polym. Sci.* **2004**, *93*, 2484–2493.
8. Hu, C.; Reddy, N.; Yan, K.; Yang, Y. *J. Agric. Food Chem.* **2011**, *59*, 1723–1728.
9. Reddy, N.; Hu, C.; Yan, K.; Yang, Y. *Appl. Energy* **2011**, *88*, 1664–1670.
10. Hu, C.; Reddy, N.; Luo, Y.; Yan, K.; Yang, Y. *Biomass Bioenergy* **2011**, *35*, 884–892.
11. Schilling, C. H.; Tomasik, P.; Karpovich, D. S.; Hart, B.; Shepardson, S.; Garcha, J. *J. Polym. Environ.* **2004**, *12*, 257–263.
12. Reddy, N.; Shi, Z.; Temme, L.; Hu, H.; Lan, X.; Hou, X.; Yang, Y. *J. Agric. Food Chem.* **2014**, *62*, 2406–2411.
13. Qin, Z.; Gao, Q.; Zhang, S.; Li, J. *Bioresources* **2013**, *8*, 5369–5379.
14. Reddy, M. M.; Mohanty, A. K.; Misra, M. *J. Mater. Sci.* **2012**, *47*, 2591–2599.
15. Reddy, M. M.; Mohanty, A. K.; Misra, M. *Macromol. Mater. Eng.* **2012**, *297*, 455–463.
16. Wu, Q.; Seike, S.; Mohanty, A. K. *Macromol. Mater.Eng.* **2007**, *292*, 1149–1157.
17. Reddy, N.; Jin, E.; Chen, L.; Jiang, X.; Yang, Y. *J. Agric. Food Chem.* **2012**, *60*, 4872–4879.

Chapter 7

3D Electrospun Fibrous Structures from Biopolymers

Helan Xu[1] and Yiqi Yang[*,1,2,3]

[1]Department of Textiles, Merchandising and Fashion Design, 234, HECO Building, University of Nebraska-Lincoln, Lincoln, Nebraska 68583-0802, United States
[2]Department of Biological Systems Engineering, 234, HECO Building, University of Nebraska-Lincoln, Lincoln, Nebraska 68583-0802, United States
[3]Nebraska Center for Materials and Nanoscience, 234, HECO Building, University of Nebraska-Lincoln, Lincoln, Nebraska 68583-0802, United States
[*]Tel: +001 402 472 5197. Fax: +001 402 472 0640. E-mail: yyang2@unl.edu.

Electrospun three-dimensional (3D) fibrous biopolymers are receiving increasing attention as tissue engineering scaffolds. 3D structures could more closely resemble the stereoscopic architectures of native extracellular matrices (ECMs), and thus could provide similar guidance to signaling and migration of cells. Furthermore, fibrous structures could provide larger surface area than non-fibrous ones to facilitate cell attachment and growth. Due to the high efficiency and broad applicability, electrospinning has become the most widely accepted method in developing ultrafine fibers from biopolymers. However, since last decade, researchers started applying electrospinning technology to produce 3D ultrafine fibrous scaffolds. Via incorporating porogens or microfibrous frames, using coagulation bath as receptors, and changing electrical properties of spinning dopes, 3D fibrous structures have been developed from natural biopolymers, including proteins (collagen, gelatin, silk fibroin, zein, soyprotein, wheat gluten, etc.), polysaccharides (chitosan, alginate, hyaluronic acid, etc.), and bio-derived synthetic polymers, mainly polylactic acid

(PLA) and polycaprolactone (PCL). These 3D fibrous scaffolds from biopolymers played increasingly important roles in tissue engineering and medical applications.

Introduction

Tissue Engineering

The ultimate target of scaffold design in tissue engineering is to produce substrates that can function as artificial extracellular matrices (ECMs) until new ECMs are resynthesized by cells cultured. To achieve this goal, the substrates should have architectures similar to the natural ECMs. Typical natural 3D ECM is shown in Figure 1 (*1*). The cell attaches on the fibrils in the ECM and spread into a stereo shape. As dynamic, mobile and flexible three dimensional (3D) sub-micron fibrous networks, natural ECMs not only provide physical support to cells, but also define cellular behaviors and final tissue functions. In natural ECMs, highly organized ultrafine collagen fibrils in 3D architectures play dominant roles in maintaining the biological and structural integrity of the tissues or organs (*2*). The 3D structures could provide better connection among cells than conventional 2D cultures, by providing another dimension for interaction, migration and morphogenesis of cells. Figure 2 showed the cross-sections of 2D fibrous and 3D fibrous scaffolds after culturing with cells for 15 days (*3*). The red color represented stained cells, indicating that more cells could be found in the interior of 3D fibrous scaffolds. Moreover, the collagen ECMs are highly dynamic because they undergo constant remodeling to maintain proper physiologic functions. Hence, optimal tissue engineering scaffolds should be able to restore both structural integrity and physiological functions of native ECMs.

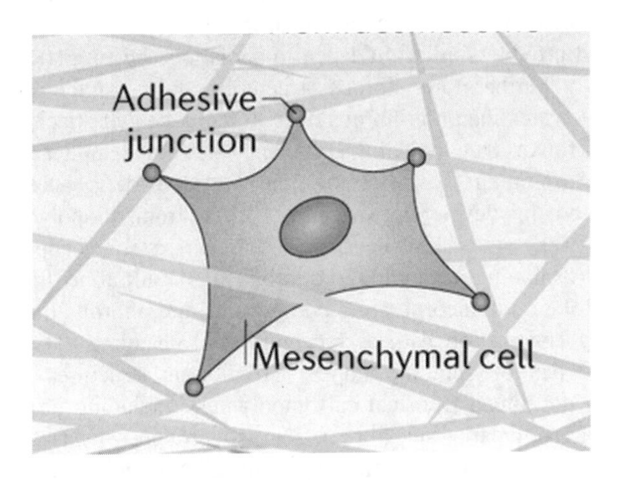

Figure 1. Scheme of a cell in 3D natural extracellular matrix. Reproduced with permission from reference (1). Copyright (2013) Nature.

Ideally, the basic building blocks of tissue engineering scaffolds should be ultrafine fibers oriented in multiple dimensional scales. Comparing to films, sponges and other types of scaffolds, fibrous architectures could more closely emulate the structures of the native ECMs, provide better guidance for cell migration and proliferation, and ultimately, determine the functions of neo-tissues. Fibers also can provide much larger surface area to facilitate adsorption of receptor proteins, such as integrin, fibronectin and vitronectin, and thus enhance cell adhesion. In general, properties and processing techniques are the two major concerns in developing 3D ultrafine fibrous scaffolds.

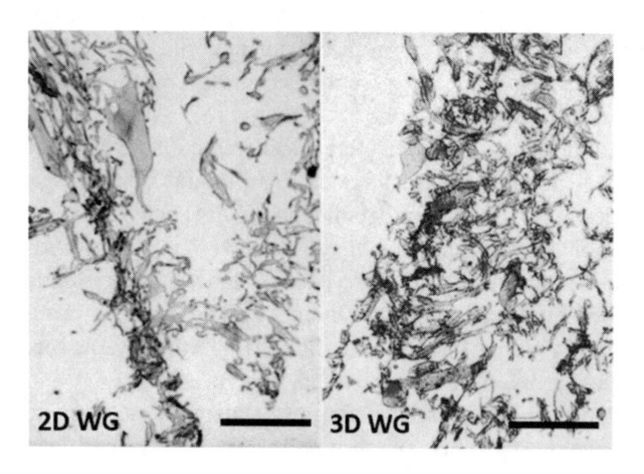

Figure 2. Histological analysis of the Oil Red O stained scaffolds with adipose derived mesenchymal stem cells 15 days after induction of adipogenic differentiation. Left, 2D wheat glutenin (WG) fibrous scaffolds; right, 3D WG fibrous scaffolds. Scale bar represents 100 μm. Reproduced with permission from reference (3). Copyright (2014) Elsevier.

Electrospinning

Electrospinning is one common method to fabricate nano- to micro-scale fibers from various biopolymers because it is applicable to a variety of biopolymers and efficient (*4*). During electrospinning, polymer solution is forced through a capillary, forming a drop of polymer solution at the tip. By applying a high voltage between the tip and a collecting surface, the polymer solution carries charges and is attracted by the collecting surface with opposite charges. As soon as the electric field strength overcomes surface tension of the droplet, the droplet travels to the collector. During flying, the polymer solution is drawn in the electric field and the solvent evaporates, subsequently, a nonwoven fibrous mat is formed on the targeted surface.

Most tissue engineering applications required for three-dimensional (3D) scaffolds with sufficient interconnected pores to allow cells to penetrate, facilitate formation of uniform tissue and transport mass. The goal could be achieved via random arrangement of fibers. Conventional electrospinning technologies generated flat fibrous composed of nanofibers randomly oriented in two dimensions (2D) (*5*). Without allowing penetration of cells into interior of the scaffold, the 2D electrospinning mats could only be developed into flat shapes instead of stereo spheroid shapes as in most native ECMs.

To fabricate ultrafine fibrous structures with 3D organization, conventional electrospinning methods have been modified via multiple approaches. Porogens, such as ice and salt crystals, have been incorporated to increase bulkiness of electrospun ultrafine fibrous scaffolds (*6*). Wet electrospinning was developed using a coagulation bath instead of a collecting board to receive fluffy fibers (*7*). Using porous structures as receptors for electrospinning was a facile approach to obtain 3D ultrafine fibrous scaffolds. Micro-fibrous scaffolds were usually used to collect electrospun fibers from various materials. In addition, to fabricate special tubular scaffolds, such as vascular prostheses, rotating drums with desired diameter were used instead of collecting boards (*8*). All the methods have their own merits and drawbacks. These methods shared a disadvantage of retained the 2D alignment of the conventional 2D electrospinning. Presence of coarse fiber frames, porogens and wet bath only enlarged the distance among deposited fibers, but did not change the alignment of the fibers. With higher volume of voids inside, the cells could penetrate more deeply. However, orientation of attached cells determined by the alignment of the fibers would not be changed. Stacking of cells oriented in planar directions could be a potential issue in regeneration of thick 3D tissues or organs, such as liver.

Adjusting the surface charge of spinning solution was another newly emerging approach to develop scaffolds composed of 3D randomly oriented ultrafine fibers (*9, 10*). Figure 3 shows that plant protein, zein and synthetic polymer polyethylene glycol could be electrospun 3D cotton ball-like structures on the left sides of Figure 3(a) and 3(b). The conventional 2D electrospun scaffolds with the same mass are shown on the right sides of Figure 3(a) and 3(b). Therefore, porosity and accessible area for cells in the 3D scaffolds could be remarkably increased.

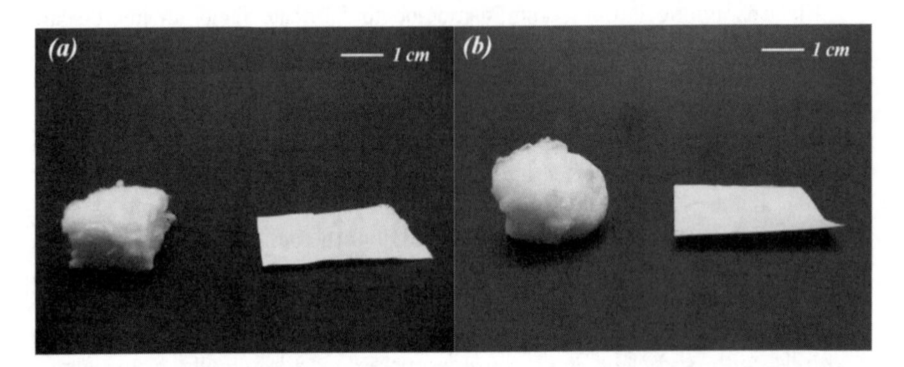

Figure 3. (a) Electrospun zein fibrous scaffolds, left: 3D; right: 2D; (b) electrospun Poly ethylene glycol, left: 3D; right: 2D. Reproduced with permission from reference (9). Copyright (2013) The American Chemical Society.

Biopolymers for Tissue Engineering

Materials for development of tissue engineering scaffolds should be biocompatible and biodegradable, as they usually perform roles as temporary therapeutic substrates to shelter cells. In long term, the neo-tissues are expected to restore physiological functions after being synthesized by the cells cultured on scaffolds. During neo-tissue regeneration, scaffolds, the artificial tissues, should degrade gradually to give room to the newly formed tissues. Proteins, polysaccharides and degradable synthesized polymers are the major biomaterials for fabrication of tissue engineering scaffolds.

Proteins have been employed to develop tissue engineering scaffolds for their biocompatibility, biodegradability, versatility in loading therapeutic agents and provide unique affinity to cells. The basic units of proteins are similar to that of collagen, the major component in natural ECMs. Proteins could be easily degraded into short peptides or amino acids, which are building blocks for native proteins in body and could be reused or metabolized in physiological environments. Hydrophilic protein scaffolds are preferred to hydrophobic ones for cell attachment (*11*). In addition, with tunable surface charges under different pHs, proteins could load substrates with different charges. Meanwhile, the hydrophobic domains in proteins could attract water insoluble substrates via hydrophobic interactions. Some animal proteins, such as collagen and keratin have bio-signaling moieties, such as tripeptide, Arg-Gly-Asp (RGD) in the molecules to promote cell adhesion (*12–14*). Proteins have been extensively electrospun into 2D fibrous structures for tissue engineering (*15*), and are attracting attention in 3D electrospinning (*10*).

Polysaccharides, polymers of monosaccharides in multiple combinations, have several merits, i.e. hygroscopicity, biocompatibility and non-toxicity for biomaterial applications. Polysaccharides have diverse origins, chemical structures, molecular weights and ionic characters, leading to different processability, stability and degradability. In addition, some polysaccharides, like hyaluronic acid existed in special tissues, like cartilages, and therefore showed intrinsic advantages in developing engineering scaffolds for these tissues. The widely used polysaccharides, i.e., chitosan, hyaluronic acid and alginate will be introduced in terms of processing methods and properties of products in this chapter.

A plethora of synthetic biopolymers, including polycaprolactone (PCL), polylactic acid (PLA) and polylactic-co-glycolic acid (PLGA), have been investigated and used as scaffold matrices (*16*). These polymers are generally biodegradable and biocompatible. Featured with highly repeated molecular compositions and structures, scaffolds from these bio-based synthetic polymers usually have highly predictable and reproducible chemical, physical and biological properties. Easy processability makes these synthetic biopolymers favorable in scaffold design. With mechanical properties superior to most natural polymers, synthetic biopolymers are especially advantageous in constructing scaffolds to repair high load-bearing defects. Degradation rates of synthetic polymers can be regulated to match tissue growth into neo-ECMs. Moreover, a variety of copolymers, polymer blends, biologically and chemically functionalized polymers significantly promoted broader applications of these synthetic polymers.

Proteins

Animal Proteins

Collagen

Collagen is a category of fibrous proteins existing in connective tissues, providing structural integrity and mechanical support to many tissues, such as skin, cartilage, tendons and blood vessels. As biomaterial, collagen has its unique advantages. Collagen has biological cues in the molecules to facilitate cell adhesion (*13*). Collagen can be degraded by enzymes in body, such as collagenases and serine proteases. However, collagen has issues of adverse immune response and risks of transmitting pathogens even after purification. Furthermore, collagen-based scaffolds usually suffer from poor mechanical properties, difficulty in spinning and fast degradation.

Collagen has been extensively electrospun into nanofibers (*17*), and shown compatibility with a number of cell lines. In Table 1, publications regarding electrospun 3D ultrafine fibers from collagen are summarized. There have been no examples of using sole collagen as scaffolds in Table 1.

As shown in Table 1, all the collagen based scaffolds combined two or more materials, usually one biodegradable synthetic polymer, such as polycaprolactone (PCL). Collagen was added to promote the initial attachment of cells. However, as collagen degraded or dissolved, the affinity to cells of the whole scaffolds might be reduced. The poor water stability and mechanical properties of collagen-based materials could be the critical reason. Solvent for collagen dissolution could be one cause, and usually crosslinking was used to compensate.

Careful selection of solvent could be critical for collagen regeneration. Native collagen fibrils in physiological conditions are water stable and mechanically robust. However, regenerated collagen fibers usually were not water stable probably due to the destroyed triple helical configuration during dissolution. 1,1,1,3,3,3-hexaflouro-2-propanol (HFIP) was the most widely used solvent for collagen. However, due to its corrosive nature, HFIP introduced apparent loss of the natural triple helical configuration, increased the solubility of regenerated collagen in physiological environments, and thus decreased the water stability of collagen based fibers (*18*). To substitute HFIP, acetic acid has been used to dissolve collagens, but still caused 70% decrease in the percentage of triple helical configuration of collagen. In addition, a water/alcohol/salt solvent system has been developed to dissolve collagen (*19*). But the salt concentration as high as 50% impaired the protein fibers, indicated by decrease in strength and water stability. Aqueous ethanol solution was found to be a more non-corrosive and non-toxic solvent system that could more efficiently preserve the triple helical configuration of collagen (*20*).

Table 1. 3D fibrous scaffolds from collagen

Material	*Solvent*	*Crosslinker*	*Diameter (nm)*	*Method*	*Cell culture*	*Reference*
Collagen I	Acetic acid	EDC/NHS	1200 ± 210	Coarse fiber incorporation	BMSCs	(*23*)
Collagen I	HFIP	EDC	160	Coarse fiber incorporation	Osteoblasts	(*24*)
Collagen II	HFIP	GTA	5200 ± 700	Coarse fiber incorporation	BMSCs	(*25*)
Collagen I	HFIP	N/A	350 ± 85	Coarse fiber incorporation	Osteoblasts	(*26*)
Collagen I/PCL/HA/PEO	HFIP	N/A	1000	PEO removed as sacrificial substrates	BMSCs	(*27*)
Collagen I/PCL/HA	Acetic acid	N/A	N/A	NaCl as porogens	Osteoblasts	(*28*)
Collagen I/PCL/PEO	HFIP	N/A	250 ± 73	PEO removed as sacrificial substrates	*In vivo* study using rats	(*29*)

EDC: 1-ethyl-3-[3-dimethylaminopropyl]-carbodiimide; NHS: N-hydroxysuccinimide; BMSCs: Bone marrow-derived mesenchymal stem cells; HFIP: 1,1,1,3,3,3 hexafluoro-2-propanol; GTA: Glutaraldehyde; N/A: Not Applicable; PCL: polycaprolactone; PEO: Polyethylene oxide.

Crosslinking was usually needed to improve water stability and performance properties of collagen fibers. Glutaraldehyde (GTA) was the primary crosslinking agent for proteins to enhance the mechanical properties and water stability (*21*). However, the cytotoxicity and calcification effect on protein materials rendered GTA not an optimal crosslinker for collagen targeting biomedical applications. Recently, non-toxic crosslinkers, such as 1-ethyl-3-(3-dimethylaminopropyl) carbodiimide (EDC)/N-hydroxysuccinimide (NHS) and genipin, have been applied to crosslink protein biomaterials (*22*). However, the crosslinking efficiency of these crosslinkers could not endow adequate water stability, as the crosslinked proteins still easily collapsed into films in high humidity or aqueous conditions. Citric acid with glycerol as extender has also been used to crosslink electrospun collagen nanofibers (*20*).

Silk Fibroin

Silk fibroin is the insoluble protein secreted by silk worms. It has been used as surgical sutures for decades due to its outstanding mechanical properties and water stability. Comparing with many other proteins, silk fibroin has higher percentage of crystalline formed by anti-parallel sheets of hydrophobic peptide chains (*30*). The highly stable configuration of silk fibroin is due to the strong hydrophobic interaction among the polypeptides, which primarily consist of glycine (Gly) (43%), alanine (Ala) (30%) and serine (Ser) (12%) (*31*). All the three amino acids have small side groups, facilitating tight packing of polypeptides to form crystals. Most regenerated fibroin fibers in nano- or micro-scales, showed good mechanical behaviors due to easy formation of beta sheet of polypeptides (*32*).

Silk fibroin showed promising potential to support cell adhesion, proliferation, growth and differentiation to promoting neo-tissue formation. Moreover, silk fibroin scaffolds were preferred in regenerating load-bearing tissues, such as ligament (*33*), cartilage (*34*) and bone (*35*) tissue engineering due to its superb mechanical properties and slow degradation rate.

Table 2 demonstrates a few examples of nano- and micro-scale silk fibroin fibers in 3D dimensions produced via electrospinning methods. Strong solvents, such as HFIP and trifluoro acetic acid (TFA) were used to dissolve silk fibroin. One example of using water as solvent for electrospinning produced micro-scale fibers, indicating worse spinnability of fibroin in water than in HFIP and TFA (*36*). Fibroin fibers with diameter on micro-scale were also generated by electrospinning fibroin dissolved in 9.3 M LiBr (*37*). The precipitation of salts and poor solubility of fibroin in water could also be the cause. Methanol bath, microfiber template and small-diameter mandrel were used to induce formation of 3D structures.

Table 2. 3D Fibrous scaffolds from silk fibroin

Material	*Solvent*	*Diameter (nm)*	*Method*	*Cell culture*	*Reference*
Silk fibroin	$CaCl_2$ /water/ethanol	100-400	Methanol bath	N/A	(*38*)
Silk fibroin/PEO	9.3 M LiBr	<1000	Mandrel with diameter of 38 mm	Human endothelial cells, smooth muscle cells	(*39*)
Silk fibroin/gelatin/PLA	$CaCl_2$ /H_2O/ethanol, formic acid, chloroform	<1000	Multiple electrospinning, post-treatment (rolling)	Fibroblasts	(*40*)
Silk fibroin	$CaCl_2$ /water/ethanol	500-3000	Template	N/A	(*41*)
Silk fibroin	Formic acid	<1000	Methanol bath and NaCl porogen	Fibroblasts and in vivo rats	(*42*)
Silk fibroin/PEO	Water	<1000	NaCl porogen	In vivo rat	(*43*)

N/A: Not applicable; PEO: Polyethylene oxide; PLA: Poly(lactic acid).

Other Animal Proteins

Recently, feather keratin has been successfully electrospun into 3D fibrous scaffolds and demonstrated satisfactory support to chondrogenic differentiation of adipose derived mesenchymal stem cells (*44*). The dissolution of highly crosslinked feather keratin was achieved via controlled disentanglement and de-crosslinking of keratin molecules using reductant and surfactant (*45*).

Plant Proteins

Plant proteins are attracting increasing attention as substrates for medical applications due to their specific advantages. In addition to satisfactory biodegradability and biocompatibility, plant proteins have lower possibility to transmit zoonotic diseases comparing to animal proteins. Plant proteins also have much larger availability since most of the proteins can be derived from byproducts from agricultural industries. Potential of plant proteins for medical applications has been reviewed elsewhere (*46*). Zein, soyprotein and wheat gluten are three major plant proteins to be electrospun into 3D ultrafine fibrous scaffolds.

Zein is a prolamin, the alcohol soluble protein that can be extracted from agricultural byproducts such as dry milled corn (DMC), corn gluten meal (CGM), and distiller's grains (DG). Zein has received considerable attention in medical applications because of its good biocompatibility, biodegradability and good support to multiple cell lines (*47*).

Zein has unique advantages over many other proteins for drug delivery. Zein is not water soluble and is more hydrophobic and water stable than animal proteins, such as collagen and gelatin. Moreover, the strong surface charge of zein under different pH conditions makes it possible to load drugs with different charges (*48*). Meanwhile, the hydrophobic domains in its compositions enable zein to attract hydrophobic substances via hydrophobic interaction. These properties make zein suitable for loading high amount of various drugs. Zein can be easily dissolved in aqueous ethanol solution (*48*, *49*).

Only a few publications reported 3D ultrafine zein fibrous scaffolds for biomedical applications, and are summarized in Table 3. Cai and coworkers invented a novel spinning method to develop citric acid crosslinked 3D ultrafine fibrous scaffolds from zein, and proved its much higher cell accessibility and proliferation than 2D zein scaffolds crosslinked under the same condition (*9*). Development of the 3D ultrafine fibrous scaffolds broadened the applications of zein scaffolds by making possible regeneration of tissues with certain thickness. However, the untreated 3D ultrafine fibrous zein scaffolds were still not water stable and needed crosslinking to enhance their dimensional integrity in aqueous environment.

Based on the similar spinning mechanism, more water stable plant proteins, soyprotein and wheat glutenin were also electrospun into 3D randomly oriented ultrafine fibrous structures (*3*, *10*). The highly crosslinked soyprotein and wheat glutenin scaffolds showed considerable water stability without any crosslinking as illustrated in Figure 4. The 3D soy protein scaffolds swelled, yet still maintained their fibrous structures after immersed in PBS under 37 °C for 28 days.

Table 3. 3D ultrafine fibrous scaffolds from plant proteins

Material	*Solvent*	*Crosslinker*	*Diameter (nm)*	*Method*	*Cell culture*	*Drug loading*	*Reference*
Zein	Carbonate buffer/SDS	Citric acid	700-2200	3D ES	Fibroblasts	N/A	(*9*)
Soyprotein	Carbonate buffer/SDS	N/A	800-1500	3D ES	Adipose mesenchymal stem cells	N/A	(*10*)
Wheat glutenin	Carbonate buffer/SDS	N/A	800-1500	3D ES	Adipose mesenchymal stem cells	N/A	(*3*)

3D: Three dimensional; SDS: sodium dodecyl sulfate; ES: Electrospinning; N/A: Not applicable.

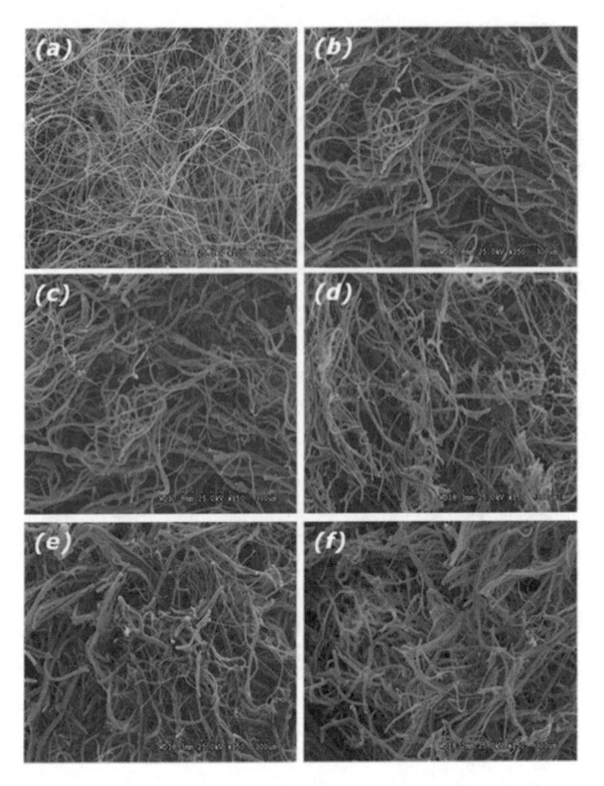

Figure 4. (a) As-spun soyprotein fibers, (b) to (f) soyprotein fibers immersed in PBS at 37 °C for (b) 3, (c) 7, (d) 14, (e) 21 and (f) 28 days. Scale bar = 300 µm. Reproduced with permission from reference (10). Copyright (2013) The Royal Society of Chemistry. (DOI: 10.1039/C3RA47547F)

Polysaccharides

Chitosan and hyaluronic acid (HA) from animals, and alginate from algal, were the three most widely studied polysaccharides as biomaterials. However, there have been not many publications reporting 3D electrospinning from all the three polysaccharides.

Chitosan, the only polysaccharide that carries positive charge, is the partially deacetylated derivative of chitin obtained from shells of crabs and shrimp. Chitosan is difficult to electrospin into fibrous structure (*50*). Surface tension of aqueous chitosan solution was high due to the polycationic character of chitosan in acidic aqueous solution due to the amine groups in the backbones.

Hyaluronic acid presents at high concentrations in hyaline articular cartilage, synovial fluid and vitreous humour, and is also a principal constituent in ECMs found in mammalian species. In cartilage, HA plays a major role in lubrication, cell differentiation and tissue hydration due to the high hydrophilicity of the backbone. HA is also very difficult to be electrospun because of the unusually high viscosity and surface tension of the aqueous solution. In addition, strong water retention ability of HA leads to the fusion of HA nanofibers on the collector due to the insufficient evaporation of the solvents during electrospinning. HA has been successfully developed after blowing was incorporated in electrospinning.

Alginate is an anionic polysaccharide derived from algal. It contains block structures of (1,4)-linked-β-D-mannuronic acid (M) and α-L-guluronic acid (G) residues into a linear copolymer. Alginate is usually non-degradable in body and could not be excreted by kidney due to the large molecular weight. However, with slight oxidization, alginate could be degraded in aqueous media, and showed potential for cartilage regeneration (*51*). However, due to lack of entanglements of the rigid and extended chains in aqueous solution, aqueous alginate solution was difficult to be electrospun into nano- or submicron fibers.

Even be electrospun into 3D fibrous structures, these polysaccharides should be crosslinked or the scaffolds might lose their fibrous forms due to swelling in aqueous environments.

Here we summarized the few publications of 3D electrospinning of the polysaccharides in Table 4. The principle involved in 3D electrospinning of chitosan, alginate and hyaluronic acid were incorporation of fibrous framework as receptor, rolling of 2D electrospun fibrous mats for chitosan scaffolds (*52–54*).

The scaffolds collected on the coarse chitosan fibrous receptors had significantly high porosity than the traditional 2D scaffolds collected directly on the aluminum foil. The tight stacking of electrospun fibers were prevented by the fibrous frameworks. However, the obtained chitosan scaffolds did not have real 3D orientation of fibers, as the fibers were still parallel to the collectors. For the 3D chitosan fibrous scaffolds obtained by rolling the 2D fibrous mats, the low porosity of the scaffolds might not be favored for cell penetration. In the work of 3D electrospinning of alginate blended with PEO, a new mechanism was proposed. The 3D fibrous scaffolds were formed without any modification on the electrospinning apparatus. Alginate with strong negative charge induced strong fiber-fiber repulsion during electrospinning of scaffolds. The fibers showed spatially random distribution and formed large 3D pores throughout the thickness of the scaffolds. For hyaluronic acid, salt microspheres were used as sacrificial cores during collection of nanofibers. Orientation of the hyaluronic fibers might not be as random as the 3D electrospun alginate scaffolds, since most of the fibers still deposited parallel to the collector.

So far, not much attention has been paid on the 3D fibrous scaffolds of polysaccharides, probably due to the technical difficulty in preparation of spinning dopes with proper spinnability. However, 3D electrospinning of HA still deserved research since HA was the major macromolecule in cartilage tissues and 3D orientation and distribution of fibers in scaffolds were necessary for proper guidance of cell growth. In addition, crosslinking might also be indispensable to prolong the residence time of scaffolds in physiological environments as regeneration of cartilage tissues was usually time-consuming.

Table 4. Fibrous scaffolds from polysaccharides

Material	*Solvent*	*Diameter (nm)*	*Method*	*Cell culture*	*Drug loading*	*Reference*
Chitosan	Acetic acid	<3000	Combination of coarse fibrous structures	Chondrocytes	N/A	(*52*)
Chitosan	Acetic acid	<300	Post-treatment (rolling)	N/A	N/A	(*53*)
Chitosan	Acetic acid	100-1000	Post-treatment (rolling)	Schwann cells	N/A	(*54*)
Alginate/PEO	Water	200-300	Electrical repulsion	N/A	N/A	(*55*)
Hyaluronic acid/collagen	NaOH/DMF	<400	Salt leaching	Chondrocytes	N/A	(*56*)

N/A: Not applicable; PEO: Polyethylene oxide; DMF: Dimethylformamide.

PLA

Poly(lactic acid) (PLA) is one of the most well recognized biopolymers derived from renewable resources, such as corn starch in the United States, starch in Asia and sugar from sugarcane in the rest of the world. Hydrolysis of ester bonds facilitated degradation of PLGA, PLA and PGA in body, while molecular weight, crystallinity and monomer ratios all influence the degradation rates (*57*). PLA has been electrospun into nanofibrous scaffolds due to its easy processability. PLA could be readily dissolved in chloroform and electrospun into nanofibers. Electrospun PLA scaffolds usually had better mechanical properties especially under wet conditions than scaffolds from natural scaffolds. In physiological environments, PLA was insusceptible to water molecules due to its hydrophobic nature, while natural polymeric scaffolds had poor wet properties as the hydrogen bonds that provided strength of scaffolds were readily interfered by water molecules.

Two major drawbacks of PLA hindered its applications as tissue engineering scaffolds. First disadvantage was the cell non-adhesive surface character due to the hydrophobicity and lacking in signaling ligands in the molecular structures of PLA. Hydrophobicity of PLA surface impeded wetting of scaffolds as well as adsorption of proteins in the culture medium, and thus limited adhesion, proliferation, and appropriate cellular functioning. Incorporation of natural chondrocyte-friendly polymers, such as fibrin, alginate and collagen, and bio-signaling molecules, such as TGF β-1 with PLA scaffolds promoted cartilage matrix synthesis and assisted to generate of hyaline-like neo-tissues (*58*, *59*). In addition, chemical and physical surface treatments of PLA, such as alkaline treatment and plasma treatment (*60*, *61*), both of which improved surface hydrophilicity, and thus promoted protein or cell adhesion on the surface. The second shortcoming is generation of acidic degradation products. Detrimental necrotic effects of the acidic degradation products of PLGA on chondrocytes might be augmented in the closed environments, such as a knee joint, where they cannot be diluted by influx of body fluid.

Table 5 showed the 3D PLA fibrous scaffolds developed via multiple mechanisms. The scaffolds developed by incorporation of porogens could improve porosity of the scaffolds, but could not change the 2D orientations of the PLA scaffolds. Laser ablation and removal of components from electrospinning of blend of PLA and other polymers (such as PEO) could create more pores in the electrospun scaffolds. In addition, coagulation was incorporated to replace boards as receptors for the electrospun fibers. Agglomeration of fibers was avoided by solvent molecules to create large pores and increase distances among fibers. Incorporation of coarse fiber templates could increase the porosity or fluffiness of the scaffolds, but could not change the 2D orientations of PLA fibers either. So far, new methods are still in need to create scaffolds with fibers oriented in all directions in space.

Table 5. 3D electrospun fibrous scaffolds from PLA

Material	*Solvent*	*Diameter (nm)*	*Method*	*Cell culture*	*Drug loading*	*Reference*
Incorporation of porogens						
PLLA	HFIP	<1500	Montmorillonite (MMT) nano-sized platelets, NH_4HCO_3/ NaCl leaching/gas forming method	N/A	N/A	(*62*)
PLGA	HFIP	<1000	Ice crystals	Fibroblasts	N/A	(*63*)
PLGA	Chloroform	<10000	Ice crystals	N/A	N/A	(*64*)
PLGA/tricalcium phosphate	Chloroform	5000	Ice crystals	In vivo study in New Zealand white rabbits	N/A	(*65*)
Removal of components						
PLLA/PEO	Dichloromethane and DMF	<2000	Removal of PEO	N/A	N/A	(*66*)
PLGA/PEO	Chloroform	4000-9000	Removal of PEO	Human dermal fibroblasts	N/A	(*67*)
PLLA	HFIP	200-3000	Femtosecond laser ablation	BMSCs	N/A	(*68*)
Wet electrospinning						
PGA	Chloroform	300-1000	Coagulation bath (pure water, 50% tertiary-butyl alcohol, and 99% t-BuOH.)	N/A	N/A	(*69*)

Material	*Solvent*	*Diameter (nm)*	*Method*	*Cell culture*	*Drug loading*	*Reference*
PLA/siloxane-doped calcium carbonate (vaterite) particles	Chloroform	20000	Methanol bath	Fibroblasts	N/A	(*70*)
Incorporation of templates						
PDLLA/PCL	DMF and THF	<500	Templates	N/A	N/A	(*71*)
PLGA	HFIP	530	Incorporation of coarse fibrous structures	Normal human epidermal keratinocytes	N/A	(*72*)

PLLA: Poly-L-lactic acid; HFIP: 1,1,1,3,3,3 hexafluoro-2-propanol; N/A: Not applicable; PLGA: poly(lactide-*co*-glycolide); PEO: Polyethylene oxide; DMF: Dimethylformamide; THF: tetrahydrofuran; PGA: Poly (glycolic acid); DMF: Dimethylformamide; PDLLA: Poly-D,L-lactic acid.

PCL

Polycaprolactone (PCL), an aliphatic linear polyester composed of hexanoate basic units, is biodegradable due to the hydrolysis of ester bonds in physiological conditions (*73*). PCL has been approved by FDA for biomedical applications, and is regarded as a soft and hard-tissue compatible material including resorbable suture, drug delivery system and bone graft substitutes. With low melting temperature (around 60 °C) and solubility in many organic solvents, PCL has been electrospun into 2D fibrous scaffolds. As a semi-crystalline polymer with a crystallinity up to 69% (*74*), PCL has good mechanical properties and shows slower degradation kinetics than other biodegradable polyesters (*75*). Therefore, PCL is preferred for long-term in vitro cell culture before implantation into the defects, since the scaffolds would preserve their structural integrity and mechanical behavior while chondrocytes or MSCs can develop properly and secrete ECM molecules (*76*).

PCL has general drawbacks as a synthetic polymer, such as absence of cell recognition sites and hydrophobicity that is also adverse to cell adhesion. PCL scaffolds have been applied in cartilage repair. Chondrocytes or MSCs were reported to attach and proliferate on PCL films (*77*), electrospun PCL mats (*78*), and cartilaginous ECMs were reported to develop in PCL scaffolds (*79*). PCL sponges fabricated via porogen leaching method for cartilage repair in vitro and in rabbit model, respectively. Based on the in vitro study, the authors did not recommend to utilize PCL for cartilage repair, since chondrocytes did not proliferate significantly in the scaffolds though the phenotypes retained (*76*).

Table 6 shows 3D PCL fibrous scaffolds fabricated via incorporation of template, sacrificial porogens and coagulation bath. Porogens, templates and coagulation bath presented physical blocking to increase the distances among fibers. Incorporation of templates and porogens could lead to better control of porosity and pore sizes of the 3D fibrous scaffolds. However, most of the fibers could not orient in the thickness direction. In terms of incorporation of coagulation bath, the fibers could have better chances to orient in the vertical directions. However, the porosity and pore sizes of the 3D fibrous scaffolds could not be controlled.

PCL also suffered the drawbacks of other synthetic biopolymers. Similar to PLA, absence of signaling motifs and hydrophobic character of PCL did not facilitate cell adhesion and proliferation during cell culture. However, it was reported that degradation products of PCL was less harmful than PLA (*80*).

Table 6. 3D electrospun fibrous scaffolds from PCL

Material	*Solvent*	*Diameter (nm)*	*Method*	*Cell culture*	*Drug loading*	*Reference*
Incorporation of porogens						
PCL	Acetone	<750	Montmorillonite (MMT) nano-sized platelets; NH_4HCO_3/NaCl leaching/gas forming method	CFK2 chondrocytes	N/A	(*81*)
PCL/collagen or PCL/PEO	HFIP	1610 ± 270	Removal of PEO; small diameter mandrel	Human fetal osteoblasts	N/A	(*82*)
PCL	Chloroform and methanol	<1000	Na_2CO_3/$CaCO_3$	N/A	N/A	(*83*)
Wet electrospinning						
PCL	DMF and methylene chloride	1200-1900	Ethanol bath	N/A	N/A	(*84*)
Incorporation of templates						
PCL	HFIP	200-600	Incorporation of coarse fibrous structures	Chondrocytes	N/A	(*85*)
PCL	Chloroform and methanol	9300 ± 600	Incorporation of coarse fibrous structures	Osteoblasts	N/A	(*86*)

Continued on next page.

Table 6. (Continued). 3D electrospun fibrous scaffolds from PCL

Material	*Solvent*	*Diameter (nm)*	*Method*	*Cell culture*	*Drug loading*	*Reference*
PCL/PEO	THF and DMF	<1000	Rotating mandrel	MSCs	N/A	(*87*)
PCL	Acetone	500	Rotating mandrel	Tendon fibroblasts	N/A	(*88*)
PCL/PLA	Chloroform and DMF	400-500	Incorporation of coarse PLA fibers	Chondrocytes	N/A	(*89*)
PCL	DMF and methylene chloride	700-3000	Incorporation of coarse fibrous structures	N/A	N/A	(*90*)
PCL	Chloroform and DMF	1500-3500	Engineered collectors	Fibroblasts	N/A	(*91*)
PCL	Chloroform	275-350	Templates	Fibroblasts	N/A	(*92*)

HFIP: 1,1,1,3,3,3 hexafluoro-2-propanol; N/A: Not applicable; DMF: dimethylformamide.

Conclusions

In medical applications, biopolymers are preferred materials due to their similar biodegradability and biocompatibility. Proteins, polysaccharides and synthetic biopolymers have been electrospun into 3D fibrous scaffolds. Primarily, incorporation of sacrificial porogens to create voids, using templates and coagulation as receptors, and adjustment of conductivity and electrical properties of spinning dopes have been employed to obtain 3D fibrous structures from electrospinning. Applications of porogens could lead to formation of non-connective pores while the other methods resulted in formation of connective pores. The methods of using porogens, templates and coagulation bath presented physical blocking to increase the distances among the deposited fibers. However, adjustment of electrical properties of spinning dopes achieved multiple directions of fiber deposition, and thus created the most randomly oriented fibrous scaffolds. Natural biopolymers showed poor processability and thus only a few examples of 3E electrospinning were reported. Incorporation of natural and synthetic polymers was common in electrospinning of natural biopolymers. Recently, a new dissolution method was employed to develop 3D fibrous scaffolds from pure proteins, and showed potential to be applied on many other hydrophilic biopolymers. On the other hand, synthetic biopolymers was much easier to be electrospun into 3D structures, but showed biocompatibility inferior to scaffolds from natural biopolymers. With advances in dissolution and spinning technologies, fibrous protein biomaterials may have promising prospect for a variety of biomedical applications in drug delivery, tissue engineering and regenerative medicine.

References

1. Watt, F.; Huck, T. S. *Nat. Rev. Mol. Cell Biol.* **2013**, *14*, 467–473.
2. Lee, J.; Cuddihy, M. J.; Kotov, N. A. *Tissue Eng., Part B* **2008**, *14*, 61–86.
3. Xu, H.; Cai, S.; Yang, Y. *J. Biotechnol* **2014**, *184*, 179–186.
4. Ng, R.; Zang, R.; Yang, K. K.; Liu, N.; Yang, S.-T. *RSC Adv.* **2012**, *2*, 10110–10124.
5. Pham, Q. P.; Sharma, U.; Mikos, A. G. *Tissue Eng.* **2006**, *12*, 1197–1211.
6. Nam, J.; Huang, Y.; Agarwal, S.; Lannutti, J. *Tissue Eng.* **2007**, *13*, 2249–2257.
7. Park, S. Y.; Ki, C. S.; Park, Y. H.; Jung, H. M.; Woo, K. M.; Kim, H. J. *Tissue Eng., Part A* **2010**, *16*, 1271–1279.
8. Inoguchi, H.; Kwon, I. K.; Inoue, E.; Takamizawa, K.; Maehara, Y.; Matsuda, T. *Biomaterials* **2006**, *27*, 1470–1478.
9. Cai, S.; Xu, H.; Jiang, Q.; Yang, Y. *Langmuir* **2013**, *29*, 2311–2318.
10. Xu, H.; Cai, S.; Sellers, A.; Yang, Y. *RSC Adv.* **2014**, *4*, 15451–15457.
11. Ito, Y. *Biomaterials* **1999**, *20*, 2333–2342.
12. Ruoslahti, E. *Annu. Rev. Cell Dev. Biol.* **1996**, *12*, 697–715.
13. Mogford, J. E.; Davis, G. E.; Platts, S. H.; Meininger, G. A. *Circ. Res.* **1996**, *79*, 821–826.
14. Rouse, J. G.; Van Dyke, M. E. *Materials* **2010**, *3*, 999–1014.
15. Khadka, D. B.; Haynie, D. T. *Nanomed. Nanotechnol.* **2012**, *8*, 1242–1262.

16. Lutolf, M. P.; Hubbell, J. A. *Nat. Biotechnol.* **2005**, *23*, 47–55.
17. Matthews, J. A.; Wnek, G. E.; Simpson, D. G.; Bowlin, G. L. *Biomacromolecules* **2002**, *3*, 232–238.
18. Zhou, J.; Cao, C.; Ma, X.; Lin, J. *Int. J. Biol. Macromol.* **2010**, *47*, 514–519.
19. Dong, B.; Arnoult, O.; Smith, M. E.; Wnek, G. E. *Macromol. Rapid Commun.* **2009**, *30*, 539–542.
20. Jiang, Q.; Reddy, N.; Zhang, S.; Roscioli, N.; Yang, Y. *J. Biomed. Mater. Res., Part A* **2013**, *101A*, 1237–1247.
21. Choi, K.-H.; Choi, B. H.; Park, S. R.; Kim, B. J.; Min, B.-H. *Biomaterials* **2010**, *31*, 5355–5365.
22. Shin, H. J.; Lee, C. H.; Cho, I. H.; Kim, Y.-J.; Lee, Y.-J.; Kim, I. A.; Park, K.-D.; Yui, N.; Shin, J.-W. *J. Biomater. Sci., Polym. Ed.* **2006**, *17*, 103–119.
23. Ahn, S.; Koh, Y. H.; Kim, G. *J. Micromech. Microeng.* **2010**, *20*, 065015.
24. Yeo, M.; Lee, H.; Kim, G. *Biomacromolecules* **2010**, *12*, 502–510.
25. Guaccio, A.; Guarino, V.; Perez, M. A. A.; Cirillo, V.; Netti, P. A.; Ambrosio, L. *Biotechnol. Bioeng.* **2011**, *108*, 1965–1976.
26. Lee, H.; Yeo, M.; Ahn, S.; Kang, D. O.; Jang, C. H.; Lee, H.; Park, G. M.; Kim, G. H. *J. Biomed. Mater. Res., Part B* **2011**, *97*, 263–270.
27. Phipps, M. C.; Clem, W. C.; Grunda, J. M.; Clines, G. A.; Bellis, S. L. *Biomaterials* **2012**, *33*, 524–534.
28. Akkouch, A.; Zhang, Z.; Rouabhia, M. *J. Biomed. Mater. Res., Part A* **2011**, *96A*, 693–704.
29. Klumpp, D.; Rudisile, M.; Kühnle, R. I.; Hess, A.; Bitto, F. F.; Arkudas, A.; Bleiziffer, O.; Boos, A. M.; Kneser, U.; Horch, R. E. *J. Biomed. Mater. Res., Part A* **2012**, *100*, 2302–2311.
30. Altman, G. H.; Diaz, F.; Jakuba, C.; Calabro, T.; Horan, R. L.; Chen, J.; Lu, H.; Richmond, J.; Kaplan, D. L. *Biomaterials* **2003**, *24*, 401–416.
31. Jin, H.-J.; Kaplan, D. L. *Nature* **2003**, *424*, 1057–1061.
32. Jiang, C.; Wang, X.; Gunawidjaja, R.; Lin, Y. H.; Gupta, M. K.; Kaplan, D. L.; Naik, R. R.; Tsukruk, V. V. *Adv. Funct. Mater.* **2007**, *17*, 2229–2237.
33. Liu, H.; Fan, H.; Wang, Y.; Toh, S. L.; Goh, J. C. H. *Biomaterials* **2008**, *29*, 662–674.
34. Shangkai, C.; Naohide, T.; Koji, Y.; Yasuji, H.; Masaaki, N.; Tomohiro, T.; Yasushi, T. *Tissue Eng.* **2007**, *13*, 483–492.
35. Li, C.; Vepari, C.; Jin, H.-J.; Kim, H. J.; Kaplan, D. L. *Biomaterials* **2006**, *27*, 3115–3124.
36. Chen, F. H.; Rousche, K. T.; Tuan, R. S. *Nat. Clin. Pract. Rheumatol.* **2006**, *2*, 373–382.
37. Jin, H.-J.; Chen, J.; Karageorgiou, V.; Altman, G. H.; Kaplan, D. L. *Biomaterials* **2004**, *25*, 1039–1047.
38. Ki, C. S.; Kim, J. W.; Hyun, J. H.; Lee, K. H.; Hattori, M.; Rah, D. K. *J. App. Polym. Sci.* **2007**, *106*, 3922–3928.
39. Soffer, L.; Wang, X.; Zhang, X.; Kluge, J.; Dorfmann, L.; Kaplan, D. L.; Leisk, G. *J. Biomater. Sci., Polym. Ed* **2008**, *19*, 653–664.
40. Wang, S.; Zhang, Y.; Wang, H.; Yin, G.; Dong, Z. *Biomacromolecules* **2009**, *10*, 2240–2244.
41. Zhou, J.; Cao, C.; Ma, X. *Int. J. Biol. Macromol.* **2009**, *45*, 504–510.

42. Park, S. Y.; Ki, C. S.; Park, Y. H.; Jung, H. M.; Woo, K. M.; Kim, H. J. *Tissue Eng., Part A* **2010**, *16*, 1271–1279.
43. Lee, O. J.; Ju, H. W.; Kim, J. H.; Lee, J. M.; Ki, C. S.; Kim, J.-H.; Moon, B. M.; Park, H. J.; Sheikh, F. A.; Park, C. H. *J. Biomed. Nanotechnol.* **2014**, *10*, 1294–1303.
44. Xu, H.; Yang, Y. *ACS Sustainable Chem. Eng.* **2014**, *2*, 1404–1410.
45. Xu, H.; Cai, S.; Yang, Y. *Langmuir* **2014**, *30*, 8461–8470.
46. Reddy, N.; Yang, Y. *Trends Biotechnol.* **2011**, *29*, 490–498.
47. Jiang, Q.; Yang, Y. *J. Biomater. Sci., Polym. Ed.* **2011**, *22*, 1393–1408.
48. Xu, H.; Jiang, Q.; Reddy, N.; Yang, Y. *J. Mater. Chem.* **2011**, *21*, 18227–18235.
49. Xu, H.; Zhang, Y.; Jiang, Q.; Reddy, N.; Yang, Y. *J. Environ. Manage.* **2013**, *125*, 33–40.
50. Lee, K. Y.; Jeong, L.; Kang, Y. O.; Lee, S. J.; Park, W. H. *Adv. Drug Delivery Rev.* **2009**, *61*, 1020–1032.
51. Bouhadir, K. H.; Lee, K. Y.; Alsberg, E.; Damm, K. L.; Anderson, K. W.; Mooney, D. J. *Biotechnol. Prog.* **2001**, *17*, 945–950.
52. Shim, I. K.; Suh, W. H.; Lee, S. Y.; Lee, S. H.; Heo, S. J.; Lee, M. C.; Lee, S. J. *J. Biomed. Mater. Res., Part A* **2009**, *90A*, 595–602.
53. Wang, W.; Itoh, S.; Matsuda, A.; Ichinose, S.; Shinomiya, K.; Hata, Y.; Tanaka, J. *J. Biomed. Mater. Res., Part A* **2008**, *84A*, 557–566.
54. Wang, W.; Itoh, S.; Konno, K.; Kikkawa, T.; Ichinose, S.; Sakai, K.; Ohkuma, T.; Watabe, K. *J. Biomed. Mater. Res., Part A* **2009**, *91A*, 994–1005.
55. Bonino, C. A.; Efimenko, K.; Jeong, S. I.; Krebs, M. D.; Alsberg, E.; Khan, S. A. *Small* **2012**, *8*, 1928–1936.
56. Kim, T. G.; Chung, H. J.; Park, T. G. *Acta Biomater.* **2008**, *4*, 1611–1619.
57. Kenley, R. A.; Lee, M. O.; Mahoney, T. R.; Sanders, L. M. *Macromolecules* **1987**, *20*, 2398–2403.
58. Bai, H. Y.; Chen, G. A.; Mao, G. H.; Song, T. R.; Wang, Y. X. *J. Biomed. Mater. Res., Part A* **2010**, *94*, 539–546.
59. Han, S. H.; Kim, Y. H.; Park, M. S.; Kim, I.; Shin, J. W.; Yang, W. I.; Jee, K. S.; Park, K. D.; Ryu, G. H.; Lee, J. W. *J. Biomed. Mater. Res., Part A* **2008**, *87*, 850–861.
60. Park, G. E.; Pattison, M. A.; Park, K.; Webster, T. J. *Biomaterials* **2005**, *26*, 3075–3082.
61. Wan, Y.; Qu, X.; Lu, J.; Zhu, C.; Wan, L.; Yang, J.; Bei, J.; Wang, S. *Biomaterials* **2004**, *25*, 4777–4783.
62. Lee, Y. H.; Lee, J. H.; An, I.-G.; Kim, C.; Lee, D. S.; Lee, Y. K.; Nam, J.-D. *Biomaterials* **2005**, *26*, 3165–3172.
63. Leong, M. F.; Rasheed, M. Z.; Lim, T. C.; Chian, K. S. *J. Biomed. Mater. Res., Part A* **2009**, *91*, 231–240.
64. Simonet, M.; Schneider, O. D.; Neuenschwander, P.; Stark, W. J. *Polym. Eng. Sci.* **2007**, *47*, 2020–2026.
65. Schneider, O. D.; Weber, F.; Brunner, T. J.; Loher, S.; Ehrbar, M.; Schmidlin, P. R.; Stark, W. J. *Acta Biomater.* **2009**, *5*, 1775–1784.
66. Whited, B. M.; Whitney, J. R.; Hofmann, M. C.; Xu, Y.; Rylander, M. N. *Biomaterials* **2011**, *32*, 2294–2304.

67. Gang, E.; Ki, C.; Kim, J.; Lee, J.; Cha, B.; Lee, K.; Park, Y. *Fibers Polym.* **2012**, *13*, 685–691.
68. Lee, B. L.-P.; Jeon, H.; Wang, A.; Yan, Z.; Yu, J.; Grigoropoulos, C.; Li, S. *Acta Biomater.* **2012**, *8*, 2648–2658.
69. Yokoyama, Y.; Hattori, S.; Yoshikawa, C.; Yasuda, Y.; Koyama, H.; Takato, T.; Kobayashi, H. *Mater. Lett.* **2009**, *63*, 754–756.
70. Kasuga, T.; Obata, A.; Maeda, H.; Ota, Y.; Yao, X.; Oribe, K. *J. Mater. Sci. Mater. Med.* **2012**, *23*, 2349–2357.
71. Zhang, D.; Chang, J. *Nano Lett.* **2008**, *8*, 3283–3287.
72. Kim, S. J.; Jang, D. H.; Park, W. H.; Min, B.-M. *Polymer* **2010**, *51*, 1320–1327.
73. Hu, Y; Zhang, L; Cao, Y; Ge, H; Jiang, X; Yang, C. *Biomacromolecules* **2004**, *5*, 1756–62.
74. Labet, M.; Thielemans, W. *Chem. Soc. Rev.* **2009**, *38*, 3484–3504.
75. Lam, C. X.; Hutmacher, D. W.; Schantz, J. T.; Woodruff, M. A.; Teoh, S. H. *J. Biomed. Mater. Res., Part A* **2009**, *90*, 906–919.
76. Garcia-Giralt, N; Izquierdo, R; Nogués, X; Perez-Olmedilla, M; Benito, P; Gómez-Ribelles, JL; Checa, MA; Suay, J; Caceres, E; Monllau, JC *J. Biomed. Mater. Res., Part A* **2008**, *85*, 1082–9.
77. Ishaug-Riley, S. L.; Okun, L. E.; Prado, G.; Applegate, M. A.; Ratcliffe, A. *Biomaterials* **1999**, *20*, 2245–2256.
78. Li, W.-J.; Tuli, R.; Okafor, C.; Derfoul, A.; Danielson, K. G.; Hall, D. J.; Tuan, R. S. *Biomaterials* **2005**, *26*, 599–609.
79. Hoque, M. E.; San, W. Y.; Wei, F.; Li, S.; Huang, M.-H.; Vert, M.; Hutmacher, D. W. *Tissue Eng., Part A* **2009**, *15*, 3013–3024.
80. Taylor, M. S.; Daniels, A. U.; Andriano, K. P.; Heller, J. *J. App. Biomater.* **1994**, *5*, 151–157.
81. Nam, J.; Huang, Y.; Agarwal, S.; Lannutti, J. *Tissue Eng.* **2007**, *13*, 2249–2257.
82. Ekaputra, A. K.; Prestwich, G. D.; Cool, S. M.; Hutmacher, D. W. *Biomacromolecules* **2008**, *9*, 2097–2103.
83. Wang, Y.; Wang, B.; Wang, G.; Yin, T.; Yu, Q. *Polym. Bull.* **2009**, *63*, 259–265.
84. Hong, S.; Kim, G. *Appl. Phys. A* **2011**, *103*, 1009–1014.
85. Park, S. H.; Kim, T. G.; Kim, H. C.; Yang, D.-Y.; Park, T. G. *Acta Biomater.* **2008**, *4*, 1198–1207.
86. Yeo, M. G.; Kim, G. H. *Chem. Mater.* **2011**, *24*, 903–913.
87. Baker, B. M.; Gee, A. O.; Metter, R. B.; Nathan, A. S.; Marklein, R. A.; Burdick, J. A.; Mauck, R. L. *Biomaterials* **2008**, *29*, 2348–2358.
88. Bosworth, L. A.; Alam, N.; Wong, J. K.; Downes, S. *J. Mater. Sci. Mater. Med.* **2013**, *24*, 1605–1614.
89. Thorvaldsson, A.; Stenhamre, H.; Gatenholm, P.; Walkenström, P. *Biomacromolecules* **2008**, *9*, 1044–1049.
90. Kim, T. G.; Chung, H. J.; Park, T. G. *Acta Biomater.* **2008**, *4*, 1611–1619.
91. Vaquette, C.; Cooper-White, J. J. *Acta Biomater.* **2011**, *7*, 2544–2557.
92. Lee, J.; Jang, J.; Oh, H.; Jeong, Y. H.; Cho, D.-W. *Mater. Lett.* **2013**, *93*, 397–400.

Chapter 8

Phase Separated Fibrous Structures: Mechanism Study and Applications

Qiuran Jiang*,[1,2,3] and S. M. Kamrul Hasan[1,2,3]

[1]Key Laboratory of Textile Science andTechnology, Ministry of Education, College of Textiles, Donghua University, Shanghai 201620, P. R. China
[2]Shanghai Key Laboratory of Advanced Micro and Nano Textile Materials, College of Textiles, Donghua University, Shanghai 201620, P. R. China
[3]Department of Technical Textiles, College of Textiles, Donghua University, Shanghai 201620, P. R. China
***Tel: +86-21-67792380. Fax: +86-21-67792627. E-mail: jj@dhu.edu.cn.**

Phase separation is one of the most important technologies to fabricate ultrafine fibers. Different from electrospinning, phase separation can produce fibers arranged in three dimension and thus has unique advantages, such as mimicking the fibrous structure in extracellular matrix more closely. However, to precisely control the architectures of phase separated scaffolds is challenging. This chapter introduces the current understanding about the mechanisms of phase separation and summarizes the factors, which can control the morphologies of resultant scaffolds. Methods to produce fibrous phase separated scaffolds are discussed in this chapter. Applications and limitations are also analyzed.

Introduction

Fibers in extracellular matrices (ECMs), mainly collagen/elastin fibers and interfibrillary matrix fibers of proteoglycan, are around 10 to 500 nm (*1*). Fiber assembling controls the structure and mechanical properties of ECMs. These fibers may locate under cells to form flat fibrous mats, such as in basement membranes, or three dimensionally (3D) distribute around cells with a random orientation, such as in interstitial matrix (*2*). The topography of ECMs is considerably important for regulating essential cellular functions, such as attachment, spreading, migration, proliferation, differentiation and morphogenesis, and affects the biomechanical behaviors and biofunctions of the tissues/organs (*2*). Hence, tissue engineering scaffolds need to be designed and constructed according to the distinct architectures of ECMs in the targeted tissues/organs. The two dimensional (2D) fibrous structure of ECMs has been successfully emulated via technologies, like electrospinning. However, to mimic the 3D structure of ECMs is challenging. Phase separation is one of the established technology, that can produce 3D fibrous substances. This method is mold-based and efficient.

This chapter aims at providing a background on the technology of phase separation. The importance to emulate the fibrous 3D ECMs is introduced first, and methods of phase separation to form 3D fibrous bulks are discussed in details. Subsequently, the mechanism to control scaffolds morphology during the phase separation process is analyzed. The information about the main applications and limitations of using phase separation to make 3D fibrous bulks is provided.

Necessity of Three Dimensional Substrates

Huge numbers of published works are related to the technology of electrospinning, which has been widely used to produce flat fibrous scaffolds for various applications. These scaffolds are able to mimic the 2D ECMs (*3*), but difficult to emulate the 3D ECMs. A number of 3D models (i.e. cells seeded within 3D matrices) have been developed for a range of tissues (*4–6*). A truly 3D environment for cells can provide cues necessary for most cells to form appropriate physiological tissue structures both *in vivo* and *in vitro* (*7*). Three dimantional architecture is favored for cell adhesion (*8–11*), and facilitates establishing solute concentration gradients in a local pericellular scale or a tissue-scale. Furthermore, a 3D environment affects morphogenetic and remodeling events, such as epithelial acinar development, mammary (*12–15*) and kidney epithelial-cell activities (*16*) and morphogenesis (*17*). In native ECMs, where cells interact with neighboring cells, fibrillar collagen and various structural and nonstructural proteins, organization of cells in 3D promotes proper transmission of mechanical cues and expressions of receptor among cells and between cells and substrates. The interaction dramatically affects integrin ligation, cell contraction and associated with intracellular signaling (*18*, *19*). Even a small change in mechanical stress can elicit changes in the local distribution of cell-secreted proteases, such as cysteine, which hydrolyzes proteins, morphogens, which enhance cell growth and migration, and chemokines, which excite cell responses to certain chemicals and directe cell to cell signaling and anisotropy in engineered tissues (*20*). The abilities to subtly transmit mechanical signals and shield cells from tensile forces are associated with the composition and 3D architecture of native ECMs.

Necessity of Fibrous Morphology

Besides a 3D structure, tissue engineering scaffolds require for an interconective porous morphology with a high porosity to ensure efficient nutrient transportation throughout scaffolds. Recently, many researches proved that along with appropriate pore size and porosity, nanofibrous structures of scaffolds are able to promote cell adhesion and proliferation due to enhanced protein adsorption for their (nanofibers) high surface area to volume ratios (*8–11*, *21–26*). In the nanofibrous environment, cells can develop more filapodia and move along fibers, known as *contact guidance*, which in turn improves cell adhesion to substrates (*27*). If the interfiber gap is relatively small, cells can adhere to adjacent fibers and reduce focal contacts and form a sail-like structure (*28*). Moreover, nanofibrous scaffolds resemble collagen fibrils in native ECMs. Therefore, current researches in tissue engineering strive towards the development of 3D nanofibrous scaffolds.

3D Fibrous Structure Produced via Phase Separation

The phase separation technique can produce particles, nanofibrous or macroporous structures. This technique is based on thermodynamic demixing of a homogeneous polymer-solvent solution into a polymer-rich phase and a polymer-lean phase, usually by either exposing the solution to another immiscible solvent or cooling the solution below a binodal solubility curve (*29*, *30*). There are mainly two types of phase separation mechanisms, which have been used to produce 3D scaffolds, namely non-solvent induced phase separation (NIPS) and thermally induced phase separation (TIPS). NIPS process uses a non-solvent to induce phase separation in a polymer/solvent/non-solvent ternary or even quaternary system. The initially homogeneous solution becomes unstable with the addition of the non-solvent, and the solution separates into a polymer-rich phase and a polymer-lean phase. The polymer-rich phase develops a continuous matrix, while the polymer-lean phase, which mainly consists of the mixture of solvents, flows through the matrix resulting in porous channels in the matrix. When the phase separation of the polymer is completed, this process gives rise to a continuous interconnective porous structure. But, there is fairly any work, that uses NIPS to produce scaffolds with fibrous morphology. Contrarily, TIPS process has been employed to produce 3D fibrous scaffolds and the interfiber gas and fiber length as well as transverse scales can be controlled.

Thermally Induced Phase Separation

Thermally induced phase separation uses thermal energy as a latent solvent to induce phase separation (*31*). The polymer solution is quenched below the freezing point of the solvent and subsequently freeze-dried to produce porous structure. There are two types of TIPS techniques available for scaffold preparation. One is liquid-liquid (L-L) TIPS and the other is solid-liquid (S-L) TIPS. In order to study the temperature dependence of L-L TIPS, a typical temperature concentration phase diagram for a binary polymer–solvent system with an upper critical solution temperature is presented in Figure 1. The upper critical solution temperature is a temperature above which no phase transition occurs at any concentration. The two curves in Figure 1 are the binodal curve, where first derivative of Gibb's free energy of mixing is equal to zero and represents the thermodynamic equilibrium of L-L demixing, and the spinodal curve, where Gibb's free energy of mixing second derivative is equal to zero. When the temperature of a solution is above the binodal curve, the polymer solution is homogeneous. The maximum point, at which both the binodal and the spinodal curves merge, is the critical point of the system. The area under the spinodal curve is the unstable region and the area located in the zone between the binodal and spinodal curves is the metastable region, where the second derivative of Gibb's free energy of mixing is larger than zero. When the solution is quenched in the metastable region for a certain solution concentration (off critical quench), a polymer-rich phase and a solvent-rich phase (polymer-lean phase) coexist in the L–L demixing region depending on the concentrations of components. L–L demixing in the metastable region can produce a poor connected stringy or beady morphology, when the polymer concentration is lower or higher than the critical point concentration, respectively (*32*). The generation of these structures is based on a nucleation and growth mechanism. If the system is quenched into the unstable region (critical quench), the L-L phase separation takes place by a spinodal mechanism, where the second derivative of Gibb's free energy of mixing is negative, resulting in a well-interconnected porous structure. Figure 2 shows morphologies of gelatin scaffolds prepared in different regions of the phase diagram (*33*). All the scaffolds were produced at the same temperature, however, from solutions with different concentrations. The solution with a high concentration of 1 % was probably in the unstable region and produced scaffolds with a sponge morphology, while lowering the concentration of solutions drove the solution system into the metastable region and produced beady scaffolds (0.1 %). Continurously decrease in solution concentration to around 0.01 % generated fibrous scaffolds.

The difference between the critical quench and off-critical quench is that, for critical quench, the system enters the unstable region and spontaneously decomposes into two equilibrium concentration phases at a certain temperature, but for off-critical quench the system enters into the metastable region and phase separation takes place by a nucleation and growth mechanism, where free energy and concentration fluctuations are needed to overcome the nucleation barrier and to initiate phase separation. In case of polydisperse polymers, cloud point curves are constructed instead of the binodal curve, which is only applied to a polymer with a monodisperse molecular weight (*34*, *35*). In a ternary system, addition of nonsolvent to the polymer-solvent system increases its cloud by reducing polymer-diluent interaction, which may results in formation of polymer droplets with a greater droplet domain (Figure 3). For L-L TIPS, quenching induces the possibility of different gelation behaviors, but S-L TIPS limits the possibility in this respect. S–L TIPS takes place, when a polymer solution is cooled quickly to freeze the solvent and there is no enough time for L–L phase separation to occur (no time for gelation). In this process, the quenching temperature is a key factor in controlling the morphology of the resultant polymer matrix, because the crystalline morphologies of the polymer and the solvent are dependent on the crystallization temperature (*36*).

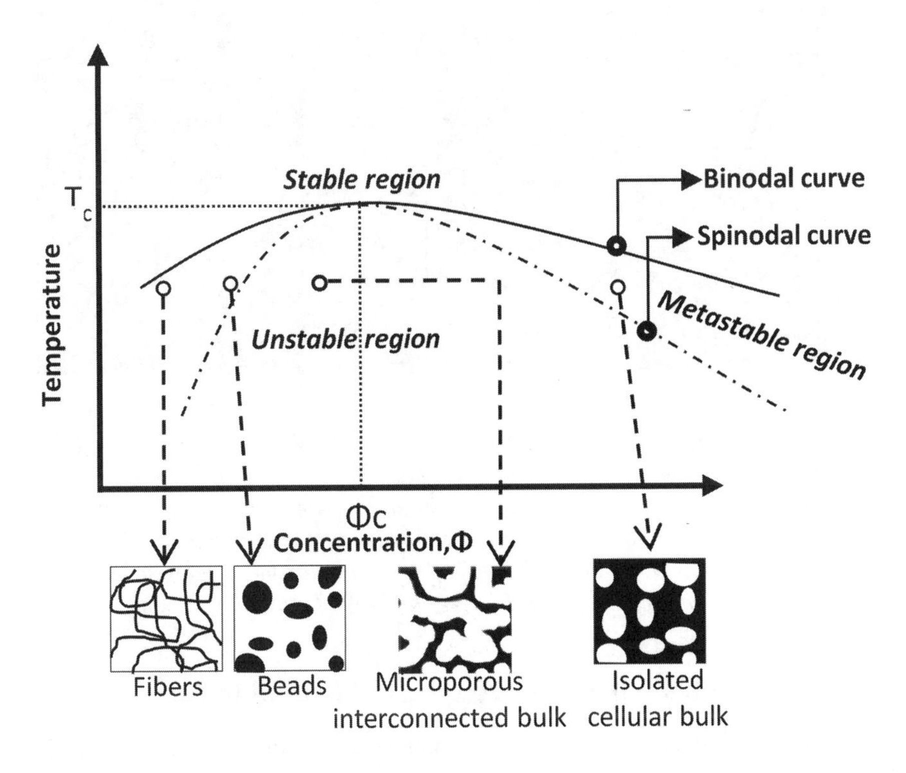

Figure 1. A schematic representation of binary phase diagram of a polymer solution.

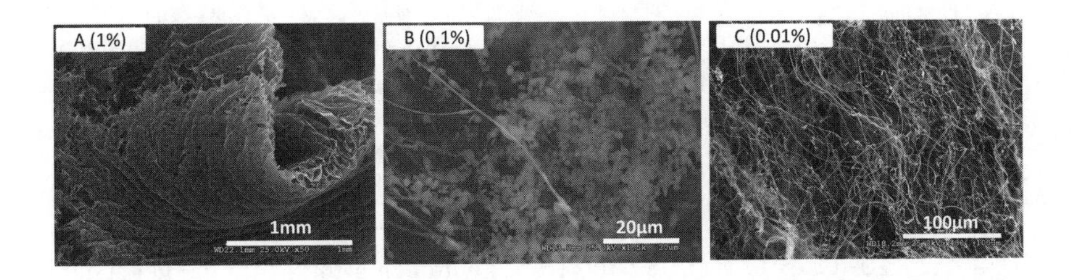

Figure 2. SEM images of gelatin scaffolds prepared via ULCPS from gelation solutions with concentrations at 1wt.% (A), 0.1wt.% (B) and 0.01wt.% (C).

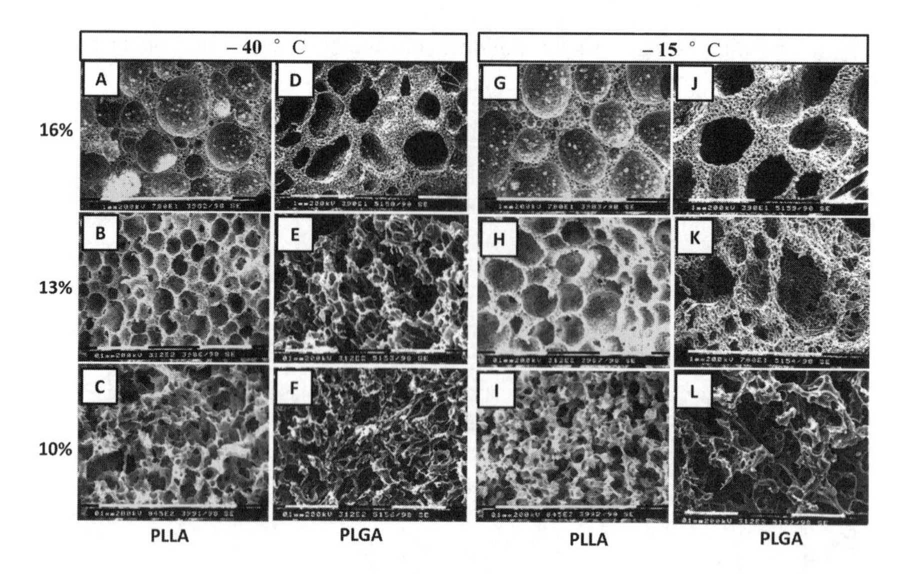

Figure 3. Influence of the non-solvent amount on scaffold structure. SEM images of the cross-sections of scaffolds produced at −40°C (A-F) and −15 °C (G-L). (A-C, G-I) scaffolds prepared from PLLA solutions with 16 w/v %, 13 w/v % and 10 w/v % of water in dioxane. (D-F, J-L) scaffolds prepared from PLGA solutions with 16 w/v %, 13 w/v % and 10 w/v % of water in dioxane. (Reproduced with permission from reference (37). Copyright (1999) John Wiley & Sons, Inc.).

Factors Controlling the Morphology of Scaffolds during TIPS

One of the advantages of TIPS is that multi-scale morphology can be obtained by adjusting various thermodynamic and kinetic parameters. Nevertheless, this advantage can also be considered as a disadvantage when a specific single structure is required. Controlling key parameters to produce scaffolds with desired architecture is chanllenging and well understanding of the phase separation mechanism is required (*37*).

Effect of Gelation Temperature

To employ L-L TIPS for fabrication nanofibrous scaffolds involves five characteristic steps: polymer dissolving, gelation, solvent exchanging, freezing and freeze-drying. During the gelation process, the gelation temperature (T_{gel}) exerts critical influence to the architecture of the resultant scaffolds (*38*, *39*). He *et. al.* observed the scaffolds fabricated from the same poly-(caprolactone)-block-poly L-lactide (PCL-*b*-PLLA) solutions in tetrahydrofuran (THF) at different T_{gel} (−40 °C, −20 °C, 4 °C, 8 °C, 12 °C, 16 °C) with aging for 2 h in the gel status. Although all the scaffolds were nanofibrous networks, only the structures of the scaffolds produced at lower temperatures (−40 °C and −20 °C) were uniform. Pieces of PCL-*b*-PLLA chips were observed in the obtained scaffolds, when T_{gel} increased concomitantly. Additionally, those fibers produced at higher T_{gels} were thicker than those formed at low T_{gels}. Shao *et al.* reported the same effect of T_{gel} on the morphology of precipitated poly L-lactide (PLLA) via TIPS (*40*).

Effect of Polymer Concentration

Polymer concentration plays a vital role in adjusting scaffold morphology. Some researchers found that there is a general trend of increase in fiber diameter with the rise of polymer concentration, but this effect is only significant, when polymer concentration changes in higher concentration ranges (*36*, *41*). During TIPS, polymer concentration has significant effect on pore size and distribution. Sara *et al.* proved that a low polymer (PLLA) concentration led to the formation of nonuniform larger pores (*42*). This phenomenon might connect to the incomplete microstructure formation at a low concentration (*43*). Sara *et al.* also found that higher PLLA concentrations increased the thickness of pore walls and could also uniform the pore size (*42*).

Base on this concept, Jiang *et al.* developed an ultralow concentration phase separation technology (ULCPS) (*33*). By this means, gelatin ultrafine fibrous scaffolds with 3D randomly orientation could be produced (Figure 4). Unlike convential TIPS scaffolds, these fiber bulks had an extremely loose structure. The specific pore volume of a gelatin electrospun mats was around 10.6 cm^3/mg (Figure 4 B), but the specific pore volume of the ULCPS gelatin scaffolds could reach 259.5 cm^3/mg (Figure 4 C). The interfiber gaps in the conventional TIPS gelatin scaffolds were around 2 μm, while interfiber gaps in the ULCPS scaffolds distributed in two regions 30-140 μm and 0.5-1.5 μm. The big pores in these scaffolds facilitated cell infiltration, and promoted cell growth.

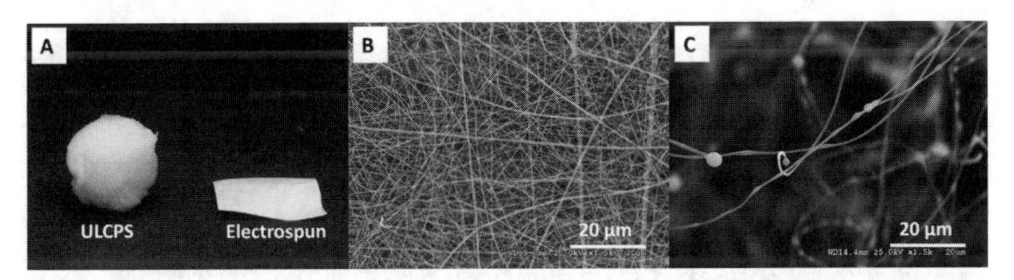

Figure 4. Morphological comparison between ULCPS and electrospun scaffolds. (A) digital image of ULCPS and electrospun gelatin scaffolds with the same weight;(B) SEM image of the electrospun gelatin scaffold; (C) SEM image of the ULCPS gelatin scaffold.

Effect of Cooling Rate

Cooling rate during phase separation has a significant effect on scaffold morphology. Fast cooling is unfavorable for single crystal nucleation and growth, and hence smaller crystals are genereated resulting in the decrease in pore size. This phenomenon has also been observed in PLLA/dioxane/water ternary system by He *et al.* (*44*). Pores near the wall of scaffolding molds were typically smaller than those in the middle, because the cooling rate is higher near the wall (*42*).

Effect of Polymer Properties

Polymer molecular structure affacts both of the scaffold morphology and the optimal processing conditions. Compared to homopolymers, the blocks in co-polymers prevents phase separation. As a result, the co-polymer systems can self-organize into different complex structures. He *et al.* reported that, with a similar fiber diameter, nano-fibrous structures were obtained at a relative higher T_{gels} in the pure PLLA system than in the PCL-b-PLLA system (*39*). Polymer crystallization behavior plays an important role in nanofiber formation. Shao *et al.* suggested that the crystallization of PLLA is a decisive step of nanofiber formation with the lateral assembly of nanofibrils in the early stage of phase separation (*45*). Additionally, the molecular weight of polymer has significant effect on the pore interconnectivity. Bernke *et al.* revealed that the high molecular weight PLLA had a low critical polymer concentration, and thus allowed for the formation of highly interconnected porous substence with nano-fibers around 50 to 500 nm in diameter (*46*). Polymers with different semi-crystalline have different crystallization behaviors, which in turn affects the final scaffold morphology (Figure 5) (*37*). PLLA, poly lactic-co-(glycolic acid) (PLGA) and poly-DL-lactide (PDLLA) scaffolds produced via TIPS at the same concentration and the same solven-nonsolvent ratio displayed different structures. In the scaffolds of PLGA and PDLLA, larger pores could be seen. Guobao *et al.* produced 3D scaffolds with nanofibers and chips via TIPS of PLLA/ PDLLA and PLLA/PLGA blends. It was possible to control the ratio of nanofibers and chips by controlling the amount of each polymer (*47*). It is necessary for adjusting the total surface area of scaffolds and the protein adsorption capability (*47*).

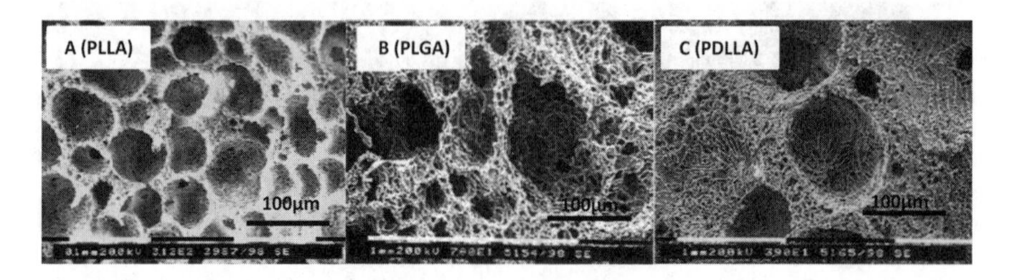

Figure 5. SEM image of PLLA (A), PLGA(B) and PDLLA(C) scaffolds produced via TIPS, where the polymer concentration and the quenching temperature are 10 w/v % and −15 °C respectively for all polymers. (Reproduced with permission from reference (37).Copyright (1999) John Wiley &Sons, Inc.)

Influence of Incorporation of Other Technology

To facilitate cell infilatration in a phase separated scaffolds, researchers have tried the combination of other technology with phase separation, and porogen leaching is one of them. By using inter-linked porogen with similar sizes can reserve the space for macropores in TIPS fibrous scaffolds (*48*). Figure 6 illustrates the process of TIPS combined with porogen leaching to make scaffolds with interconnective macropores. The porogens are put into molds and bound together during, for example, heating (*49*). The polymer solution is then cased into the mold and phase separated to generate fibrous scaffolds. Porogens are dissolved and removed from the scaffolds and leave large interconnective pores. With the combination of porogen leaching method, slightly increase in fiber diameter can be observed. Fibers diameters are still in the range of 50 to 500 nm, which are similar to the size of collagen fibers found in native ECMs (*47*).

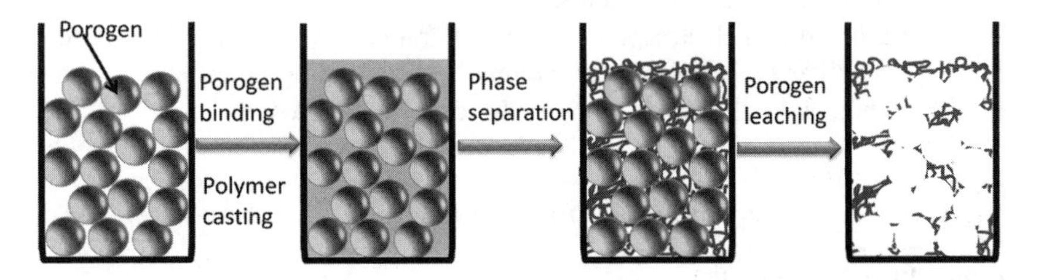

Figure 6. Schematic illustration of the process to generate interconnective pores in the thermally induced phase separated scaffolds via porogen leaching technique.

Researchers also developed fibrous scaffolds via combination of TIPS with electrospinning (*50*). PLGA solution was cased onto stacked electrospun poly–(caprolactone) (PCL) fiber mats, and phase separated.The scaffolds possessed a hierarchically porous 3D structure.

Fibrous Structure Produced via Phase Separation in an Electric Field

Ding *et al.* discovered special "fishnet-like nano-nets", when they were developing polyamide-6 (PA-6) electrospun mats (*51*). Pieces of nets were three dimetionally randomly distributed in the electrospun mats, though each piece of nets was 2D. Fibers in the nano-nets was around 20 to 30 nm in diameter, much smaller than normal electrospun fibers ranging from 100 nm to several micrometers. They called this structure as nano-fiber/net (NFN). According to Ding's studies, the polymer solution jet was instable and broken up via electrospinning into small charged polymer droplets, which then formed the nano-nets (*52*). Up to now, four possible mechanisms for the formation of nano-nets have been proposed (*53*). One of them was phase separation of charged droplets, which was the only theory raised based on the experimental observation. In the electrostatic field, the charged droplets driven by the electrostatic force depart from the tip of the spinning needle and are deformed by the drag force between the air and droplets followed by generating thin liquid films, which undergo quick phase separation to form polymer-rich domains as fiber nets and solvent-rich domains. Pores are left at positions of solvent-rich domains after the rapid solvent evaporation.

With this electrospinning/netting (ESN) technology, a combinated bulk of electrospun fiber mats at around hundreds nanometers and a large amount of nano-nets at about tens of nanometers with a controllable density and coverage rate can be obtained via a single step process. The brunauer-emmett-tellersurface area of resultant materials is increased for potential applications, such as adsorption of drugs, filtration, and sensor fabrication (*54*). Currently, the nano-nets can be fabricated from various materials, for instance, PA-6 (*54*), polyacrylic acid (*55*), poly(vinyl alcohol) (*51*), polyurethane (PU) (*56*), poly(trimethylene terephthalate) (*57*), gelatin (*58*) chitosan (*59*) and silk (*60*).

However, normal electrospun mats are the frame of NFN bulks. The compact structure is still an obstacle for penetration of relatively large substances, such as cells, and thus limits the applications of these NFN scaffolds.

Applications

Biomedical Applications

Currently, the TIPS nanofibrous materials are applied mainly in the biomedical field. The compact structure provides TIPS fibrous bulks with required mechanical properties for potential applications as bone repair scaffolds (*61*). To mimic the nanostructure and chemical composition of the native bone ECMs, Xiaohua *et al.* (*62*) and Bo *et al.* (*61*) developed TIPS gelatin scaffolds combined with silica for bone tissue engineering. Combination of several materials in TIPS scaffolds can promote cell growth. The presence of poly lactic-co-(glycolic acid) (PLGA) in a PU scaffold developed via TIPS gave rise to improved mechanical properties, cell attachment and viability (3T3 fibroblasts), when compared to the scaffolds fabricated from pure PLGA and PU (*63*). Huang *et al.* produced PLGA/nanohydroxyapatite (NHA) composite TIPS scaffolds. They found that with the increase of the PLGA concentration and hydroxyapatite (HA) content the mechanical properties and the capability in water absorption of the composite scaffolds were enhanced greatly (*64*). This reason might be the introduction of NHA. Significantly promoted cell growth and higher alkaline phosphatase activity were observed on the PLGA/NHA scaffolds, especially those with more NHA contents, rather than the pure PLGA scaffolds (*64*). The quaternary system (polymer/co-solvents/coagulant) may be applied to construct different nano-structured porous matrices from different biodegradable polymers such as PCL, PU and PLGA via TIPS. The nanofibrous polylactide (PLA) matrices (< 100 nm) fabricated by this method were found to promote the differentiation of mouse pre-osteoblasts or specific osteogenic differentiation of human bone marrow-derived mesenchymal stem cells without osteogenic induction additives (*65*). The collagen production and calcium deposition were 2.5 to 6 times greater on nano-structured PLA. The PLA nanofibers could be further grafted with chitosan or its derivatives after air plasma activation and the modified nanofibers were still osteoinductive with extra antimicrobial activity (*65*).

The NFN fibers composed of normal electrospun frame and nano-nets could also be used for biomedical applications. The researchers in Kim's group demonstrated the feasibility to culture osteoblasts on NFN scaffolds (chitosan/PA-6). NFN scaffolds provided better support for cell growth then normal electrospun control scaffolds (*66*).

However, both of the TIPS scaffolds and the NFN scaffolds share the same problem: limited cell infiltration ability due to their compact structures. Therefore, it is hard to reconstruct tissues emulating the native ones using these technologies. To solve this problem, Xiaohua *et al.* first prepared 3D nano-fibrous gelatin scaffolds with well-defined interconnetive pores via TIPS and porogen leaching techniques (*62*). The resultant gelatin nanofibrous scaffolds exhibited high porosity, interconnectivity and good mechanical strength and it also showed excellent biocompatibility and mechanical stability. They further incorporated bone-like apatite onto the surface of gelatin scaffolds through a stimulated body fluid (SBF) incubation method. The addition of apatite enhanced the differentiation of osteoblastic cells and the mechanical strength of the scaffold (*62*). By this way, cells could penetrate into the inner domains of scaffolds, but the surface attached by cells was still 2D, and it was hard to embed cells in the bulk of fibers.

To enhance cell infiltration to phase separated (PS) fiber bulks, Jiang *et al.* employed the technology of ULCPS (*33*). With 23.5 times higher specific pore volume, the ULCPS scaffolds allowed enhanced cell infiltration. Nevertheless, the mechanical perperites were sacrificed. The low resistance to compression of these scaffolds restricted the applications for bone reconstruction, but the scaffolds could be used for soft tissue repair. Thus, scaffolds with sufficient mechanical properties and cell infiltration property have yet to be developed.

Other Applications

Currently, limited records could be found on employing TIPS or ULCPS fibers for nonbiomeidal applications. However, the NFN fibers have been used in many areas. Since the NFN fiber bulks possess a compact structure with a comparatively large surface area, they have been employed as filters for gas and water filtrations. Ding's group developed a two-tier filter with PA-66 NFN deposited on a nonwoven polypropylene (PP) scaffold (*67*). The filtration efficiency to NaCl aerosols (300 nm in diameter) could achieve 99.9 % after filtration for 90 min. Due to the presence of the phase separted nano-nets, the filtration efficiency could be enhanced and the pressure drop was lowered comparing to the performance of the filter with normal electrospun fibers. Besides, the potential of using NFN to fabric ultrasensitive quartz crystal microbalance (QCM) and colorimetric sensors has been proved. Polyacrylic acid (PAA) NFN membranes was coated on QCM to produce a high sensitive humidity sensor with long term stability (*68*). Sensors for the detection of trimethylamine was developed via coating QCM with PAA/NaCl composite NFN membrances (*69*). The high sensitivies of NFN membrances coated sensors were attributed to their structure with a high surface area that provides abundant absorbing sites, and enable obvious responses.

Limitations

Although highly porous fibrous scaffolds from various materials have been developed via phase separation, thorogh understanding about the mechanism of phase separation has not be clearly revealed. Thus, it is still difficult to effectively control the process and the architechure of the resultant materials, especially when the component of a phase saparation system is complex. Additionally, the scaffolds produced by TIPS and ESN showed compact structures, though they had required mechanical properties, penetration of, such as, cells into these scaffolds were difficult. ULCPS technology could solve the infiltration problem, but sacrificed mechanical properties. Up to now, to produce scaffolds with sufficient mechanical properties and highly permeable structures is still a challenge. Meanwhile, comparing with tranditional fabrication technologies of fibers in industry, the productivity of phase separation is still too low for large scale industrial applications, and the cost of processing is too high. It is also difficult to execute a continuous production.

Summary

Besides electrospinning and molecular self-assambly, phase separation is another effective way to produce nanofibrous bulks, especially in 3D. The mechanism of phase separation is complex and has not been thoroughly revealed. In general, it can be induced by invation of non-solvent, temperature change, or forces in the electrostatic field. By adjusting processing parameters, such as the types of polymers, solvent to non-solvent ratio, polymer concentration, quenching temperature and cooling rate, it is possible to control the morpholoy of resultant phase separated scaffolds. Phase separation alone or with combination of other methods, such as porogen leaching or electrospinning, can produce nanofibrous porous substrates, that meet the intrinsic requirements for many appliations, for instance, tissue engineering scaffolds, filters, and sensors. Limiations of phase separation in, for instance, difficulties in continurous production with a high productivity, fabrication of bulks with high mechanical properties and large pores, hindered future applications of this technology. Further modification of this technology is still in need.

References

1. Liu, X.; Ma, P. X. *Biomaterials* **2009**, *30*, 4094–4103.
2. Dvir, T.; Timko, B. P.; Kohane, D. S.; Langer, R. *Nat. Nanotechnol.* **2011**, *6*, 13–22.
3. Chen, V. J.; Smith, L. A.; Ma, P. X. *Biomaterials* **2006**, *27*, 3973–3979.
4. Abbott, A. *Nature* **2003**, *424*, 870–872.
5. Langer, R.; Tirrell, D. A. *Nature* **2004**, *428*, 487–492.
6. Lee, J.; Cuddihy, M. J.; Kotov, N. A. *Tissue Eng., Part B* **2008**, *14*, 61–86.
7. Harrison, K. Introduction to Polymeric Scaffolds for Tissue Engineering. In *Biomedical Polymers*, 1st ed.; Jenkins, M., Ed.; Woodhead Publishing Limited: Cambridge, U.K., 2007; Vol. 1, pp 1–32.

8. Woo, K. M.; Chen, V. J.; Ma, P. X. *J. Biomed. Mater. Res., Part A* **2003**, *67*, 531–537.
9. Thomas, J. W.; Michael, C. W.; Janice, L. M.; Rachel, L. P.; Jeremiah, U. E. *Nanotechnology* **2004**, *15*, 48–54.
10. Nair, L. S.; Bhattacharyya, S.; Laurencin, C. T. Nanotechnology and Tissue Engineering: The Scaffold Based Approach. In *Nanotechnologies for the Life Science: Tissue,Cell and Organ Engineering*; Kumar C., Ed.; Wiley-VCH Verlag GmbH & Co. KGaA: Weinheim, 2007; Vol. 9, pp 1−56.
11. Gupta, D.; Venugopal, J.; Prabhakaran, M. P.; Dev, V. R.; Low, S.; Choon, A. T.; Ramakrishna, S. *Acta Biomater.* **2009**, *5*, 2560–2569.
12. Bissell, M. J.; Rizki, A.; Mian, I. S. *Curr. Opin. Cell. Biol.* **2003**, *15*, 753–762.
13. Debnath, J.; Brugge, J. S. *Nat. Rev. Cancer* **2005**, *5*, 675–688.
14. Paszek, M. J.; Weaver, V. M. *J. Mammary Gland Biol. Neoplasia* **2004**, *9*, 325–342.
15. Wozniak, M. A.; Desai, R.; Solski, P. A.; Der, C. J.; Keely, P. J. *J. Cell Biol.* **2003**, *163*, 583–595.
16. Zegers, M. M.; O'Brien, L. E.; Yu, W.; Datta, A.; Mostov, K. E. *Trends Cell Biol.* **2003**, *13*, 169–176.
17. Langer, R. *Chem. Eng. Sci.* **1995**, *50*, 4109–4121.
18. Knight, B.; Laukaitis, C.; Akhtar, N.; Hotchin, N. A.; Edlund, M.; Horwitz, A. R. *Curr. Biol.* **2000**, *10*, 576–585.
19. Roskelley, C. D.; Desprez, P. Y.; Bissell, M. J. *Proc. Natl. Acad. Sci. U.S.A.* **1994**, *91*, 12378–12382.
20. Griffith, L. G.; Swartz, M. A. *Nat. Rev. Mol. Cell Biol.* **2006**, *7*, 211–224.
21. Shanmugasundaram, S.; Griswold, K. A.; Prestigiacomo, C. J.; Arinzeh, T.; Jaffe, M. *Proceedings of the IEEE 30th Annual Northeast Bioengineering Conference*, Springfield, MA, 2004; pp 140−141.
22. Bondar, B.; Fuchs, S.; Motta, A.; Migliaresi, C.; Kirkpatrick, C. J. *Biomaterials* **2008**, *29*, 561–572.
23. Li, W. J.; Jiang, Y. J.; Tuan, R. S. *Tissue Eng.* **2006**, *12*, 1775–1785.
24. Sun, T.; Norton, D.; McKean, R. J.; Haycock, J. W.; Ryan, A. J.; MacNeil, S. *Biotechnol. Bioeng.* **2007**, *97*, 1318–1328.
25. Kwon, I. K.; Kidoaki, S.; Matsuda, T. *Biomaterials* **2005**, *26*, 3929–3939.
26. Tuzlakoglu, K.; Bolgen, N.; Salgado, A. J.; Gomes, M. E.; Piskin, E.; Reis, R. L. *J. Mater. Sci. Mater. Med.* **2005**, *16*, 1099–1104.
27. Curtis, A.; Riehle, M. *Phys. Med. Biol.* **2001**, *46*, 47–65.
28. Martin, I.; Vunjak-Novakovic, G.; Yang, J.; Langer, R.; Freed, L. E. *Exp. Cell. Res.* **1999**, *253*, 681–688.
29. Vandeweerdt, P.; Berghmans, H.; Tervoort, Y. *Macromolecules* **1991**, *24*, 3547–3552.
30. Williams, J. M.; Moore, J. E. *Polymer* **1987**, *28*, 1950–1958.
31. Mulder, M. *Basic Principles of Membrane Technology: Preparation of Synthetic Membranes*, 2nd ed.; Kluwer Academic Publishers Group: Dordrecht, The Netherlands, 1996; Vol. 3, pp 71−156.
32. Sperling, L. H. *Introduction to Physical Polymer Science*, 4th ed.; John Wiley & Sons, Inc.: Hoboken, NJ, 2005; Vol. 4, pp 145−195.

33. Jiang, Q.; Xu, H.; Cai, S.; Yang, Y. *J. Mater. Sci. Mater. Med.* **2014**, *25*, 1789–800.
34. Tsai, F. J.; Torkelson, J. M. *Macromolecules* **1990**, *23*, 775–784.
35. van deWitte, P.; Dijkstra, P. J.; vandenBerg, J. W. A.; Feijen, J. *J. Membr. Sci.* **1996**, *117*, 1–31.
36. Ma, P. X.; Zhang, R. Y. *J.Biomed. Mater. Res.* **1999**, *46*, 60–72.
37. Nam, Y. S.; Park, T. G. *J.Biomed. Mater. Res.* **1999**, *47*, 8–17.
38. P.X. Ma, R. J. Z. *J. Biomed. Mater. Res.* **1999**, *46*, 60–72.
39. He, L.; Liu, B.; Xipeng, G.; Xie, G.; Liao, S.; Quan, D.; Cai, D.; Lu, J.; Ramakrishna, S. *Eur. Cell. Mater.* **2009**, *18*, 63–74.
40. Shao, J.; C., C.; Wang, Y.; Chen, X.; Dua, C. *Appl. Surf. Sci.* **2012**, *258*, 6665–6671.
41. Yang, F.; Murugan, R.; Ramakrishna, S.; Wang, X.; Ma, Y. X.; Wang, S. *Biomaterials* **2004**, *25*, 1891–1900.
42. Molladavoodi, S.; Gorbet, M.; Medley, J.; Kwon, H. J. *J. Mech. Behav. Biomed.* **2013**, *17*, 186–197.
43. Schugens, C.; Maquet, V.; Grandfils, C.; Jerome, R.; Teyssie, P. *J. Biomed. Mater. Res.* **1996**, *30*, 449–461.
44. He, L. M.; Y., Q. Z.; Zeng, X. *Polymer* **2009**, *50*, 4128–4138.
45. Shao, J.; Chen, C.; Wang, Y.; Chen, X.; Du, C. *Appl. Surf. Sci.* **2012**, *258*, 6665–6671.
46. Papenburg, B. J.; Bolhuis-Versteeg, L. A. M.; Grijpma, D. W.; Feijen, J.; Wessling, M.; Stamatialis, D. *Acta Biomater.* **2010**, *6*, 2477–2483.
47. Wei, G.; Ma, P. X. *Biomaterials* **2009**, *30*, 6426–6434.
48. Chen, V. J.; Ma, P. X. *Biomaterials* **2004**, *25*, 2065–2073.
49. Wei, G.; Ma, P. X. *J. Biomed. Mater. Res., Part A* **2006**, *78*, 306–315.
50. Vaquette, C.; Cooper-White, J. *Acta Biomater.* **2013**, *9*, 4599–4608.
51. Ding, B.; Li, C. R.; Wang D.; Shiratori, S. Fabrication and Application of Novel Two-Dimensional Nanowebs via Electrospinning. In *Nanotechnology: Nanofabrication, Patterning, and Self Assembly*; Charles J., Dixon C. J., Curtines, O. W., Eds.; Nova Science Publishers, Inc.; New York, 2010; pp 51−69.
52. Ding, B.; Li, C. R.; Miyauchi, Y.; Kuwaki, O.; Shiratori, S. *Nanotechnology* **2006**, *17*, 3685–3691.
53. Wang, X.; Ding, B.; Sun, G.; Wang, M.; Yu, J. *Prog. Mater. Sci.* **2013**, *58*, 1173–1243.
54. Ding, B.; Wang, X.; Yu, J.; Wang, M. *J. Mater. Chem.* **2011**, *21*, 12784–12792.
55. Parajuli, D. C.; Bajgai, M. P.; Ko, J. A.; Kang, H. K.; Khil, M. S.; Kim, H. Y. *ACS Appl. Mater. Interfaces* **2009**, *1*, 750–757.
56. Barakat, N. A. M.; Kanjwal, M. A.; Sheikh, F. A.; Kim, H. Y. *Polymer* **2009**, *50*, 4389–4396.
57. Wu, D.; Shi, T.; Yang, T.; Sun, Y.; Zhai, L.; Zhou, W.; Zhang, M.; Zhang, J. *Eur. Polym. J.* **2011**, *47*, 284–293.
58. Wang, X.; Ding, B.; Yu, J.; Yang, J. *Colloids Surf., B* **2011**, *86*, 345–352.
59. Wang, X. F.; Ding, B.; Yu, J. Y.; He, J. H.; Sun, G. *Int. J. Nonlinear Sci. Numer. Simul.* **2010**, *11*, 509–515.

60. Ayutsede, J.; Gandhi, M.; Sukigara, S.; Ye, H.; Hsu, C. M.; Gogotsi, Y.; Ko, F. *Biomacromolecules* **2006**, *7*, 208–214.
61. Bo, L.; Kwan-Ha, S.; Da-Young, N.; In-Hwan, J.; Young-Hag, K.; Won-Young, C.; Hyoun-Ee, K. *J. Mater. Chem.* **2012**, *22*, 14133–14140.
62. Jiang, H.; Xiaohua, L.; Ma, P. X. *Biomaterials* **2008**, *29*, 3815–3821.
63. Cooper-White, J. J.; Rowlands, A. S.; Lim, S. A.; Martin, D. *Biomaterials* **2007**, *28*, 2109–2121.
64. Huang, Y. X.; Ren, J.; Chen, C.; Ren, T. B.; Zhou, X. Y. *J. Biomater. Appl.* **2008**, *22*, 409–432.
65. Hsu, S. h.; Huang, S.; Wang, Y. C.; Kuo, Y. C. *Acta Biomater.* **2013**, *9*, 6915–6927.
66. Nirmala, R.; Navamathavan, R.; Kang, H.-S.; El-Newehy, M. H.; Kim, H. Y. *Colloids Surf., B* **2011**, *83*, 173–178.
67. Na, W.; Xianfeng, W.; Bin, D.; Jianyong, Y.; Gang, S. *J. Mater. Chem.* **2012**, *22*, 1445–1452.
68. Xianfeng, W.; Bin, D.; Jianyong, Y.; Moran, W.; Pan, F. *Nanotechnology* **2010**, *21*, 055502(6 pp).
69. Wang, X.; Ding, B.; Yu, J.; Si, Y.; Yang, S.; Sun, G. *Nanoscale* **2011**, *3*, 911–915.

Chapter 9

Applications of Lightweight Composites in Automotive Industries

Da Zhao[*,1] and Zhou Zhou[2]

[1]DeepFlex, Inc., Houston, Texas 77082, United States
[2]Department of Materials Science and Engineering, North Carolina State University, Raleigh, North Carolina, 27606, United States
[*]E-mail: daniel.zhao@deepflex.com

Lightweight composites produced from renewable and biodegradable materials have been very promising to find wide applications in automotive industries and to lessen the dependence on exhausting petroleum resource. In this chapter, various natural fibers, traditional polymer and bio-polymer matrix systems are presented. Potentials and limitations of modification methods on natural fiber and manufacturing techniques for lightweight composites are reviewed in details. Major technical concerns and challenges regarding natural fiber reinforced composites are addressed. Examples of recent automotive industrial applications are given and future trends are outlined.

1. Introduction

During the past decades, the high depletion rate of petroleum resources, the growing environmental and societal awareness and the demand for sustainability have stimulated the effort to develop bio-materials which are more environmental compatible during the whole cycle of production, usage and removal. In automotive industries, natural fiber reinforced composites, or bio-composites, deriving from renewable resources, have therefore attracted extensive attention as promising alternatives to replace traditional fiber (i.e., glass, carbon and aramid) reinforced petroleum-based composites (*1–7*). Moreover, strong regulations from European Union (E.U.) and Asia countries like Japan have addressed the waste disposal issues and demanded high recyclability of vehicles (*8*), which has expedited the use of bio-materials. Natural fibers such as jute, flax, hemp, sisal and kenaf have already been used as reinforcements in composites for automotive interior and exterior parts like door panels, dashboards, headliners, seat cushions and floor mats (*5*). Natural fibers offer enormous environmental advantages including worldwide availability, biodegradability, low pollutant emissions and low greenhouse gas emissions (*9–11*). Their superior environmental performance is the strong driver for their increasing applications.

Besides numerous environmental advantages, natural fiber reinforced composites also bring excellent properties and economic benefits. In order to reduce weight, the automotive industry has already shifted from steel to aluminum and now is shifting towards fiber-reinforced composites for many components (*12*). The use of lightweight materials becomes even more vital for further electrical vehicles in order to offset the added weight of batteries (*4*). With even lower density than glass fibers (about 40% lower), the addition of natural fibers into vehicles can further reduce the weight and give high specific mechanical properties. Comparing to synthetic fibers, natural fibers also offer significant cost advantages due to wide availability and benefits of easy processing. Moreover, energy consumption of the production of natural fibers for composites is about 60 percent lower than the manufacture of glass fibers (*13*).

The incorporations of natural fibers with polymer matrix derived from both non-renewable (traditional thermoplastic or thermoset) and renewable sources (biopolymers) have been developed (*4*, *6*). Natural fiber reinforced polypropylene (PP) composites have obtained commercial success for automotive applications with their unique advantage of recyclability (*14*). The low moisture resistance of natural fibers and their relatively poor interface with matrix remain to be the major negative issues affecting the properties of resultant composites. A lot of efforts have focused on different natural fiber modification techniques to improve their compatibility with polymer matrix (*15*, *16*). Injection molding, compression molding and resin transfer molding (RTM) techniques have been widely adopted in the industry to manufacture lightweight composites. Bio-composites produced from natural fibers and renewable biopolymers can be fully environmental friendly and have been attracting increased attention (*11*, *17*).

2. Reinforcing Natural Fibers

In natural fiber reinforced composites, the fibers serve as reinforcements to give strength and stiffness in the same way as synthetic fibers. For automotive applications, most interests have been given to bast fibers such as jute (*18–21*), ramie (*22*), hemp (*23*), flax (*24*) and kenaf (*25–27*), leaf fibers such as sisal (*28–30*), pineapple (*31*) and banana fibers (*32*, *33*), as well as fibers from wood sources or crop residues (*34–36*). Among them, jute is one of the most abundant fiber plants worldwide and ramie is of much attention due to its high stiffness. In general, the bast fibers exhibit superior flexural strength and elastic modulus, while the leaf fibers have better impact properties. The physical and mechanical properties of different natural fibers in comparison to important synthetic fibers were given in Table 1. Specific strength and stiffness are the most important performance indicators for automotive applications, in which weight is a critical factor. The tensile strengths of natural fibers are generally inferior to E-glass fiber. However, with lower densities, the specific strengths of natural fibers are quite comparable to glass fiber. The specific stiffness of natural fibers can be even superior to glass fiber.

Table 1. Properties of important natural fibers and synthetic fibers (*1, 3*)

Fiber	*Density (g/cm^3)*	*Elongation (%)*	*Tensile Strength (MPa)*	*Young's Modulus (GPa)*
Jute	1.3	1.5–1.8	393–773	26.5
Hemp	1.5	1.6	690	27.6
Flax	1.5	2.7–3.2	345-1035	70
Kenaf	1.5-1.6	1.6	350-930	40-53
Ramie	1.5	1.2–3.8	400–938	61.4-128
Sisal	1.5	2.0–2.5	511–635	9.4-22.0
Cotton	1.5-1.6	7.0–8.0	400	5.5-12.6
Softwood kraft pulp	1.5	4.4	1000	40
Glass	2.5	2.5	2000–3500	70.0
Aramid	1.4	3.3–3.7	3000–3150	63.0-67.0
Carbon	1.4	1.4–1.8	4000	230.0-240.0

Moreover, natural fibers present many advantages over synthetic fibers such as lower tool wear, cheaper cost, safer to handle, wider availability and biodegradability for their acceptance in high volume automotive industry. However, there are also shortcomings like low moisture resistance, low thermal stability, low microbial resistance, a lack of quality consistency, as well as seasonal and regional quality variations (*1, 3, 4, 6*). A multi-step manufacturing process is required in order to produce high quality natural fibers, which will contribute to the improved mechanical properties of their composites.

3. Modification of Natural Fibers

Natural fibers are mainly composed of cellulose, which contains a lot of strongly polarized hydroxyl (–OH) groups forming intramolecular and intermolecular bonds, thus causing the fibers to be hydrophilic. The incompatibility between polar-hydrophilic fibers and non-polar-hydrophobic matrix leads to the non-uniform wetting of the fibers in the matrix and their poor interface adhesion. Load transfer is therefore insufficient between the fiber reinforcement and matrix, which results in the poor mechanical properties of the composites and reduced life span when exposing to environmental attacks.

Moreover, natural fibers are moisture sensitive and their moisture content can reach as high as 20%. This high water and moisture sorption can cause swelling and a plasticizing effect resulting in dimensional instability. Presence of water can create voids in the matrix and also lead to poor adhesion between fiber and matrix. The hydrophilic nature of natural fibers can be a problem in the finished composites as well. These aspects render incorporation of natural fibers for exterior surface and body applications of vehicles complicated to fulfill all the performance requirements although they are more widely used for interior components.

Main disadvantages of natural fibers as reinforcements in composites are the poor compatibility with matrix and their relative high moisture absorption. Therefore, surface modifications of natural fibers are imperative to improve the mechanical properties of the resultant composites and there have been extensive studies addressing this issue during the past years (*9*, *15*, *16*, *37–40*). Pre-treatments of the natural fibers can increase the surface roughness, chemically modify the surface to mitigate moisture absorption. Both physical and chemical methods have been investigated and applied.

In general, physical treatments change structural and surface properties of the fibers without extensively affecting the bulk chemical composition. Physical methods like steam explosion and thermo-mechanical processes are used to separate natural fiber bundles into individual filaments, while plasma, corona and dielectric barrier techniques are to modify the surface structure of the fibers (*41*). Plasma treatment is a clean and environmentally friendly method, allowing significant surface modifications of fibers such as cleaning, etching, deposition and polymerization (*42*, *43*). The etching effect of plasma can increase the surface roughness of fibers and enhance the mechanical interlocking between fiber and matrix. Reactive free radicals and groups are produced and surface cross-linking can also be introduced. Depending on the type and nature of the gases used, different varieties of surface modifications can be achieved. Improved interfacial adhesion and mechanical properties of bio-composites have been reported using plasma techniques (*43–48*). More recently, the combination of chemical pretreatments, like alcohol with plasma treatment has been found to be effective to reduce the surface hydrophilicity of natural fibers and thus enhance the interface strength of the composites (*49–51*). However, it should be noted that process control is the critical aspect that final morphology and chemical modification strongly depend on the treatment conditions. Similarly, corona treatment, also as air plasma treatment without gas injection, can induce surface oxidation of cellulose fibers and have been found to be effective for the improvement of compatibility between fibers and matrix (*52*, *53*).

Chemical methods mainly include mercerization, acetylation, silane treatment and maleated coupling (*37*, *38*). Mercerization or alkaline treatment is one of the most widely used chemical methods to disrupt fiber bundles and hydrogen bonding in the network structure, thereby increasing the surface roughness and improving fiber wetting (*54–57*). Acetylation is another method of making natural fiber surface more hydrophobic by introducing an acetyl functional group to coat the hydroxyl groups of natural fibers. Its effects on various fiber structures and properties have been illustrated (*58*, *59*). Both alkali treatment and acetylation seem to best improve the thermal stability of natural fibers. Silane treatment involves the introduction of silanes to react with the hydroxyl groups and improve the surface wetting. It has been reported to give excellent mechanical improvement results (*60–62*). Some chemical modification techniques can be very complicated, thus have drawbacks in cost and may not be suitable for industrial applications. Nowadays, coupling agents are preferred and widely used to strengthen natural fiber reinforced composites in industrial applications (*63–65*). Compared to other chemical treatments, the fundamental difference is that the addition of adhesion promoter, which contains both a hydrophilic part and hydrophobic part, can not only modify fiber surface but also the polymeric matrix to achieve better interfacial bonding. The use of coupling agent containing malefic anhydride is the most common and has been shown to immensely improve the properties of natural reinforced composites by numerous studies (*66–71*).

4. Matrix Polymers

Main polymer matrices used for natural fiber-reinforced composites include traditional petro-based thermoplastics and thermosets. Variety of thermoplastics that can be successfully used with natural fibers is limited due to the low thermal stability of natural fibers (processing temperature up to 230 °C). For automotive applications, thermoplastic polypropylene (PP) has became the most commonly used system and obtained much commercial success due to its low density, easy processability, good mechanical properties, as well as its recyclability (*14*, *19*, *27*, *72*). Several other utilized thermoplastics include polyethylene (PE) (*73*, *74*), polystyrene (PS) (*30*) and polyvinylchloride (PVC) (*75*). However, there are still processing limitations like high melt viscosity, which is a serious problem in the case of injection molding process. The primary thermoset resins used in natural-fiber composites are unsaturated polyester (*76*, *77*), epoxy (*29*, *33*) and vinyl ester resins (*20*, *78*). They exhibit higher mechanical properties than thermoplastics, however render the overall product not easily recyclable.

While bio-composites made from natural fibers and traditional thermoplastics or thermosets are not fully eco-friendly since the matrices are still petro-based and non-biodegradable, biopolymers derived from renewable resources have attracted increasing interest to be developed into more eco-compatible bio-composites (*11*, *17*, *21*, *79–81*). Biopolymers can be synthesized from raw materials derived from a variety of sources including soybean, castor beans, corn and sugar cane. Just like their traditional counterparts, biopolymers can be extruded, blown, injection-molded and thermoformed. Due to large availability and outstanding mechanical properties, polylactic acid (PLA) has been most thoroughly investigated in biopolymer research aera (*26*, *82–87*). PLA can be produced via lactic acid from fermentable sugar source like corn, and belongs to thermoplastic aliphatic polyesters family (*88*). Other biopolymers like polyhydroxybutyrate (PHB) (*89*, *90*) and polyhydroxybutyrate-co-valerate (PHBV) (*91*) have also been investigated. Table 2 presents the properties of both traditional polymers and bio-polymers that mostly used for automotive applications. Like the synthetic thermosets, bio-based thermosets are difficult to recycle but can be manufactured to be biodegradable or partially biodegradable according to specific application demands. While bio-polymers bring environmental benefits, there are still drawbacks like poor durability during long time use, high cost, brittleness, low melt viscosity for further processing which limit their use in large scale production at current development stage (*92*).

Table 2. Properties of important polymers for natural fiber reinforced composites (*1, 4*)

Polymers	*Density (g/cm^3)*	*Elongation (%)*	*Tensile Strength (MPa)*	*Young's modulus (GPa)*
PP	0.9	200	35	0.83
HDPE	0.95	30	28	1.04
Nylon	1.12	29	66	3.5
PS	1.05	15	35	2.76
Epoxy	1.02-1.61	6.2	32	0.5
PLA	1.21-1.25	4	65.5	0.35-3.5
PLLA	1.25-1.29	3-4	15.5-65.5	0.83-2.7
PHB	1.18-1.26	5-8	24-40	3.5-4
PHBV	1.23-1.25	17.5-25	20-25	0.5-1.5

5. Manufacturing Techniques

In general, natural fiber reinforced composites are processed by traditional manufacturing techniques, including compounding, extrusion, injection molding, compression molding, and resin transfer molding (RTM). However, due to the non-uniformity of nature fibers, relatively poor interface with matrix and their different thermal behaviors compared to synthetic fibers, optimization of manufacturing techniques, as well as innovative process development become key factors to ensure high quality of bio-composites (*93*). For thermoplastic matrix, injection molding and extrusion have been well developed in the automobile industry for bio-composites. For thermoset materials, there are a few techniques that are widely adopted such as compression molding, resin transfer molding (RTM) and vacuum assist resin transfer molding (VARTM).

5.1. Injection Molding

Injection molding is a mature and widely used processing technique for making thermoplastic composite products, which is especially suitable to form complex shapes with fine details. The advantages include consistency in quality, excellent surface finish, good dimensional accuracy, high production output rate and low labor cost due to full automation (*94*). Many researchers have reported production of high quality bio-composites using injection molding process (*27, 35, 95–97*) .

The original thermoplastic polymer used by injection molding was designed for plastic pellets. For fiber reinforced composites, the pellets with chopped short fibers with critical fiber length are fed into a heated compression barrel with a rotating screw. One of the biggest challenges to adopt injection molding and other thermoplastic based processes is to produce pellets with highly consistent quality. Both North American and European companies have done extensive work to produce better pellets for bio-composites parts. For example, Green Dot's Terratek® WC wood-plastic composites pellets with smaller and more uniform size have been developed to ensure superior processing properties for injection molding and sheet extrusion. It allows designers and OEMs to produce more high quality parts using existing equipments without extensive mold modifications. Another example is NCell™ natural fiber composites pellet produced by GreenCore Composites Inc. (Canada). It consists of PP or PE matrix reinforced with up to 40% natural cellulosic microfibers.

5.2. Compression Molding

Compression molding using thermoset matrix is an efficient technique to manufacture large parts especially for light, strong and thin structures in automotive mass production. Many studies have been conducted on the process feasibility of bio-composites through compression molding (*98–100*).

Sheet molding compound (SMC) and bulk molding compound (BMC) are traditional initial charges for compression molding. The precut fibers (chopped, mat or stitched) can be mixed with compound and placed inside a pre-heated mold cavity. Both heat and pressure are applied to the charge, which are molten to form cavity shape (*101*). As the fibers are placed inside the mold and no shear stress and violent motion are applied, fiber damages can be kept minimal. Therefore, long fibers can also be used to produce bio-composites with higher volume fraction using this technique (*93*).

As short natural fibers are the primary reinforcements for bio-composites, short fiber processes from textile and paper industries combined with compression molding have been adopted. For example, PLA/kenaf bio-composites were fabricated by a nonwoven based carding process. The fibers were carded and consolidated by needle punching. The perform was then cut and hot-pressed (*102*). NafPur Tec process consists of a spray system which applies resins on both sides of nonwoven preform of natural fibers (*13*). The prepreg was then cured by compression molding process. During these processes, the requirements of raw materials can be relatively low. The process is simple but has high production rate. Thus, the products can be excellent candidates for low end applications such as trunk liner, automobile floor mats, etc.

5.3. Resin Transfer Molding (RTM)

RTM is one of liquid composite molding processes and has been widely used in composites industry. During RTM process, dry fibers or porous fibrous preform is placed into mold, in which two matching mold halves are clamped together to hold the preform. Molten plastic is injected into the heated mold and then post-cured and cooled (*103*). With high volume production capability and cost effectiveness, it is also widely investigated for the potential to produce bio-composites for automotive applications.

Back to 1997, O'Dell presented an experimental study of RTM process for a nonwoven jute fiber and polyester-styrene composite system (*104*). This study showed that jute fibers could be processed just as well as glass fibers using RTM. Sèbe et al. (*105*) prepared hemp fiber reinforced polyester composites using RTM technique. There was no major problem during the process and good flexural properties of composites were obtained. As the first attempt to use predominant natural materials for composites, Williams and Wool (*79*) also adopted RTM technique to manufacture flax and hemp fibers reinforced soy oil resin composites. Oksman manufactured unidirectional high-quality flax fiber reinforced epoxy composites using RTM process (*106*). Good flow properties were observed in the mold and the composites were found to have very promising mechanical properties. A recent work from Francucci et al. modeled the infiltration of reinforcements during RTM process. Conventionally, Darcy's law is widely used in RTM and other resin infusion processes. However, due to polarity, natural fibers have swelling behavior, which does not exist in synthetic fibers such as glass and carbon fiber. Swelling will lead to decrease in both saturated and unsaturated permeability. Unlike classic models, a non-constant permeability value was adopted in the model. Therefore, this proposed new model gave a more accurate prediction of the flow front of the infused resin (*107*).

5.4. Vacuum Assist Resin Transfer Molding (VARTM)

VARTM is a low pressure molding process and the main difference to RTM is that one of the solid mold faces is replaced by a flexible vacuum bag. It has lower tooling cost and the set-up is relatively easy. The low pressure will result in low and inconsistent composites quality, and thickness of the parts along with the fiber volume fraction cannot be precisely controlled. However, it is still a clean and cost-effective manufacturing process to produce automobile bio-composites.

O'Donnell et al. (*81*) used VARTM process to make composites using nature fiber mats from flax, cellulose, pulp, hemp and soybean oil-based resin. Recycled paper as a cheap resource of cellulose fibers was also found to work well with this resin system under the same process. Francucci et al. (*78*) made natural fiber reinforced vinyl ester composites via VARTM process and found that the capillary pressure was significantly higher in natural fiber than in synthetic fiber fabrics. Most recently, Schuster et al. (*108*) fabricated composite panels made of flax fiber and bio-based thermoset resins using VARTM and found out that bio-based resins can be used successfully to impregnate natural fibers using VARTM process without further chemical improvements such as dilutants to reduce the viscosity.

5.5. Processing Concerns and Challenges

Although there are processing techniques promising to manufacture high quality bio-composites, extensive studies are still needed to address various technical concerns and challenges. One drawback of raw natural fibers is that their properties vary from different harvest seasons and regions. By blending fiber batches from multiple suppliers and harvests are utilized to address this issue, which can ensure the relatively consistent performance of the finished products (*13*). There are also improvements to increase the consistency of raw materials by using smaller pellets or microfibers for injection molding as mentioned before.

With consistent raw materials, optimal processing conditions are the next key determinant to acquire ideal properties and quality consistency in the products. Several factors specifically regarding bio-composites must be considered. Adequate compatibility between fiber and matrix need to be achieved through modification processes. Moisture control in the fibers and process environment is of primary concern. To maintain the moisture at a desired low level can minimize swelling effect and part distortion problems. Fiber attritions and damages need to be minimized during the processing to ensure reinforcement properties. Moreover, due to low thermal stability of natural fibers, the range of processing temperature and processing time has to be limited and controlled. Some other process factors, such as permeability and resin viscosity, also exist in conventional composites manufacturing. However, the behavior and impact of those factors are quite different for bio-composites (*109*).

To understand the interrelationship between all the process factors and accurately control them is very crucial to obtain a robust and repeatable manufacturing process. Design of experiment (DOE), analysis of variance analysis (ANOVA), linear optimization and other statistical analysis and mathematic optimization tools should be implemented to improve the processing conditions. Zarrinbakhsh et al. (*110*) conducted statistical analysis to optimize the processing condition for bio-composites. A 3^2-full factorial design of experiment and regression modeling were implemented to attain a distillers' grains-filled bio-composite with balanced mechanical properties. ANOVA was also implemented to develop least square regression models containing statistically significant main effects and interaction effect. Their models showed good predictability for the testing results. Finally, response surface was used to find optimized window for process and bio-composites performance optimization. This study shows that statistical approaches can greatly benefit the process development of bio-composites in automotive industry. Furthermore, when the processing conditions of raw materials such as pellet, fiber, preform and prepreg are also considered into the entire process optimization system, a multi-stage optimization concept will be important and effective (*111*).

6. Automotive Applications

The use of natural fiber reinforced composites for automotive application is not a new idea. Back to early 1940, Henry Ford began the experimenting by adopting soybeans to produce composite components. Although the importance of the idea was not fully realized due to the cheap price of petroleum over that time, it showed the early interests and feasibility for the new generation of bio-composites for automobile application. Bio-composites were later successfully introduced in several vehicles, first in Europe and later in North America (*5*). To date, almost all major automotive manufacturers including Daimler Chrysler, Volkswagen, BMW Ford, General Motors, Toyota and Mitsubishi have made large effort to incorporate bio-composites in interior and exterior components. The stringent legislation in large automotive markets such as in E.U. has been a strong drive force. More importantly, major auto manufacturers have set the priority in global sustainability, which is closely related to corporate responsibility to pursue a more environmental benign automotive supply chain. The most common components that have been made with bio-composites include door panels, dashboards, seat backs, package trays, headliners, trunk liners and other interior trim parts.

In Europe, DaimlerChrysler has done the most development work on bio-composites through its Mercedes-Benz brand. Back in 1996, Mercedes-Benz used jute/epoxy composites for the door panels in its E-class vehicles (*112*). Later for many years, DaimlerChrysler have integrated polypropylene and a wide variety of natural fibers including flax, hemp, sisal and coconut for over 50 different interior components in the Mercedes-Benz A-, C-, E-, and S-class models (*3*). Other European auto manufacturers have also used bio-composites in their products. Volkswagen's Audi launched the A2 midrange car in which door trim panels were made of polyurethane reinforced with mixed flax/sisal in 2000 (*11*). BMW used prepreg sisal fiber mats reinforced thermoset acrylic copolymer for the inner door panels in the 7 Series sedan in 2010 (*113*). Fiat used fuel lines made of castor oil derived nylon in several vehicle models.

In North America, Ford adopted wheat straw reinforced materials for storage bin and inner lid in its 2010 Flex crossover vehicle and brought cellulose fiber reinforced thermoplastic to 2014 Lincoln MKX for the floor console armrest substrate. Ford is also using soy-based foam seating in all of its North American models, while kenaf-reinforced polypropylene composites are used in interior door panels for the Focus, Fiesta, and Mondeo models. General Motors has used flax reinforced polypropylene composites for rear shelf components in the Chevrolet Impala. A mixture of kenaf and flax fibers is used in package trays and door panel inserts for Saturn L300 and European-market Opel Vectras, while wood fibers are used in seatbacks of the Cadillac Deville and in cargo floor of the GMC Envoy and Chevrolet TrailBlazer (*3*).

Toyota is the leader in bio-composites development in Asia. Toyota Raum 2003 model adopted 100% bioplastics with PLA matrix reinforced by kenaf fibers for the spare tire cover. The Toyota Prius and SAI use corn-based plastic in interior components such as instrument panels, air conditioning system outlets and ceiling surface (*5*). In 2011, Toyota developed new bio-based polyethylene terphthalate (PET) from sugar cane and began using this new material in the luggage compartment liner of the new Lexus CT200h hybrid-electric compact car. Other Asian auto companies like Mitsubishi developed a new automobile floor mat using bio-polyethylene (bio-PE) manufactured from sugar cane molasses with PP fiber on surface, while Honda used wood fiber in floor area parts of the Pilot SUV.

There have been fewer exterior applications of bio-composites since the components must be able to withstand extreme conditions. DaimlerChrysler has used abaca banana fibers to replace fiberglass in the standard underbody cover for the spare-wheel of the Mercedes-Benz A-Class vehicle in 2004, which was the first component for exterior use. This novel mixture of abaca fibers and PP thermoplastic was patented in 2002 (*4*). Another example is that Lotus manufactured the Eco Elise body panels which contains hemp fibers as the primary constituent in the high quality 'A' class polyester composites in 2008. In recent years, there have been also attempts to incorporate nanomaterials in natural fiber reinforced composites for potential structural applications (*6*).

Automotive manufacturers are constantly looking for material solutions to reduce weight and cost, enhance performance and mitigate environment impact. The use of bio-composites made from renewable and environmentally benign materials are expected to continually increase in automotive applications. However, it should be mentioned that most of the mentioned "bio-composites" use natural fibers but the polymer matrix are still petro-based synthetic materials. They are used mostly in non-structural applications and not applied in large surfaces of vehicle's body.

7. Conclusions

Automotive industries have already witnessed the revolution of light weight composites. The incorporation of bio-composites into vehicles can lead to further weight reduction and less fuel consumption, which has already been embraced by many automotive manufacturers on a global scale. With enormous environmental and economic benefits, bio-composites have been expected to replace glass fiber reinforced composites for various automotive applications. However, several major technical concerns must be addressed before their wide adoption, such as low moisture stability of natural fibers, incompatibility between natural fibers and polymer matrices, inconsistency of raw materials and immature manufacturing techniques. This chapter has demonstrated state-of-the-art surface treatment and manufacturing techniques to address these issues. Recent technology advances offer significant opportunities for bio-composites to expand their role in automotive industries.

References

1. Ashori, A. *Bioresour. Technol.* **2008**, *99*, 4661–4667.
2. Joshi, S. V.; Drzal, L. T.; Mohanty, A. K.; Arora, S. *Composites, Part A* **2004**, *35*, 371–376.
3. Holbery, J.; Houston, D. *JOM* **2006**, *58*, 80–86.
4. Koronis, G.; Silva, A.; Fontul, M. *Composites, Part B* **2012**, *44*, 120–127.
5. Hill, K.; Swiecki, B.; Cregger, J. *The Bio-Based Materials Automotive Value Chain* ; Center for Automotive Research, Sustainability & Economic Development Strategies Group: Ann Arbor, MI, 2012.

6. Njuguna, J.; Wambua, P.; Pielichowski, K.; Kayvantash, K. In *Cellulose Fibers: Bio- and Nano-Polymer Composites*; Springer: Berlin, 2011; p 661–700.
7. Puglia, D.; Biagiotti, J.; Kenny, J. M. *J. Nat. Fibers* **2005**, *1*, 23–65.
8. Directive 2000/53/EC of the European Parliament and of the Council of 18 September 2000 on end-of life vehicles; *Official Journal of the European Union*, 2000.
9. Faruk, O.; Bledzki, A. K.; Fink, H.-P.; Sain, M. *Prog. Polym. Sci.* **2012**, *37*, 1552–1596.
10. John, M. J.; Thomas, S. *Carbohydr. Polym.* **2008**, *71*, 343–364.
11. Mohanty, A. K.; Misra, M.; Drzal, L. T. *Natural Fibers, Biopolymers, and Biocomposites*; CRC Press: Boca Raton, FL, 2005.
12. Mouti, Z.; Westwood, K.; Kayvantash, K.; Njuguna, J. *Materials* **2010**, *3*, 2463–2473.
13. Brosius, D. *Compos. Technol.* **2006**, *12*, 32–37.
14. Malkapuram, R.; Kumar, V.; Negi, Y. S. *J. Reinf. Plast. Compos.* **2009**, *28*, 1169–1189.
15. Mohanty, A. K.; Misra, M.; Drzal, L. T. *Compos. Interfaces* **2001**, *8*, 313–343.
16. Kalia, S.; Kaith, B. S.; Kaur, I. *Polym. Eng. Sci.* **2009**, *49*, 1253–1272.
17. Mohanty, A. K.; Misra, M.; Hinrichsen, G. *Macromol. Mater. Eng.* **2000**, *276*, 1–24.
18. Alves, C.; Ferrão, P. M. C.; Silva, A. J.; Reis, L. G.; Freitas, M.; Rodrigues, L. B.; Alves, D. E. *J. Cleaner Prod.* **2010**, *18*, 313–327.
19. Rana, A. K.; Mandal, A.; Bandyopadhyay, S. *Compos. Sci. Technol.* **2003**, *63*, 801–806.
20. Ray, D.; Sarkar, B. K.; Rana, A. K.; Bose, N. R. *Composites, Part A* **2001**, *32*, 119–127.
21. Behera, A. K.; Avancha, S.; Basak, R. K.; Sen, R.; Adhikari, B. *Carbohydr. Polym.* **2012**, *88*, 329–335.
22. Goda, K.; Sreekala, M. S.; Gomes, A.; Kaji, T.; Ohgi, J. *Composites, Part A* **2006**, *37*, 2213–2220.
23. Müssig, J.; Schmehl, M.; von Buttlar, H. B.; Schönfeld, U.; Arndt, K. *Ind. Crops Prod.* **2006**, *24*, 132–145.
24. Bos, H. t. L. *The Potential of Flax Fibres as Reinforcement for Composite Materials*; Technische Universiteit Eindhoven: Eindhoven, The Netherlands, 2004.
25. Akil, H. M.; Omar, M. F.; Mazuki, A. A. M.; Safiee, S.; Ishak, Z. A. M.; Abu Bakar, A. *Mater. Des.* **2011**, *32*, 4107–4121.
26. Ochi, S. *Mech. Mater.* **2008**, *40*, 446–452.
27. Zampaloni, M.; Pourboghrat, F.; Yankovich, S. A.; Rodgers, B. N.; Moore, J.; Drzal, L. T.; Mohanty, A. K.; Misra, M. *Composites, Part A* **2007**, *38*, 1569–1580.
28. Li, Y.; Mai, Y.-W.; Ye, L. *Compos. Sci. Technol.* **2000**, *60*, 2037–2055.
29. Oksman, K.; Wallström, L.; Berglund, L. *J. Appl. Polym. Sci.* **2002**, *84*, 2358–2365.

30. Antich, P.; Vazquez, A.; Mondragon, I.; Bernal, C. *Composites, Part A* **2006**, *37*, 139–150.
31. Arib, R. M. N.; Sapuan, S. M.; Ahmad, M.; Paridah, M. T.; Zaman, H. M. D. *Mater. Des.* **2006**, *27*, 391–396.
32. Liu, H.; Wu, Q.; Zhang, Q. *Bioresour. Technol.* **2009**, *100*, 6088–6097.
33. Sapuan, S. M.; Leenie, A.; Harimi, M.; Beng, Y. K. *Mater. Des.* **2006**, *27*, 689–693.
34. Stark, N. M.; Rowlands, R. E. *Wood Fiber Sci.* **2003**, *35*, 167–174.
35. Yu, Y.; Yang, Y.; Murakami, M.; Nomura, M.; Hamada, H. *Adv. Compos. Mater.* **2013**, *22*, 425–435.
36. Dai, D.; Fan, M. In *Natural Fibre Composites: Materials, Processes and Properties*; Woodhead Publishing Limited: Cambridge, 2014; pp 3−65.
37. John, M. J.; Anandjiwala, R. D. *Polym. Compos.* **2008**, *29*, 187–207.
38. Li, X.; Tabil, L. G.; Panigrahi, S. *J. Polym. Environ.* **2007**, *15*, 25–33.
39. Valadez-Gonzalez, A.; Cervantes-Uc, J. M.; Olayo, R.; Herrera-Franco, P. J. *Composites, Part B* **1999**, *30*, 309–320.
40. George, J.; Sreekala, M. S.; Thomas, S. *Polym. Eng. Sci.* **2001**, *41*, 1471–1485.
41. Mukhopadhyay, S.; Fangueiro, R. l. *J. Thermoplast. Compos. Mater.* **2009**, *22*, 135–162.
42. Strobel, M.; Lyons, C. S.; Mittal, K. *Plasma Surface Modification of Polymers: Relevance to Adhesion*; Vsp: Utrecht, 1994.
43. Chen, X.; Yao, L.; Xue, J.; Zhao, D.; Lan, Y.; Qian, X.; Wang, C. X.; Qiu, Y. *Appl. Surf. Sci.* **2008**, *255*, 2864–2868.
44. Marais, S.; Gouanvé, F.; Bonnesoeur, A.; Grenet, J.; Poncin-Epaillard, F.; Morvan, C.; Métayer, M. *Composites, Part A* **2005**, *36*, 975–986.
45. Martin, A. R.; Manolache, S.; Mattoso, L. H. C.; Rowell, R. M.; Denes, F. In *Proceedings from the Third International Symposium on Natural Polymers and Composites-ISNaPol*; Embrapa−Empresa Brasileira de Pesquisa Agropecuaria, Embrapa Instrumentacao Agropecuaria: São Pedro, SP, 2000; pp 431−436.
46. Seki, Y.; Sever, K.; Sarikanat, M.; Gülec, H. A.; Tavman, I. H. In *Proceedings of the 5th International Advanced Technologies Symposium (IATS'09)*; The Scientific and Technological Research Council of Turkey: Karabük, Turkey, 2009.
47. Yuan, X.; Jayaraman, K.; Bhattacharyya, D. *Composites, Part A* **2004**, *35*, 1363–1374.
48. Kafi, A. A.; Magniez, K.; Fox, B. L. *Compos. Sci. Technol.* **2011**, *71*, 1692–1698.
49. Wang, J.; Zhou, Z.; Huang, X.; Zhang, L.; Hu, B.; Moyo, S.; Sun, J.; Qiu, Y. *J. Adhes. Sci. Technol.* **2013**, *27*, 1278–1288.
50. Zhou, Z.; Liu, X.; Hu, B.; Wang, J.; Xin, D.; Wang, Z.; Qiu, Y. *Surf. Coat. Technol.* **2011**, *205*, 4205–4210.
51. Zhou, Z.; Wang, J.; Huang, X.; Zhang, L.; Moyo, S.; Sun, S.; Qiu, Y. *Appl. Surf. Sci.* **2012**, *258*, 4411–4416.
52. Gassan, J.; Gutowski, V. S. *Compos. Sci. Technol.* **2000**, *60*, 2857–2863.

53. Ragoubi, M.; Bienaimé, D.; Molina, S.; George, B.; Merlin, A. *Ind. Crops Prod.* **2010**, *31*, 344–349.
54. Bachtiar, D.; Sapuan, S. M.; Hamdan, M. M. *Mater. Des.* **2008**, *29*, 1285–1290.
55. Bisanda, E. T. N. *Appl. Compos. Mater.* **2000**, *7*, 331–339.
56. Gomes, A.; Matsuo, T.; Goda, K.; Ohgi, J. *Composites, Part A* **2007**, *38*, 1811–1820.
57. Gassan, J.; Bledzki, A. K. *Compos. Sci. Technol.* **1999**, *59*, 1303–1309.
58. Tserki, V.; Zafeiropoulos, N. E.; Simon, F.; Panayiotou, C. *Composites, Part A* **2005**, *36*, 1110–1118.
59. Bledzki, A. K.; Mamun, A. A.; Lucka-Gabor, M.; Gutowski, V. S. *Express Polym. Lett.* **2008**, *2*, 413–22.
60. Abdelmouleh, M.; Boufi, S.; Belgacem, M. N.; Dufresne, A. *Compos. Sci. Technol.* **2007**, *67*, 1627–1639.
61. Xie, Y.; Hill, C. A. S.; Xiao, Z.; Militz, H.; Mai, C. *Composites, Part A* **2010**, *41*, 806–819.
62. Ismail, H.; Abdul Khalil, H. P. S. *Polym. Test.* **2000**, *20*, 33–41.
63. Araujo, J. R.; Waldman, W. R.; De Paoli, M. A. *Polym. Degrad. Stab.* **2008**, *93*, 1770–1775.
64. Gauthier, R.; Joly, C.; Coupas, A. C.; Gauthier, H.; Escoubes, M. *Polym. Compos.* **1998**, *19*, 287–300.
65. Yang, H.-S.; Kim, H.-J.; Park, H.-J.; Lee, B.-J.; Hwang, T.-S. *Compos. Struct.* **2007**, *77*, 45–55.
66. Park, J.-M.; Quang, S. T.; Hwang, B.-S.; DeVries, K. L. *Compos. Sci. Technol.* **2006**, *66*, 2686–2699.
67. Fuqua, M. A.; Ulven, C. A. *J. Biobased Mater. Bioenergy* **2008**, *2*, 258–263.
68. Keener, T. J.; Stuart, R. K.; Brown, T. K. *Composites, Part A* **2004**, *35*, 357–362.
69. Mutjé, P.; Vallejos, M. E.; Girones, J.; Vilaseca, F.; López, A.; Lopez, J. P.; Méndez, J. A. *J. Appl. Polym. Sci.* **2006**, *102*, 833–840.
70. Mohanty, S.; Nayak, S. K.; Verma, S. K.; Tripathy, S. S. *J. Reinf. Plast. Compos.* **2004**, *23*, 625–637.
71. Gassan, J.; Bledzki, A. K. *Appl. Compos. Mater.* **2000**, *7*, 373–385.
72. Bourmaud, A.; Baley, C. *Polym. Degrad. Stab.* **2007**, *92*, 1034–1045.
73. Torres, F. G.; Cubillas, M. L. *Polym. Test.* **2005**, *24*, 694–698.
74. Mulinari, D. R.; Voorwald, H. J. C.; Cioffi, M. O. H.; da Silva, M. L. c. C. P.; Luz, S. M. *Carbohydr. Polym.* **2009**, *75*, 317–321.
75. Wang, H.; Chang, R.; Sheng, K.-c.; Adl, M.; Qian, X.-q. *J. Bionic Eng.* **2008**, *5*, 28–33.
76. Idicula, M.; Neelakantan, N. R.; Oommen, Z.; Joseph, K.; Thomas, S. *J. Appl. Polym. Sci.* **2005**, *96*, 1699–1709.
77. Aziz, S. H.; Ansell, M. P.; Clarke, S. J.; Panteny, S. R. *Compos. Sci. Technol.* **2005**, *65*, 525–535.
78. Francucci, G.; Vázquez, A.; Ruiz, E.; Rodríguez, E. S. *Polym. Compos.* **2012**, *33*, 1593–1602.
79. Williams, G. I.; Wool, R. P. *Appl. Compos. Mater.* **2000**, *7*, 421–432.

80. Torres, F. G.; Arroyo, O. H.; Gomez, C. *J. Thermoplast. Compos. Mater.* **2007**, *20*, 207–223.
81. O'Donnell, A.; Dweib, M. A.; Wool, R. P. *Compos. Sci. Technol.* **2004**, *64*, 1135–1145.
82. Bax, B.; Müssig, J. *Compos. Sci. Technol.* **2008**, *68*, 1601–1607.
83. Bledzki, A. K.; Jaszkiewicz, A.; Scherzer, D. *Composites, Part A* **2009**, *40*, 404–412.
84. Hu, R. H.; Jang, M. H.; Kim, Y. J.; Piao, Y. J.; Lim, J. K. *Adv. Mater. Res.* **2010**, *123*, 1151–1154.
85. Oksman, K.; Skrifvars, M.; Selin, J. F. *Compos. Sci. Technol.* **2003**, *63*, 1317–1324.
86. Wu, C.-S. *Polym. Degrad. Stab.* **2009**, *94*, 1076–1084.
87. Mathew, A. P.; Oksman, K.; Sain, M. *J. Appl. Polym. Sci.* **2005**, *97*, 2014–2025.
88. Le Duigou, A.; Davies, P.; Baley, C. *Compos. Sci. Technol.* **2010**, *70*, 231–239.
89. Macedo, J. d. S.; Costa, M. F.; Tavares, M. I. B.; Thire, R. M. S. M. *Polym. Eng. Sci.* **2010**, *50*, 1466–1475.
90. Barkoula, N. M.; Garkhail, S. K.; Peijs, T. *Ind. Crops Prod.* **2010**, *31*, 34–42.
91. Singh, S.; Mohanty, A. K.; Sugie, T.; Takai, Y.; Hamada, H. *Composites, Part A* **2008**, *39*, 875–886.
92. van Dam, J. E. G.; de Klerk-Engels, B.; Struik, P. C.; Rabbinge, R. *Ind. Crops Prod.* **2005**, *21*, 129–144.
93. Ho, M.-p.; Wang, H.; Lee, J.-H.; Ho, C.-k.; Lau, K.-t.; Leng, J.; Hui, D. *Composites, Part B* **2012**, *43*, 3549–3562.
94. Rosato, D. V.; Rosato, D. V.; Rosato, M. G. *Injection Molding Handbook*; Springer: Norwell, 2000.
95. Panthapulakkal, S.; Sain, M. *J. Appl. Polym. Sci.* **2007**, *103*, 2432–2441.
96. Subyakto, E. H.; Masruchin, N.; Ismadi, K. W. P.; Kusumaningrum, W. B.; Subiyanto, B. *Wood Res.* **2011**, *2*, 21–26.
97. Panthapulakkal, S.; Sain, M. *J. Appl. Polym. Sci.* **2007**, *103*, 2432–2441.
98. Fifield, L. S.; Simmons, K. L. In *Proceedings of the 10th Annual Automotive Composites Conference Exhibition, SPE Automotive Composites*; Society of Plastics Engineers: Troy, MI, 2010; Vol. 504, pp 15–16.
99. Mehta, G.; Mohanty, A. K.; Thayer, K.; Misra, M.; Drzal, L. T. *J. Polym. Environ.* **2005**, *13*, 169–175.
100. Barone, J. R.; Schmidt, W. F.; Liebner, C. F. E. *Compos. Sci. Technol.* **2005**, *65*, 683–692.
101. Mallick, P. K. *Fiber-Reinforced Composites: Materials, Manufacturing, and Design*; CRC Press: New York, 1993.
102. Kim, H.; Lee, B.; Choi, S. In *The 18th International Conference on Composite Materials*; The Korean Society of Composite Materials: Jeju Island, South Korea, 2011.
103. Seemann Iii, W. H. Plastic Transfer Molding Techniques for the Production of Fiber Reinforced Plastic Structures. U.S. Patent 4,902,215, February 20, 1990.

104. O'Dell, J. L. In *The Fourth International Conference on Woodfiber−Plastic Composites*; Forest Products Society: Madison, WI, 1997; pp 280−285.
105. Sèbe, G.; Cetin, N. S.; Hill, C. A. S.; Hughes, M. *Appl. Compos. Mater.* **2000**, *7*, 341–349.
106. Oksman, K. *J. Reinf. Plast. Compos.* **2001**, *20*, 621–627.
107. Francucci, G.; Rodríguez, E. S.; Morán, J. *J. Compos. Mater.* **2014**, *48*, 191–200.
108. Schuster, J.; Govignon, Q.; Bickerton, S. *Open J. Compos. Mater.* **2014**, *4*, 1–11.
109. Society of Plasitics Engineers Research Online.http://www.4spepro.org/ (accessed December 10, 2012).
110. Zarrinbakhsh, N.; Defersha, F. M.; Mohanty, A. K.; Misra, M. *J. Appl. Polym. Sci.* **2014**, *131*.
111. Zhou, S.; Chen, Y.; Shi, J. *IEEE Trans. Autom. Sci. Eng.* **2004**, *1*, 73–83.
112. Suddell, B. C.; Evans, W. J. In *Natural Fibers, Biopolymers and Biocomposites*; CRC Press Taylor & Francis Group: Boca Raton, FL, 2005.
113. Stewart, R. *Reinf. Plast.* **2010**, *54*, 22–28.

Chapter 10

Novel Cardiac Patch Development Using Natural Biopolymers

P. Dubey,[1] A. R. Boccaccini,[2] and I. Roy*,[1]

[1]Applied Biotechnology Research Cluster, Faculty of Science and Technology, University of Westminster, London W1W 6UW, United Kingdom
[2]Department of Material Science and Engineering, University of Erlangen-Nuremberg, Cauestr. 6., 91058 Erlangen, Germany
***E-mail: royi@westminster.ac.uk**

Cardiovascular diseases (CVD) are a major cause of death worldwide. Biomaterials including synthetic and natural polymers have been fabricated by combining the chemical, biological, mechanical and electrical aspects of material for potential tissue engineering applications. Synthetic and natural polymers help in tailoring the mechanical properties of the scaffold and are advantageous in providing cell adhesion and proliferation by offering cell recognition sites. Cardiac tissue engineering aims for the development of a bioengineered cardiac patch that can provide physical support to the damaged cardiac tissue by replacing certain functions of the damaged extracellular matrix and prevent adverse cardiac remodeling and dysfunction after myocardial infarction. Hence, cardiac patches aim to facilitate the normal functioning of the heart muscle by providing repair and support to the infarcted region of the heart. This Chapter will review application of different biodegradable natural polymeric materials and their composites in the development of cardiac patches for cardiac tissue engineering.

Introduction

Cardiovascular diseases are the major cause of death worldwide. In US, approximately 2.5 million deaths, while in United Kingdom around 1.8 million deaths each year is caused by cardiovascular disease (*1*, *2*). Myocardial infarction (MI) and congestive heart failure (CHF) form a major part of cardiovascular disease which account for 40% of the annual mortality rate in industrial and developing nations. Myocardial infarction occurs due to necrosis of the cardiac tissue leading to inflammation, apoptosis and finally CHF (*2*). Heart failure results from improper pumping efficiency of heart. Weakening of the collagen extracellular matrix causes the heart wall thinning and ventricular dilation. This results in a permanent damage to heart wall muscle as the regenerative capability of the heart tissue is permanently lost (*3*). The ventricular dilation causes the structural and functional changes in the ventricles commonly known as ventricular remodeling. This ventricular remodeling results in the impairment of the pumping efficiency, and thus leading towards the last stage congestive heart failure (CHF). In this stage, the heart is unable to pump blood to fulfill the metabolic requirements of the body (*4*, *5*).

Heart failure results in the release and activation of various toxic humoral factors, such as catechol amines, angiotensin-converting enzyme and aldosterone (*6*). Various pharmacological therapies focus on reducing the adverse effects by blocking off these humoral factors. Interventional therapy, such as surgery or implantation of pacing devices are gradually receiving more widespread applications, in particular for patients with marked limitation in activity. However, both these therapies are not useful in controlling the progression of the disease to the end stage (*7*).

Previously, there has been use of many surgical strategies, such as cardio-myoplasty, whereby prepared skeletal muscles were wrapped around the heart and contracted with the heart, which helped in improving the cardiac pumping power (*8*, *9*). It was found from clinical trials that this strategy helped in improving ventricular performance, reducing cardiac dilation and inhibiting disease progression. However, a high mortality rate during the operation posed a drawback to this approach. Studies were further conducted using biomaterial supports like polypropylene, knitted polyester to prevent negative remodeling of the left ventricle (*10*, *11*). These biomaterials were proved beneficial in animal model but were not approved by the Food and Drug Administration (FDA) (*12*).

There have been different strategies used for cardiac transplantation. In cell-based therapy, cells are injected to the infarct region via the pericardium, coronary arteries, or endocardium. In a separate approach, an *in vitro* 3D heart model was developed by seeding cells onto a matrix which was also supplied with drugs and other factors. This model facilitated delivery of cells along with drug delivery to the infarct tissue site. (*13*, *14*). In an upcoming approach, cardiac patches using different materials were prepared and served two functions of delivering cells and providing mechanical strength at the site. These patches are a circular sheet of biomaterial which was seeded with cells *in vitro* and later implanted *in vivo* at the necrotic site. It has been found that such cardiac patches could have beneficial effects on cardiac mechanics and would also help in improving the myofibril stress (*15*).

Cardiac Tissue Engineering

Cardiac tissue engineering is based on regeneration of cardiac tissue by implanting biocompatible and biodegradable scaffolds cultured with cells that are capable of forming cardiomyocytes in the infarcted region of the heart (*16*).

Cardiac Patch Development

Cardiac patches are sheets or circular structures made of polymeric materials seeded with cells. The main aim of these patches is to provide mechanical support and replace the injured heart tissue after myocardial infarction. They are small and thick structures capable of mimicking the natural ECM structure. There are various biodegradable materials that are used widely for these purposes. Biodegradable polymers are degraded naturally by the physiological system of the host and hence a further surgery is avoided for the removal of the implanted construct. This attribute is essential from the tissue engineering point of view (*17*, *18*). Depending on the method of degradation and the response of degradation product on cells, they are categorized as synthetic and biologically derived (natural) polymers. As a cardiac patch needs to withstand the diastolic pressure it should not be too rigid such that it hampers the diastolic functioning. The patch should also be able to integrate well with the electrical rhythm of the heart (*19*). Hence a tissue engineered cardiac patch with biological properties would integrate well with the cardiac cell environment. It would also mimic and replace the necrotic cardiac tissue developed after myocardial infarction. This approach would prove to be a promising for cardiac tissue engineering and myocardial regeneration.

Natural and Bacterial Derived Polymers

Bacterial-derived polymers and natural polymers are good candidates for fabrication of cardiac tissue engineering scaffolds. The vast diversity in bacterial metabolism has led to the availability of various metabolites which can be used in several medical applications and industry. The new technologies involving tissue engineering and drug delivery have helped in many therapeutics and diagnosis (*20*). As bacterial derived polymers have properties like biodegradability, non-toxicity and biocompatibility, they are very good alternatives to petroleum and oil-based polymers. Most of those biopolymers tend to internalize in mammalian cells and are rapidly degraded upon medical administration (*21*), sup- porting their suitability in clinics. There are many bacteria derived polymers which are gaining importance in tissue engineering and drug delivery because of their biodegradability and biocompatibility (*22*). With the help of new methodologies and approaches the bacterially produced polymers can be modified to various structures and scaffolds and hence used in advanced technologies and biomedical applications. There are various bacterial derived polymers including polyhydroxyalkanoates (PHA) (*23–26*), γ-polyglutamic acid (γ-PGA) (*30*, *31*), dextran (*32–35*), xanthans (*36*, *37*), gellan gum (*38*) are some which are produced in different bacterial species. The different bacterial derived polymers their bacterial sources, monomeric units, carbon source and applications are consolidated in Table 1. Polyhyroxyalkanoates (PHA) are natural polymers which have a great potential to be used in cardiac tissue engineering. They are well known biopolyesters of 3-, 4-, 5- and 6-hydroxyalkanoic acids. They are synthesized by a wide range of bacteria such as *Bacillus subtilis, Bacillus cereus, Pseudomonas mendocina, Ralstonia sps* and *Commamonas sps* etc, as an intracellular energy source, under nutrient limiting conditions with excess of carbon (*23–25*). PHAs have various properties which make them useful in different applications. They are biodegradable and exhibit elastomeric and thermoplastic properties which make them useful in the packaging industry. They are non-toxic, water insoluble and recyclable and therefore make an appropriate substitute to hydrocarbon based plastics (*26*). The molecular weight of PHAs is found to be in the range of 50,000–1,000,000 Da, depending on the microorganism used for the production and the growth conditions used. The monomer units are all in R configuration owing to the stereospecificity of biosynthetic enzymes. PHAs accumulate in cells as discrete granules, the number per cell and size of which can vary among the different species (*27*, *28*). PHAs can be divided into two main classes: Short-chain-length PHAs (SCL-PHAs), that have monomers consisting of 3-5 carbons, are partially crystalline, thermoplastic in nature. They generally lack toughness and are brittle polymers and have high melting points. Medium-chain-length PHAs (MCL-PHAs) have monomers consisting of 6–14 carbons and these polymers are elastomeric in nature with low crystallinity, low tensile strength, low melting point and high elongation at break. MCL PHAs have been extensively explored in various medical and industrial applications due to their elastomeric nature (*29*, *30*). The main enzymes responsible for PHA biosynthesis are the PHA synthases which use 3-hydroxyacyl-CoA as the substrates and polymerize them to produce PHAs, followed by the release of CoA. PHA synthases can be categorized into four different classes based on the subunit composition and substrate specificity. Class I and II enzymes comprise of one type of subunit, PhaC, whereas Class III and IV comprise of PhaE and PhaC in the form of a PhaEC complex and PhaR (molecular mass of approx. 22 kDa) and PhaC in the form of a PhaRC complex respectively. Class I, III and IV synthases polymerize short–chain-length (SCL) monomers (C_3-C_5), whereas Class II synthases polymerize medium chain length (MCL) monomers (C_6-C_{14}). The PHA synthases of *Ralstonia eutropha, Pseudomonas aeruginosa, Allochromatium vinosum* and *Bacillus megaterium* represent Classes I, II, III, and IV respectively (*24*). γ-polyglutamic acid (γ -PGA) is an anionic, water-soluble polyamino acid of D- and L-glutamic acid units linked by amide bonds that is produced by *Bacillus licheniformis* and *Bacillus subtilis, Bacillus amyloliquefaciens*, *Bacillus megatherium*, *Bacillus halodurans*, *Staphylococcus epidermidis*, *Fusobacterium nucleatum* and *Natrialba aegyptiaca* (*31–34*). γ-PGA filaments synthesized by different bacterial species vary in molecular weight and length. The molecular weight of γ-PGA from *B. subtilis* varies from 160 kDa to 1500 kDa (about

1240–11,630 glutamate residues). Polyglutamate filaments therefore consist of more than 1000 glutamate residues (*32*). Free polymer, hydrogels or nanoparticles of γ-PGA are used as hydrogels, vaccine carriers, contrast agents, on-site drug delivery vehicles, scaffolds in tissue engineering and carriers for gene therapy (*35*, *36*).

Most of the commercial dextran is produced by *Leuconostoc mesenteroides* when grown on sucrose, but also by *Streptococcus, Lactobacillus* and *Gluconobacter* sp. Dextrans are polymers of n-glucopyranose units; which comprise chains of α-l,6-linked units with α-l,4 branching points and are synthesized from the substrate (sucrose) by the enzymatic action of dextransucrase (*37–40*). Clinical dextran was primarily produced as a blood plasma extender and blood flow improving agent (*37*). Later, dextrans and their derivatives have also been used in the sustained release in nasal and colon drug delivery, bone regeneration and tissue engineering (*40–42*). Xanthan is an extracellular heteropolysaccharide with a very high molecular weight that is an ionic and a naturally acetylated cellulose derivative produced from the Gram-negative bacterium *Xanthomonas campestris* (*43*). Xanthan is an acidic polymer made up of pentasaccharide subunits, forming a cellulose backbone with trisaccharide side-chains composed of mannose, glucuronic-acid and mannose attached to alternate glucose residues in the backbone by α-1,3 linkages. It consists of D-glucosyl, D-mannosyl, and D-glucuronyl acid residues in a molar ratio of 2:2:1 and variable proportions of O-acetyl and pyruvyl residues. Because of its physical properties, it is widely used as a thickener in both food and non-food industries. Xanthan gum is also used as a stabilizer for a wide variety of suspensions, emulsions, and foams. It has been used as a drug carrier and has been used extensively in sustained release delivery formulations (*44*, *45*). Xanthan has also been applied as *in situ* gelling bioadhesive nasal inserts for drug delivery in the nasal context (*45*). Gellan gum is an anionic exopolysaccharide secreted by *Sphingomonas paucimobilis* (formerly *Pseudomonas elodea*) (*46*) and has a characteristic gelling property, which is temperature and ionic concentration dependent (*47*). The repeating unit of gellan polysaccharide is composed of β-D-glucose (D-Glc), L-rhamnose (L-Rha), and D-glucuronic acid (D-GlcA). The composition is approximately: glucose 60 %, rhamnose 20 % and glucuronic acid 20 %. In addition, considerable amount of non-polysaccharide material is found in gellan gum (cell protein and ash) that can be removed by filtration or centrifugation. The molecular weight of the polymer is around 500,000 Da. The media used for production of gellan gum are simple media containing carbon source, nitrogen source and inorganic salts. Carbohydrates such as glucose, fructose, maltose, sucrose and mannitol can be used either alone or in combination as carbon source. The amount of carbon source usually varies between 2–4 % by mass. Gellan precursors were detected by enzyme assays, and were found to be nucleotide diphosphate sugars, UDP-glucose, TDP-rhamnose and UDP-glucuronic acid. There are three types of gellan gum polymer: native, deacetylated and clarified. The acetyl groups in native gellan gum are removed by alkaline treatment to produce deacetylated gellan gum. Clarified gellan gum results from filtration of hot, deacetylated gellan gum for enhanced removal of cell protein residues (*47*, *48*). Gellan gum has gained importance in various applications in the food or pharmaceutical industries, gel electrophoresis, immobilization of cells and enzymes, and bioremediation. Gellan gum is also suitable for biomedical applications such as controlled drug release, in nasal formulations and as an *in situ* gelling agent in ophthalmic formulations (*46*).

Natural polymers are obtained from natural sources which possess properties such as biocompatibility, degradability and are easily solubilized in physiological fluids (except chitosan which is soluble in mild acidic conditions) (*22*, *49*). Various polymers used in cardiac tissue engineering are collagen, chitosan, fibrin, silk, hyaluronic acid. These polymers alone or in combination with other materials like synthetic polymers and inorganic materials have been extensively used in cardiac patch formation.

In this chapter we will concentrate on the various natural biodegradable materials that have been used to make cardiac patches and their application in cardiac tissue engineering and myocardial regeneration.

Table 1. Bacterial source, monomeric units, carbon source and applications of natural polymers

Polymer	*Produced by*	*Monomeric units*	*Substrate (carbon source used for production)*	*Applications*	*References*
Polyhydroxyalkanoates (PHAs)	*Bacillus subtilis, Bacillus cereus, Pseudomonas mendocina, Ralstonia sps* and *Commamonas sps*	3-Hydroxyalkanoates of 3 to 14 carbon atoms with a large variety of saturated or unsaturated and straight-or branched-chain containing aliphatic or aromatic side groups.	Long chain fatty acids, vegetable oils.	Drug delivery, tissue engineering, wound dressing in surgery, wound management, bio-imaging, biosensors and diagnostics	(*23–30*)
γ-polyglutamic acid (γ -PGA)	*Bacillus licheniformis* and *Bacillus subtilis, Bacillus amyloliquefaciens*, *Bacillus megatherium*, *Bacillus halodurans*, *Staphylococcus epidermidis*, *Fusobacterium nucleatum* and *Natrialba aegyptiaca*	Only D- or only L- or both enantiomers of glutamate	Glutamic acid Glycerol	Drug delivery, vaccine carriers, bio-imaging, biosensors, diagnostics, tissue engineering	(*31–36*)

Continued on next page.

Table 1. (Continued). Bacterial source, monomeric units, carbon source and applications of natural polymers

Polymer	*Produced by*	*Monomeric units*	*Substrate (carbon source used for production)*	*Applications*	*References*
Dextran	*Leuconostoc mesenteroides*, *Streptococcus*, *Lactobacillus* and *Gluconobacter* sp.	polymers of n-glucopyranose units; comprise chains of α-l ,6-linked units with α-l, 4 branching points	Sucrose	Vascular and blood applications, drug delivery, tissue engineering, dental applications	(*37–42*)
Xanthans	*Xanthomonas campestris*	D-glucosyl, D-mannosyl, and D-glucuronyl acid residues in a molar ratio of 2:2:1 and variable proportions of O-acetyl and pyruvyl residues	glucose, sucrose, starch, acid or enzymatic hydrolysates of starch, molasses or corn syrup as the major carbon source in batch cultures	Drug delivery thickener or viscosizer in both food and non-food industries, stabilizer for a wide variety of suspensions, emulsions, and foams	(*43–45*)
Gellan gum	*Sphingomonas paucimobilis* (formerly *Pseudomonas elodea*)	Native gellan gum consists of a backbone of repeaing unit of β-1,3-D-glucose, β-1,4-D-glucuronic acid, β-1,3-D-glucose, α-1,4-L-rhamnose, and two acyl groups, acetate and glycerate, bound to glucose residue adjacent to glucuronic acid .	carbohydrates such as glucose, fructose, maltose, sucrose and mannitol can be used either alone or in combination.	Drug delivery, food or pharmaceutical industries, biomedical applications.	(*46–48*)

Biomaterials for Cardiac Tissue Engineering

A few biomaterial-based cardiac patches pre-seeded with adult stem cells have been developed and implanted in the necrotic heart tissue for regeneration of the myocardium. The scaffolds provide a well distributed arrangement and enhance proper engraftment of adult stem cells along with their optimal retention, while providing support to the infarcted tissue (*50*). Natural protein-based matrices (e.g. collagen and fibrin), bacterial derived polymers polyhydroxyalkanoates (PHA), poly(L-lactic acid) acid (PLLA) and synthetic polymers (e.g. polyglycolic acid (PGA), poly-lactic-co-glycolic acid (PLGA) and poly-ε-coprolactone (PCL) are the most common biomaterials used to fabricate 3D scaffolds for tissue enginerring applications. In addition, biological biomaterials such as autologous muscle and decellularized biological matrices seeded with adult stem cells or cell sheets have also been used. The important properties required for a scaffold are biocompatibility, sterilizability, non-immunogenicity and non-toxicity of degradation products (*51*, *52*).

Natural Biodegradable Polymers for Cardiac Patch Development

Natural polymer-based biomaterials have also been widely used in cardiac regeneration through cardiac patches. Natural polymers mainly comprise of polysaccharide and protein based matrices which are made of biomolecules that mimic and interact with the components in the extracellular matrix (ECM). This helps in a better interaction of the scaffold with the internal tissue environment. It promotes cell adhesion and proliferation and hence does not induce any immunogenic response. Unlike the synthetic biodegradable polymers, the natural polymers are degraded into biomolecules which are completely non-toxic to the tissue and get easily resorbed in the body (*50*). These characteristics hence offer a better cell delivery and enhance the improvement of heart function for which scaffolds are implanted during early stage of MI.

In one of the earlier approaches of cardiac regeneration, the patch was prepared using the fibrin polymer and seeded with mesenchymal stem cells (MSCs). Fibrin is a natural polymer derived from blood clots that has been formulated into different constructs and is used regularly in tissue engineering in a variety of animal models (*53*). This patch was implanted into the infarcted myocardial tissue of pigs and an increased vascularisation was observed in the necrotic region of the scarred pig heart along with growth of cells with myocyte like characteristics (*54*). In another study fibrinogen was combined with polyglycerol sebacate to form a cardiac patch. A core/shell fiber scaffold (PGS/fibrinogen) was fabricated by Ravichandran *et al.*, using poly(glycerolsebacate) which forms the core and provides suitable mechanical support and fibrinogen which forms the shell and enhances the cell-scaffold attachment. This scaffold was formed by first spinning PGS fibers and then covering the core by spinning fibrinogen fibers forming the core/shell fibers which was seeded with neonatal cardiomyocytes and was further analyzed for cardiac applications. The scaffold was found to have a Young's modulus value of

3.28 ±1.7 MPa, quite similar to that of native myocardium. The cell proliferation studies were conducted using MTS assay after 5, 10 and 15 days and it was observed that the absorbance of cells on PGS/fibrinogen fiber scaffold was recorded to be significantly higher as compared to fibrinogen scaffold confirming the structure to be suitable for cardiac applications (*55*).

Poly (1,8-octanediol-co-citric acid) (POC) is another biodegradable elastomer which has been used to study the effect of porosity modification on the Young's modulus in the scaffold suitable for cardiac tissue engineering (*56*, *57*). In this study POC patches were prepared and coated with fibronectin, laminin and collagen and further seeded with HL-1 cardiomyocytes. There was a higher cell growth recorded in all the three scaffolds as compared to the control uncoated films. But, a maximum total coverage of the HL-1 cardiomyocytes was observed in fibronectin coated POC films due the protein- cell interaction in comparision to laminin and collagen films. This was explained as fibronectin was the optimal ECM protein to enhance cell adhesion and proliferation on POC films (*56*). In another study, a patch was prepared using a collagen matrix seeded with human umbilical cord blood mononuclear cells (HUCBCs). Collagen a natural biopolymer is an abundant protein found in the human body and has been extensively used in cardiac applications (*58*). This patch was implanted in the infarcted myocardium of mice and was studied for improvement in LV functioning. It was observed that the mice implanted with the scaffold with seeded cells combined with cell injections showed a better recovery of the scarred tissue as compared to when implanted with collagen matrix alone or collagen matrix with cells (*59*).

Yang *et al.* fabricated a hybrid cardiac patch using silk fibrion (SF) with polysaccharide microparticles of chitosan or hyaluronic acid (HA). Rat bone marrow mesenchymal stem cells (rBM-MSCs) were seeded on these scaffolds and their differentiation was induced using 5-azacytidine. It was observed that the hybrid scaffolds of SF/chitosan and SF/chitosan-hyaluronic acid demonstrated higher cell growth and higher expression of genes important in cardiac differentiation such as Gata4, Nkx2.5, Tnnt2 and Actc1, as well as cardiac proteins cardiotin and connexin 43 when compared to induced rBM-MSCs cultured on SF patches alone. The higher growth rate and expression in the hybrid scaffolds can be explained by the fact that the combination of proteins and polysaccharides mimic the cellular environment and enhance the interaction of the cells with the infarct cardiac tissue (*60*). In another study conducted by Xiang *et al.*, type I collagen–glycosaminoglycan (GAG) was used to fabricate a three dimensional scaffold which was seeded with bone marrow-derived mesenchymal stem cells (BM-derived MSCs) and was implanted in the infarcted region of the rat heart. It was observed that these scaffolds induced neovascularization in the implanted scaffold and also delivered the BM-derived MSCs to the infarct cardiac region in the rat model (*61*). In a similar study, collagen type I was used to fabricate a patch which was seeded with bone marrow derived human mesenchymal cells (hMSCs). After one week of the epicardial implantation of this patch in the rat infarcted region, there was a 23% donor cell engraftment observed, whereas it was not detectable at 4 weeks after the restorative procedure. It was also reported that the collagen I-hMSC patch induced neovascularization and showed improvent in LV remodelling. This observation was not recorded in the rat models with implanted acellular collagen type I patches, which led to the conclusion that the hMSCs play a critical role in the improvement of LV remodelling and resuming the normal functioning of the scarred cardiac tissue (*62*).

Another study was conducted by Chachques *et al.* to evaluate the clinical feasibility and potential of the biodegradable three dimensional collagen type I scaffold seeded with bone marrow cells (BMC) and implanted in a necrotic ventricle (*63*). They also wanted to inspect the combined study involving cell therapy i.e. intramyocardial injections followed by the epicardial implantation of a collagen scaffold seeded with autologous bone marrow cells in patients undergoing coronary artery bypass surgery. Phase I of the MAGNUM clinical trial involved 20 patients, 10 patients underwent surgery and received bone marrow mononuclear cells injections while the other 10 patients underwent surgery and got the BMC injections along with an implantation of epicardial collagen-BMC patch on the infarct cardiac region. It was observed that after one year of the implantation there was no immunogenic or inflammatory response to the scaffold. The thickness of the necrotic myocardium was found to increase and there was also improvement in LV remodeling and diastolic function. It was also found that the patients given BMC injections along with collagen type I-BMC implant showed higher rate of improvement as compared to the ones given only BMC injections (*63*). Another patch was prepared by Chi *et al.* using silk fibroin/hyaluronic acid (SH) and bone marrow stem cells (BMSC) were seeded onto them. These patches were implanted onto myocardial infarct (MI) rat hearts. It was found that the SH and SH/BMSC patches were intact in the MI region and there was no or minor immunological reaction observed on them when assayed using the CD68 marker (Figure 1 a,b,c). The LV wall thickness was also recorded to be higher in BMSC/SH (2.4 ± 0.10 mm) as compared to SH scaffolds (1.7±0.07 mm) implying that the BMSC/SH scaffold proved to be a better scaffold than SH alone for cardiac tissue recovery and engineering (Figure 1 d,e). TUNNEL assay was also performed to measure the apoptotic status of the cardiomyocytes and it was found that BMSC/SH showed a lower number of apoptotic cells (0.07± 0.01 %) as compared to the SH scaffold (0.10 ± 0.02). Hence it was proven that the BMSC/SH scaffold significantly prevented apoptosis of cardiomyocytes. VEGF, HGF and bFGF expressions were also studied within the constructs and it was found that the BMSC/SH scaffolds expressed a relatively higher number of each (26.97 ± 1.05, 11.14 ± 0.55 and 7.32 ± 0.43) as compared to SH scaffolds (4.27± 0.32, 5.71±0.38and 5.14±0.29) respectively. These studies hence concluded that the BMSC/SH scaffolds were much better for cardiac tissue recovery as compared to the SH scaffolds (*64*).

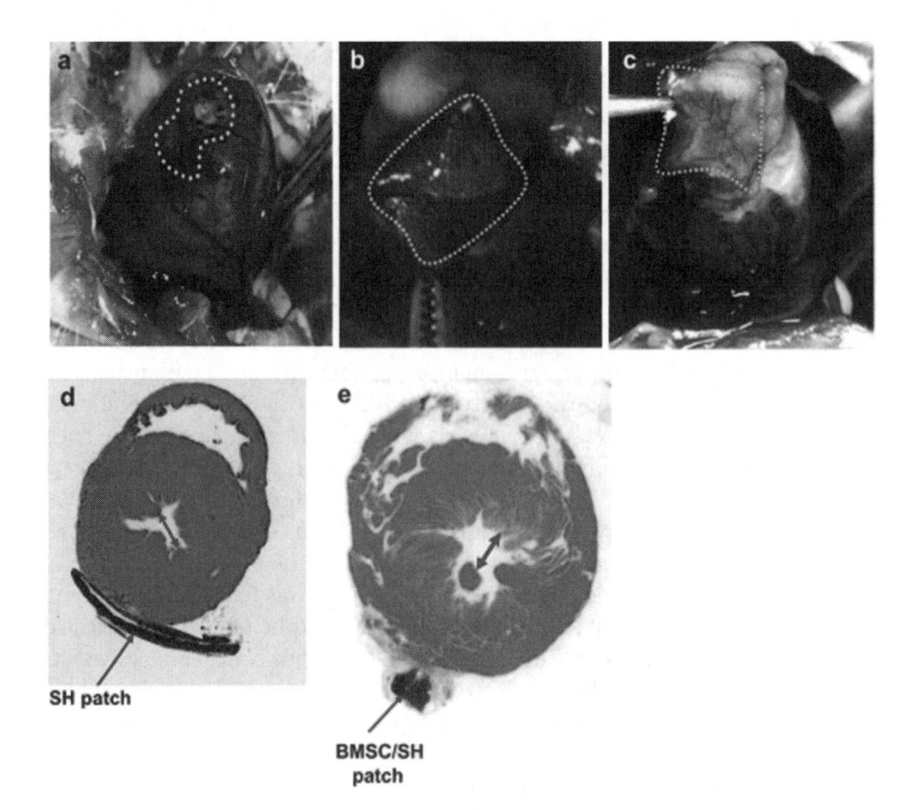

Figure 1. (a) Images of Myocardial infarcted rat heart as shown using the cryo-injury technique. The heart images are shown post patch implantation after 8 weeks of infarction (b) SH and (c) BMSC/SH. The patches, indicated by the green line, are shown to still adhere to the infarcted heart tissue. Hematoxylin and eosin staining of the heart shows the patch is attached to the heart (d) SH and (e) BMSC/SH. The images also indicate the LV inner diameter after 8 weeks of implantation which is indicated by blue arrows. Reproduced with permission from reference (64). Copyright (2012) Elsevier Limited.

Three dimensional scaffolds were also prepared using random and aligned fibers of albumin. Albumin is the most abundant protein found in blood. Albumin has some unique properties which distinguishes it from other globular proteins in the blood. It is an acidic protein with high aqueous solubility and stablity. Stability of albumin against thermal denaturation and low pH conditions is due to 17 disulfide bonds. It has high aqueous solubility, attributed to the polar surface of the molecule (*65*). Cardiomyocytes were grown on them and the beating rate and contraction amplitude was measured at intervals. These were compared with scaffolds prepared using polycaprolactone (PCL), a synthetic polymer. The cells seeded on aligned fibers were found to take the structure of the fibers while those on the random cells were not organized. Cell morphology studies conducted on Day 3 also showed that the aspect ratio of the cells grown on random and aligned (6.0±0.5 and 7.6±0.7) fibers was found to be significantly higher as compared to those grown on PCL (5.3±0.4) scaffolds. The contraction amplitude were also found to be much higher in aligned and random albumin fibers as compared to the PCL treated scaffolds. This thus indicates that the albumin scaffolds can provide support to the tissue structurally and proves to be a better scaffold for cardiac tissue engineering as compared to PCL (*66*). Chitosan/carbon nanofibers conducting scaffold was prepared and was seeded with neonatal rat heart cells and tested for cardiac tissue engineering applications for 14 days. Chitosan is a linear polysaccharide composed of β-(1-4)-linked D-glucosamine and N-acetyl-D-glucosamine units as monomers. It is produced by the deacetylation of chitin, which is mainly present in the exoskeleton of crustaceans and insects (*20*, *53*, *67*). The elastic modulus of the scaffold was measured to be 2.8 ±3.3 which was close to the rat myocardium. The metabolic activity was also recorded at 7 days and 14 days by conducting the MTS assay and both the values were found to be higher in chitosan/carbon scaffolds as compared to chitosan scaffold. Gene expression profiling showed upregulation of Tnnc1 and Cx43 in chitosan/carbon/ cell constructs after 7 days of culture and all measured cardiac- specific genes after 14 days of culture (*68*). These scaffolds demonstrated structural integrity and the ability to withstand arterial pressures in a similar mode to native vessels. To address challenging mechanical environments, while fostering slow degradation and remodeling into native tissue, researchers have been using silk fibroin-based biomaterials for cardiac tissue engineering. Silks offer unique mechanical properties in different material formats, excellent biocompatibility, controlled degradability, and versatile processability, thus offering potential for tissue engineering applications. Moreover, the ability to process silk into different structural formats using an all-aqueous process render it useful for the delivery of bioactive components via this biomaterial matrix, as well as avoiding concerns for residual organic solvents in the devices (*69*).

Three dimensional scaffolds were prepared using silk protein fibroin of Indian tropical tasar silkworm *Antheraea mylitta* (AM) and compared against silkworm *Bombyx mori* silk fibroin (BM), gelatin and fibronectin. Three days post- natal cardiomyocytes were seeded and analyzed for various parameters to investigate the potential of the scaffolds for application in cardiac tissue engineering. The metabolic activity of the cells were assayed using MTT on the different scaffolds and it was observed that cells seeded on AM and fibronectin exhibited a higher activity as compared to gelatin and BM after 2, 4 and 6 days. Fibronectin and AM scaffolds have been reported to naturally have RGD sequences which help in cell attachment and promote cellular activity. The involvement of RGD sequences on the scaffolds for cell attachment was analyzed by incubating the cells with RGD blocking agent RGDS peptides. The MTT assay was performed after 24 h and 48 h after these cells were seeded on all the scaffolds. It was found there was a significant reduction in activity on AM (100 ± 10.2% to 66 ± 10.5%) and fibronectin (100 ± 7.8% to 73.4 ± 6.0%) at 24 h. At 48 h RGDS peptide incubation reduced the metabolic activity in cultures grown on AM (100 ± 23.1% to 59.7 ± 8.9%) but not in cultures grown on fibronectin. There was no reduction observed in gelatin and BM scaffolds confirming the involvement of RGD sequences on the cardiomyocyte attachment. Immunofluorescence studies also suggested that the AM and fibronectin scaffolds exhibited formation of aligned sarcomeres while the sarcomeres on BM and gelatin appeared immature and not aligned (*70*).

Constructs prepared by *Bombyx mori* silk fibroin were seeded with menstrual blood derived stem cells (MeSC) and analyzed for its suitability in cardiac tissue engineering. Immunofluorescent staining of cardiac troponin T showed that the TNNT2, a protein in adult heart that regulates the velocity of myocardial contraction, was expressed in differentiated MeSCs. These results suggest that the interaction between stem cells and silk fibroin scaffolds might modulate and facilitate cardiomyogenic differentiations of induced stem cells during cultivation (*71*).

Conclusion

Cardiovascular tissue engineering is a very promising area for the regenerative therapies combining biomaterials and living cells. The use of this strategy has been used extensively for developing cardiac patches using various synthetic and natural biomaterials, but there have been many limitations. Current biomaterials are limited to provide appropriate biochemical, structural, and biomechanical microenvironments for cells to survive/grow, differentiate, and function. Future works should thus concentrate on the development and use of biomaterials which can overcome such limitations. Cardiac cell therapy also faces challenges on the use of appropriate cell lines in terms of availability in large amounts and ease of culturing.

Future work can be focused on the use of bacterial derived, biodegradable, biocompatible, non-immunogenic polyhydroxyalkanoates (PHAs) in cardiac tissue engineering for preparing constructs. This could bring about the use of a variety of PHAs in the field of myocardial tissue engineering, leading to novel innovative approaches. Meanwhile, the discovery of human iPSCs offers the promise of generating millions of autologous cardiomyocytes required for engineering of a clinically relevant heart patch. With the discovery of new cell types for cardiac regeneration, the development of myocardium-specific biomaterials, and improvement in the understanding of the mechanisms associated with remodeling and regeneration, it is expected that cardiac cell therapy should lead to significant clinical success.

References

1. Taylor, D. A.; Zenovich, A. G. *Diabetes, Obes. Metab.* **2008**, *10*, 5–15.
2. Rosamond, W.; Flegal, K.; Friday, G.; Furie, G. A.; Greenlund, K. *Circulation* **2008**, *117*, 25–146.
3. Taylor, D. A. *Int. J. Cardiol.* **2004**, *95*, S8–S12.
4. Vacanti, C. A. *Tissue Eng.* **2006**, *12*, 1137–1142.
5. Mills, R. M.; Young, J. B. *Clinical Management of Heart Failure*, 2nd ed.; Professional Communications: Caddo, OK, 2004; pp 1–384.
6. Packer, M. *J. Card. Failure* **2002**, *8*, 193–196.
7. Zammaretti, P.; Jaconi, M. *Curr. Opin. Biotechnol.* **2004**, *15*, 430–434.
8. Walsh, R. G. *Heart Failure Rev.* **2005**, *10*, 101–107.

9. Moainie, S. L.; Gorman, J. H.; Guy, T. S.; Bowen, F. W.; Jackson, B. M.; Plappert, T. *Ann. Thorac. Surg.* **2002**, *74*, 753–760.
10. Lorusso, R.; Milan, E.; Volterrani, M.; Giubbini, R.; vanderVeen, F. H; Schreuder, J. J.; Picchioni, A.; Alfieri, O. *Eur. J. Cardio-Thorac. Surg.* **1997**, *11*, 363–371.
11. Oh, J. H.; Badhwar, V.; Chiu, R. C. J. *J. Cardiovasc. Surg.* **1996**, *11*, 194–199.
12. Sabbah, H. N. *Ann. Thorac. Surg.* **2003**, *75*, S13–S19.
13. Eschenhagen, T.; Fink, C.; Remmers, U.; Scholz, H.; Wattchow, J.; Weil, J.; Zimmermann, W.; Dohmen, H. H.; Schäfer, H.; Bishopric, N.; Wakatsuki, T.; Elson, E. L. *FASEB J.* **1997**, *11*, 683–694.
14. Chen, Q. Z.; Harding, S. E.; Ali, N. N.; Boccaccini, A. R. *Tissue Engineering Using Ceramics and Polymers*, 1st ed.; Woodhead Publishing Limited: Boca Raton, FL, 2008; pp 335–357.
15. Wall, S. T.; Walker, J. C.; Healy, K. E; Ratcliffe, M. B; Guccione, J. M. *Circulation* **2006**, *114*, 2627–2635.
16. Jawad, H.; Ali, N. N.; Lyon, A. R.; Harding, S. E.; Boccaccini, A. *Br. Med. Bull.* **2008**, *87*, 31–47.
17. Chen, Q. Z; Harding, S. E.; Ali, N. N.; Lyon, A. R.; Boccaccini, A. R. *Mater. Sci. Eng., R* **2008**, *59*, 1–37.
18. Choi, D.; Hwang, K. C.; Lee, K. Y; Kim, Y. H. *J. Controlled Release* **2009**, *140*, 194–202.
19. Vitali, E.; Colombo, T.; Fratto, P.; Russo, C.; Bruschi, G.; Frigerio, M. *Am. J. Cardiol.* **2003**, *91*, 88–94.
20. Lee, S.-H.; Shin, H. *Adv. Drug Delivery Rev.* **2007**, *59*, 339–59.
21. Riehemann, K.; Schneider, S. W.; Luger, T. A.; Godin, B.; Ferrari, M.; Fuchs, H. *Angew. Chem., Int. Ed. Engl.* **2009**, *48*, 872–897.
22. Escarlata, R. C.; Villaverd, A. *Trends Microbiol.* **2010**, *18*, 423–430.
23. Valappil, S. P.; Rai, R.; Bucke, C.; Roy, I. *J. Appl. Microbiol.* **2008**, *104*, 1624–1635.
24. Valappil, SP.; Boccaccini, A.; Bucke, C.; Roy, I. *Antonie van Leeuwenhoek* **2007**, *91*, 1–17.
25. Hyakutake, M.; Saita, Y.; Tomizawa, S.; Mizuno, S.; Tsuge, T. *Biosci. Biotechnol. Biochem.* **2011**, *75*, 1615–1617.
26. Halami, P. M. *World J. Microbiol. Biotechnol.* **2008**, *24*, 805–812.
27. Lee, S. Y. *Biotechnol. Bioeng.* **1996**, *49*, 1–14.
28. Anderson, A. J.; Dawes, E. A. *Microbiol. Rev* **1990**, *54*, 450.
29. Rai, R.; Boccaccini, A. R.; Knowles, J. C.; Mordon, N.; Salih, V.; Locke, I. C.; Torbati, M.; Keshavarz, T.; Roy, I. *J. Appl. Polym. Sci.* **2011**, *122*, 3606–3617.
30. Rai, R.; Keshavarz, T.; Roether, J. A.; Boccaccini, A. R.; Roy, I. *Mater. Sci. Eng., R* **2010**, *72*, 29–47.
31. Candela, T.; Moya, M.; Haustant, M.; Fouet, A. *Can. J. Microbiol.* **2009**, *55*, 627–632.
32. Candela, T.; Fouet, A. *Mol. Microbiol.* **2006**, *60*, 1091–1098.
33. Buescher, J. M.; Margaritis, A. *Crit. Rev. Biotechnol.* **2007**, *27*, 1–19.

34. Hezayen, F. F.; Rehm, B. H.; Tindall, B. J.; Steinbuchel, A. *Int. J. Syst. Evol. Microbiol.* **2001**, *51*, 1133–1142.
35. Murakami, S.; Aoki, N. *Biomacromolecules.* **2006**, *7*, 2122–2127.
36. Matsusaki, M.; Akashi, M. *Biomacromolecules.* **2005**, *6*, 3351–3356.
37. Naessens, M.; Cerdobbel, A.; Soetaert, W.; Vandamme, E. J. *J. Ind. Microbiol. Biotechnol.* **2005**, *32*, 323–334.
38. Brondsted, H.; Andersen, C.; Hovgaard, L. *J. Controlled Release* **1998**, *53*, 7–13.
39. Hehre, E. J.; Hamilton, D. *J. Biol. Chem.* **1951**, *192*, 161–174.
40. Robyt, J. F.; Yoon, S. H.; Mukerjea, R. *Carbohydr. Res.* **2008**, *343*, 3039–3048.
41. Pereswetoff-Morath, L. *Adv. Drug Delivery Rev.* **1998**, *29*, 185–194.
42. Maire, M.; Chaubet, F.; Mary, P.; Blanchal, C.; Meunier, A.; Avramoglou, D. L. *Biomaterials* **2005**, *26*, 5085–5092.
43. Becker, A.; Vorhoelter, F. J.; Rehm, B.; *Microbial Production of Biopolymers and Polymer Precursors: Applications and Perspectives*; Caister Academic Press: Norfolk, U.K., 2009; pp 1–11.
44. Becker, A.; Katzen, F; PuÈhler, A; Ielpi, L *Appl. Microbiol. Biotechnol.* **1998**, *50*, 145–152.
45. Bertram, U.; Bodmeier, R. *Eur. J. Pharm. Sci.* **2006**, *27*, 62–71.
46. Bajaj, I. B.; Survase, S. A.; Saudagar, P. S.; Singhal, R. S. *Food Technol. Biotechnol.* **2007**, *45*, 341–354.
47. Hamcerencu, M.; Desbrieres, J.; Khoukh, A.; Popa, M.; Riess, G. *Carbohydr. Polym.* **2008**, *71*, 92–100.
48. Rajinikanth, P. S.; Mishra, B. *Chem. Pharm. Bull.* **2009**, *57*, 1068–1075.
49. Seeherman, H.; Wozney, J. M. *Cytokine Growth Factor Rev.* **2005**, *163*, 29–45.
50. Coutu, D. L.; Yousefi, A. M.; Galipeau, J. *J. Cell. Biochem.* **2009**, *15.108*, 537–46.
51. Leor, J.; Amsalem, Y.; Cohen, S. *Pharmacol. Ther.* **2005**, *105*, 151–63.
52. Leor, J.; Cohen, S. *Ann. N. Y. Acad. Sci.* **2004**, *1015*, 312–319.
53. Bessa, P. C.; Casal, M.; Reis, R. *J. Tissue Eng. Regener. Med.* **2008**, *2*, 81–96.
54. Liu, J.; Hu, Q.; Wang, Z. *Am. J. Physiol.: Heart Circ. Physiol.* **2004**, *287*, 501–511.
55. Ravichandran, R.; Sundarrajan, J. R. V. S.; Mukherjee, S.; Sridhar, R.; Ramakrishna, S. *Int. J. Cardiol.* **2013**, *167*, 1461–1468.
56. Bastida, L. A. H.; Barry, J. J. A.; Everitt, N. M.; Rose, F. R. A. J.; Buttery, L. D.; Hall, I. P.; Claycomb, W. C.; Shakesheff, K. M. *Acta Biomater.* **2007**, *3*, 457–462.
57. Rezwan, K.; Chen, Q. Z.; Blaker, J. J.; Boccaccini, A. R. *Biomaterials.* **2006**, *27*, 3413–3431.
58. Habraken, W. J. E. M.; Wolke, J. G. C.; Jansen, J *Adv. Drug Delivery Rev.* **2007**, *59*, 234–48.
59. Cortes-Morichetti, M.; Frati, G.; Schussler, O.; Huyen, J. P. D. V.; Lauret, E.; Genovese, J. A.; Carpentier, A. F.; Chachques, J. C. *Tissue Eng.* **2007**, *13*, 2681–2687.

60. Yang, M. C.; Wang, S. S.; Chou, N. K.; Chi, N. H.; Huang, Y. Y.; Chang, Y. L. *Biomaterials.* **2009**, *30*, 3757–65.
61. Xiang, Z.; Liao, R.; Kelly, M. S.; Spector, M. *Tissue Eng.* **2006**, *122*, 467–78.
62. Simpson, D.; Liu, H.; Fan, T. H.; Nerem, R.; Dudley, S. C., Jr *Stem Cells* **2007**, *25*, 2350–2357.
63. Chachques, J. C.; Trainini, J. C.; Lago, N.; Cortes-Morichetti, M.; Schussler, O.; Carpentier, A. *Ann. Thorac. Surg.* **2008**, *85*, 901–908.
64. Chi, N. H.; Yang, M. C.; Chung, T. W.; Chen, J. Y.; Chou, N. K.; Wang, S. S. *Biomaterials.* **2012**, *33*, 5541–5551.
65. Amiji, M.; Park, K. *J. Biomater. Sci., Polym. Ed.* **1993**, *4*, 217–234.
66. Fleischer, S.; Shapira, A.; Regev, O.; Nseir, N.; Zussman, E.; Dvir, T. *Biotechnol. Bioeng.* **2014**, *111*, 1246–1257.
67. Bhattarai, N.; Gunn, J.; Zhang, M. *Adv. Drug Delivery Rev* **2010**, *62*, 83–99.
68. Martins, A. M.; Eng, G.; Caridade, S. G.; Mano, J. F.; Reis, R. L.; Novakovic, G. V. *Biomacromolecules.* **2014**, *15*, 635–643.
69. Zhang, X.; Baughman, C. B.; Kaplan, D. L. *Biomaterials* **2008**, *29*, 2217–2227.
70. Patra, C.; Talukdar, S.; Novoyatleva, T.; Velagala, S. R.; Mühlfeld, C.; Kundu, B.; Kundu, S. C.; Engel, F. B. *Biomaterials.* **2012**, *33*, 2673–2680.
71. Rahimi, M.; Kouchesfehani, H. M.; Zarnani, A. H.; Mobini, S.; Nikoo, S.; Somaieh *J. Biomater. Appl.* **2014**, *29*, 1–10.

Chapter 11

Advanced Protein Composite Materials

Fang Wang,[1,2] Catherine Yang,[3,4] and Xiao Hu*,[1,3]

[1]Department of Physics and Astronomy, Rowan University, Glassboro, New Jersey 08028, United States
[2]Center for Analysis and Testing, Nanjing Normal University, Nanjing 210023, China
[3]Department of Biomedical and Translational Sciences, Rowan University, Glassboro, New Jersey 08028, United States
[4]Department of Chemistry and Biochemistry, Rowan University, Glassboro, New Jersey 08028, United States
***E-mail: hu@rowan.edu. Phone: 1-856-256-4860. Fax: 1-856-256-4478.**

Proteins are important biological macromolecules that have been used as materials for centuries, not only because they are environmentally friendly, renewable, and non-toxic, but also due to their excellent strength, elongation, toughness, slow degradability and great biocompatibility. Combining proteins with other essential biological materials can generate novel composite materials with diverse properties, such as suitable mechanical and chemical properties, favorable electrical and optical features or other excellent characteristics. This chapter will begin with a brief introduction of the most important natural fibrous proteins such as various silks, elastins, collagens, keratins, resilins, and their unique repeats and structures that provide distinguished physical properties. We will then discuss their mechanisms of interaction with other biomolecules, using the traditional phase diagram and glass transition theories to understand their molecular interactions and miscibility. Lastly, we will focus on different advanced protein-based composite materials developed in recent years and their related applications. This part of the chapter will be divided into four sections: protein-natural biopolymer composites, protein-synthetic polymer composites, protein-inorganic composites, and protein-small molecule composites for drug delivery and release studies. These composite materials would be broadly useful in multiple fields due to their highly tunable structures and reliable functions in the future.

Introduction

Proteins are multi-block biopolymers that play an important and indispensable role in tissue compositions and biological activities. Typical fibrous proteins include elastins and collagens from different tissues, silks from different worm and spider species, keratins from hairs and wools, and zeins from corns, *etc.* These natural structural proteins are produced by various insects, animals or plants that have evolved for millions of years with critical structural and bioactive properties (Figure 1) (*1–4*). Their properties can be used to generate a variety of protein-based composite materials which could accomplish a specific or multifunctional tasks (*3*, *4*). For example, collagens and elastins are often found together in the body to provide the combination of strength and flexibility required for specific tissue functions (*5*, *6*). The cocoons and fibers produced by silkworms, spiders and bees often form silk fibroin-sericin composites which protect insects from predators during their transmutation (*7*, *8*). Therefore, many approaches have been explored to blend the natural proteins or genetic engineered proteins with other materials to produce special characteristics, and generate multifunctional and biodegradable composite biomaterials. These composite materials meet the needs of various cellular or tissue regeneration and therapy applications for humans, which could provide suitable mechanical and chemical properties, favorable electrical and optical features or other excellent biophysical properties for various biomedical applications (*1*, *8*).

This chapter will begin with a brief explanation of the natural protein structures and their interaction mechanisms, and then followed by a detailed review of protein-based composite materials and their various applications in recent years.

Structure of Fibrous Proteins

Silks, elastins, collagens, keratins and resilins are some of the most common fibrous proteins used for protein-based biomaterials (Figure 1). These natural proteins have distinguished properties, and can be designed into materials for various chemical, mechanical, electrical, electromagnetic, or optical applications. In general, the unique repetitive amino acid sequences of these proteins are the keys for their molecular chains to form long range ordered secondary structures (*9*, *10*), such as beta-sheets in silks, coiled-coils in keratins, and triple-helix in collagens. These secondary structures are then stabilized by inter-chain hydrogen bonding in which every amino acid residue is laterally bonded to its nearest neighbor chain segments. These features reflect their roles as mechanically important structures at biological interfaces, prompting their utility as a treasured resource of biopolymers materials (*9*).

Figure 1. Natural proteins for composite materials, including various silkworm and spider silks (beta-pleased sheet structure), elastins (beta-spiral structure), wool and hair keratins (alpha-helical conformation and cross-linked by cysteines), zeins (alpha-helix structure), as well as collagens (triple-helix structure), resilins (beta-turn structure), and bee silks (coiled-coil structures) etc. (1).

Silk Proteins

Specifically, silk proteins are fibrous proteins synthesized by the larva of insects, and the best known ones are silkworms and spiders. The cocoon silks from the domesticated silk-moth, *Bombxy mori*, have been used in clothing, furnishing and upholstery fields (*11*) for thousands of years, due to its lustre, light weight, tear resistance, and great economic importance. In nature, cocoon silks protect the silkworms against microbes and predators (*12*, *13*). And one silk fiber is composed of two core silk fibroin protein fibers, along with an outer adhesive glycoprotein coats (See Figure 2). The glue-like glycoproteins are known as sericins (a set of serine-rich glycoproteins), which ensure the cohesion of the cocoon by sticking the twin filaments together. Sericin constitutes 25–30% of the weight of the fiber (*14*), and can be removed by heat or alkaline treatments, resulting in more than 300 meters of fibroin fiber from each cocoon. Silk fibroin fiber is estimated to have 70 ~ 75 % crystallinity with corresponding 25 ~ 30% amorphous region (*15*). The fibroin molecular chains are composed of a complex of three components: a large protein (H-chain) fibroin (MW~350 kDa), a second small protein (L-chain) fibroin (MW~25 kDa); and possibly a third small glycoprotein, (P25 protein, MW~30 kDa) (*15*). H-chain fibroin (which is relatively hydrophobic and can form anisotropic beta-sheet-rich nanocrystals), is linked to the L-chain fibroin (which is hydrophilic and relatively elastic) via disulfide bonds, and is often associated with a P25 protein (which plays a role in maintaining the integrity of the complex) via non-covalent hydrophobic interactions (*5*, *14–16*).

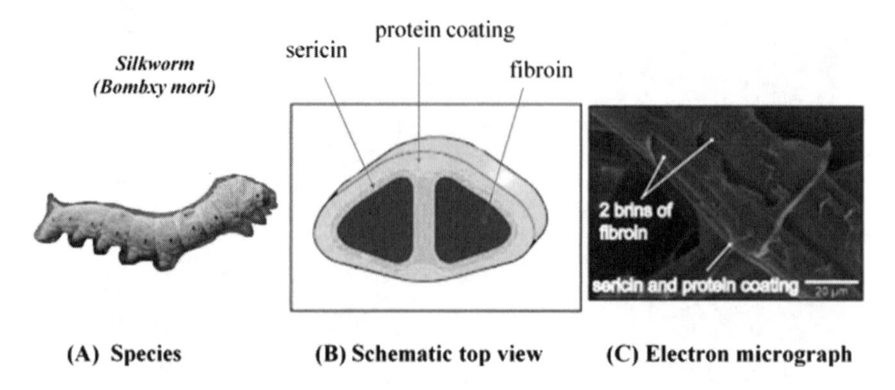

Figure 2. A) A Bombyx mori silkworm. B) Schematic illustration of the composite structure of a cocoon fiber, in which the two fibroin fibers are coated with sericins and other proteins to protect the cocoon against microbes and predators. C) Scanning electron micrograph of a single B. mori silkworm cocoon fiber (contains two fibroin fibers with a sericin coat). (Reproduced with permission from Reference (5). Copyright (2010) Elsevier Ltd.).

In contrast, spider silk proteins are synthesized from specialized abdominal glands of spiders, which function as biofactories to produce large quantities of silk fibroins. Spider silks have different properties, compositions and morphologies in comparison to the silkworm silks (*16*, *17*). The Major Ampullate (MA) spider silk fibers (of ca. 250–350 kDa) have a diameter between 1 and 20μm, and are comprised of four different layers: the core (major ampullate spidroin), the skin (minor ampullate like protein), a glycoprotein coat and a lipid coat (See Figure 3) (*18*, *19*). Koski *et al.* (*20*) measured the stiffness of some Major Ampullate spider silk webs by a non-invasive technology. The results indicated that spider silks have higher elastic stiffness, and supercontract ability than most synthetic biomaterials. Besides, it is found that the spider silk's toughness is 10 times greater than other biological materials, due to their crystal structures in polypeptide chains, whereas its super extensibility is attributed from noncrystalline regions in the protein structure (*21*). Under the same ultraviolet (UV) irradiation conditions, it is found that the degradation resistance of *Nephila clavata* spider silks was 1.7 times higher than the silkworm silks (*22*, *23*). Decomposition of silks was due to the cleavage of protein chain molecules under UV irradiation. Molecular weight of the *Nephila clavata* spider dragline was determined to be 272 ± 14 kDa using a sodium salt-polyacrylamide gel electrophoresis method (*24*). In addition, most spider species produce silk from a spinneret, which usually locates on their lower abdomens. However, it has been found that tarantula spiders secrete gluey silks from their feet (*25*). Table I lists the strength, toughness and other mechanical properties of silkworm and spider dragline silks, with comparisons to several widely used biomaterial fibers and tissues.

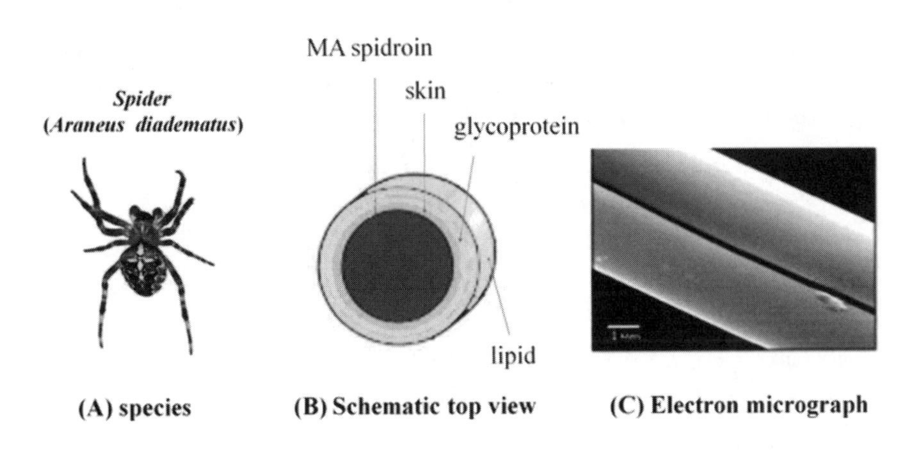

Figure 3. A) An Araneus diadematus spider. B) Schematic illustration of the composite structure of a major ampullate (MA) fiber. The core spidroin fiber is coated by a layer of minor ampullate-like (MI-like) spidroin skin, a layer of glycoprotein and a layer of lipids. C) Scanning electron micrograph of major ampullate fibers. (Reproduced with permission from Reference (5). Copyright (2010) Elsevier Ltd.).

In addition, according to the evolution studies of natural fibrous proteins, there are at least 23 independently evolved lineages that can produce silks. Different secondary structures were found in these lineages including beta-pleated sheets, coiled coils, or twisted-hellices *etc.* (*24*) Compared with the beta-sheet crystallites in the silkworm cocoons and spider dragline silks, some insect silk structures are essentially different. For example, the silks of bees, ants, and hornets from the social *Hymenoptera* (over 144,000 species) usually have a tetrameric coiled-coil conformation in their protein structures (*24*, *26*). The coiled-coil structure is composed of four alpha-helical protein chains with a similar repetitive polypeptide sequence $(abcdefg)_n$. The first residues (a), the fourth residues (d), and possibly the fifth residue (e) are highly hydrophobic, and others are hydrophilic, polar or charged, which promote formation of coiled-coil structures (*26*, *27*). Coiled-coil structures provide honeybee silks excellent extensibility (~200%), whereas the beta-sheet structures in the polyalanine motif regions of the bee silk chains provide them strength. In addition, the glycine-rich regions in the honeybee silk are dominated with beta-turns and 3_{10} helix structures, which contribute to their elasticity (*25*, *26*). Similar to the structure of honeybee silk, high-resolution ^{13}C solid-state NMR experiments also indicated that alpha-helix structures are the major conformation in the central portion of hornet silk, a silk produced by the larvae of *Vespinae mandarinia*, whereas beta-sheet conformations dominated on the ends of their protein chains (*26*, *27*).

Table I. Comparison of mechanical properties of silks from silkworm and spider dragline to other commonly used biomaterial fibers and tissues. (Partially reproduced with permission from Reference (*143*). Copyright (2003) Elsevier).

Material	*UTS* [a] *(MPa)*	*Modulus (GPa)*	*Strain at break (%)*	*References*
B. mori silk	740	10	20	(*223*, *224*, *226*)
Spider silk	875–972	11–13	17–18	(*223*, *224*)
Collagen[b]	0.9–7.4	0.0018–0.046	24–68	(*223*, *225*)
Collagen X-linked[c]	47–72	0.4–0.8	12–16	(*223*, *225*)
PLA[d]	28–50	1.2–3.0	2–6	(*223*, *225*)
Tendon (comprised of mainly collagen)	150	1.5	12	(*223*, *227*)
Bone	160	20	3	(*223*, *227*)
Kevlar (49 fiber)	3600	130	2.7	(*223*, *227*)
Synthetic Rubber	50	0.001	850	(*223*, *227*)

[a] UTS - Utimate Tensile Strength. [b] Type I Rat-tail collagen fibers were tested during stretching from 0% to 50% of their original lengths. [c] Rat-tail collagens were dehydrothermally cross-linked and tested during stretching from 0% to 50%. [d] Polylactic acid (PLA) has a molecular weight ranging from 50,000 to 300,000.

Other Natural Proteins

Several other natural proteins have been actively used as biomedical materials in recent decades. Elastin is the essential protein component in native tissues and is dominated with the amino acids valine (V), glycine (G), alanine (A) and proline (P). Elastin can provide elasticity to tissues, thus exists primarily in arteries, lung tissues, skins, ligaments, and tendons (*28*, *29*). The hydrophobic domains of elastin are rich in nonpolar amino acids, with repeating motifs such as $[GVGVP]_n$. Its hydrophilic domains contain a high content of lysines, which are involved in elastin cross-linking (*1*, *2*), that stabilize the entire protein structure in tissues.

Resilin proteins were discovered in fleas, dragonflies and other flight or jumping insects that exhibiting super jumping and flying abilities. Resilin is a natural super elastic "rubber" which can extend and retract cyclically millions of times, and can be used as an energy buffer register during the lifetime of insects. Resilin is connected via cross-linking of tyrosines in the protein side chains and can provide over 95 % resilience during the high-frequency motions (*3*, *4*), with a more than 300 % elongation ratio before its breakage (*3*). Full length resilin in *Drosophila melanogaster* consists of three domains: (GGRPSDSYGAPGGGN) repeats are enriched in the first exon, while (GYSGGRPGGQDLG) repeats are enriched in the third exon. A chitin binding domain is encoded in the second exon, which was believed to lock the chitin component of exoskeleton (*4*). A new recombinant technique to produce resilin-like proteins was also developed recently by researchers in Australia CSIRO (Commonwealth Scientific and Industrial Research Organization) (*30*), which can be used for reproducing rubber-like resilin protein biomaterials in the future. Through this new method, the protein rubber materials extracted from the resilin sequences can exhibit a high resilience of up to 98 % (*30*).

Collagens (Type I ~ XXVIII) are the most abundant and ubiquitous proteins found in connective tissues of metazoan animals, and constituted 25 ~ 35 % of total proteins in our body (*5*, *6*). The basic molecular unit of collagen network is tropocollagen, in which three tropocollagen peptide chains are twisted together to form a right-handed triple helix structure. The most common amino acid sequence of a tropocollagen chain is $(GPX)_n$, where X is any amino acid other than glycine (G), proline (P) or hydroxyproline (*5*). Collagen is the most important water-insoluble vital component in extracellular skeleton matrix of tissues, which has excellent mechanical properties to support and protect the animals. Besides, collagen has played an important role in *in vivo* cell migration and development (*29*).

Keratin is another broad category of insoluble fibrous proteins found in horns, hair, nails and wools. It consists of many parallel polypeptide chains with alpha-helix or beta-helix conformations. 'Soft' hair keratins contain a high content of cysteines in the non-helical domains, which can be crosslinked via intermolecular disulfide bonds and provide tough and durable properties to hairs (*27*, *31*).

Furthermore, Yang *et al.* (*32*) reported that they discovered a novel silk-like protein (named aneroin) from the sea anemone *Nematostella vectensis* recently, and have successfully fabricated the aneroin-based materials by wet-spinning and electro-spinning methods. The sequence of decapeptide repeats in aneroin is highly similar to those from other fibrous proteins such as silkworm silks and spider silks. And it primarily locates in the nematocysts of sea anemone, together with other proteins. Moreover, the mechanical properties of aneroin are also similar to those of spider silks, which indicated that aneroin is an excellent candidate for designing protein-based biomaterials in the future.

In addition, some other plant proteins have also been used to fabricate medical materials that can control the signaling, immunoreactions, adhesion, periodic modulation and other functions of cells (*28*, *29*, *33*). In summary, proteins are perfect candidates for designing different biomaterials with various applications. The unique sequences and structures of proteins can provide special functions and a broader range of properties in the materials, which are biocompatible, biodegradable, and tenable, and can be used for designing new implant devices and tissue regeneration materials in the future (*34*).

Mechanism of Molecular Interactions

The biological functions of tissues are often stimulated by molecular interactions between proteins. In fact, most living activities are controlled by the regulation of protein interactions in the body. For example, antibody-antigen interactions involve a variety of proteins to protect the body from harm, while interactions between actin and myosin proteins are essential in controlling muscle contractions and heart beats. Therefore, researchers have been intensively studying the protein interactions since last century in order to reveal their natural interaction mechanisms and imitate their bio-functions for health care. Various protein-biomolecule combinations or blends are also able to gain desirable features for novel biomaterials. In general, protein-biomolecule interactions include hydrophobic interaction, *van der* Waals forces, hydrogen bonds, ionic bonds, and disulfide bonds between the specific domains of proteins. From the view of polymer-molecule interactions, the free energy of mixing is the key factor to govern the miscibility of protein composites (*35–37*). And there are many techniques to detect the interactions between proteins and other biomolecules, including yeast two-hybrid system (*38*, *39*), surface plasmon resonance technology (*40*, *41*), fluorescence resonance energy transfer technology (*42*) and pull-down technology (*43*) *etc*. Among these methods, differential scanning calorimetry (DSC) is one of the most important technologies to acquire evidence for the miscibility of multivariate proteins by investigating their glass transition temperatures (T_g) or phase transition regions.

For a binary component system, the Flory-Huggins equation (*44*) can be expressed in the following form:

$$\Delta G_m = RT[\varphi_1 \ln\varphi_1/r_1 + \varphi_2 \ln\varphi_2/r_2 + \chi\varphi_1 \varphi_2] \quad (1)$$

where ΔG_m is the Gibb's free energy; R is the gas constant; r_i is the number of protein segments (i = 1, 2); φ_i is the volume fraction of the component and χ is the Flory-Huggins binary interaction parameter. The last term in the right side of equation (1), $\chi\varphi_1\varphi_2$, is assigned to the enthalpy of mixing, while the first two terms are related to the entropy of mixing. Figure 4a shows the phase diagram for a blend with two polymer components. As the two polymers blend together, miscible phase, metastable phase and immiscible phase could occur in the blending system. The solid line separates miscible and metastable regions, while the dashed line separates metastable and immiscible regions. For a fully miscible polymer system (Figure 4b), a single T_g normally appeared between the T_gs of the two individual polymer components, T_{g1} and T_{g2} (*45*, *46*). Full phase separation can be judged by the presence of both T_{g1} and T_{g2} at their original positions. Micro-heterogeneous phase might be formed if the two polymer components can be partially mixed together, where their individual T_gs will migrate close to each other [T_g (a) and T_g (b)] in the DSC scans. Based on this theory, Hu *et al.* (*47*) investigated the T_gs of silkworm silk and tropoelastin blend films by modulated DSC, and found silk proteins are well miscible with elastin proteins at different blending ratios.

Simultaneously, the variation of T_gs in the mixture of polymer-polymer blend systems can be predicted by Fox equation (*48*), Gordon-Taylo equation (*49*) or Kwei equation (*50*). The Kwei Equation is expressed as below,

$$T_g = (W_1T_{g1} + kW_2T_{g2})/(W_1 + kW_2) + qW_1W_2 \quad (2)$$

where T_g is the glass transition temperature of the final blend; T_{g1} and T_{g2} are the glass transition temperatures of their individual components, respectively; W_1 and W_2 are the weight percentages of these two components, respectively; k is a parameter representing the strength of molecular interaction between blend components; q is a parameter for the specific intermolecular interactions in the mixture, accounting for the effects of rearrangements between the molecules. The values of q can be either positive or negative. The first term on the right side of equation (2) is identical with the Gordon-Taylor equation (*51–54*), while the last term on the right side represents the effect of interaction, such as the hydrogen bonding in the binary system.

Furthermore, there are many other factors that will affect the miscibility of protein-biomolecule blends, including environmental temperature, pH, salinity, as well as protein concentration *etc*. For a protein composite system, the bounded solvents such as bound water molecules might generate a new glass transition due to its plasticization in the protein structure, which has been found in wheat proteins (*55*) and soy proteins (*56, 57*).

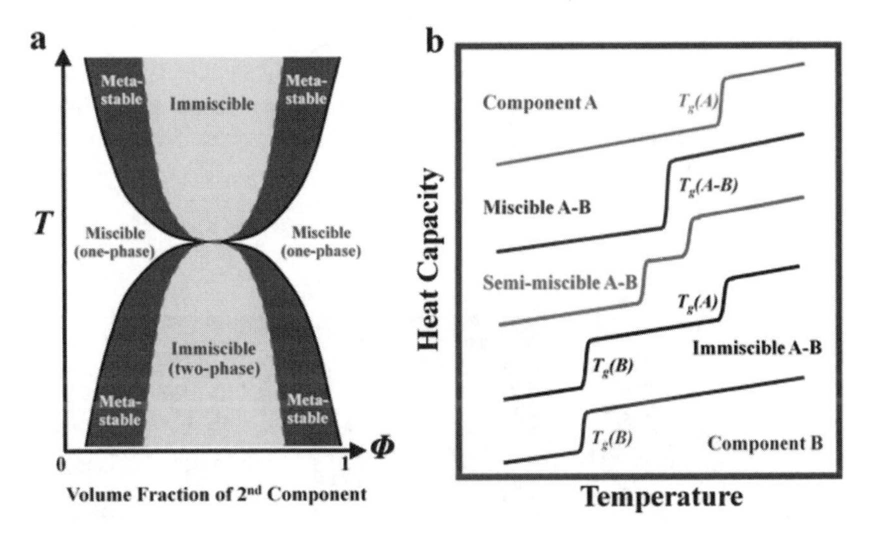

Figure 4. (a) Schematic phase diagram for a two component blend system (temperature T vs. volume fraction Φ of the second component). (b) Schematic DSC heat capacity traces for a two component blends with different miscibility (1).

Protein-Based Composite Materials

Protein fibers are natural materials that have been used for centuries in textile industry, not only because they are environmentally friendly, low cost, renewable, non-toxic and lustrous, but also due to their characteristic strength, elongation, toughness, moisture absorbance, slow degradability and great biological compatibility. Through a variety of aqueous or organic solvent processing methods, natural protein composites can be fabricated into the formats of hydrogels, tubes, sponges, composites, fibers, microspheres and thin films (Figure 5), which can be used for a wide range of applications (*58*). For example, silk particles and microcapsules are envisioned as drug carriers, whereas silk films, foams, and fabrics could be used as scaffolds in tissue engineering (*59*). However, the natural fibers also have many shortcomings such as their poor rub resistance and wrinkle recovery, as well as their low mechanical strength in water (*5*). Besides, there are many other problems that need to be solved and improved for their practical applications, such as how to control the degradation rate of protein-based biomaterials *in vivo*. In the past decade, many protein composite materials have been developed for various applications in biomedical science, tissue engineering, drug release, and bio-optics (*60–64*). A number of articles have discussed the unique advantages of these protein composite materials (*5*, *17*, *32*, *47*). These composite materials can be used to regenerate cartilages, soft tissues, corneals, vascular tissues, cervical tissues and skins, as well as to replace ligaments, repair bone defects and deliver cancer drugs (*58*, *65*). Table II lists a summary of different applications of these protein composite materials. Table III highlights the use of silk composite materials for different biomaterial property improvements.

In the following, we will discuss a variety of topics related to the applications of these proteins composites, including protein-based composite materials with biopolymers, synthetic polymers, inorganic particles as well as small molecules for drug delivery studies.

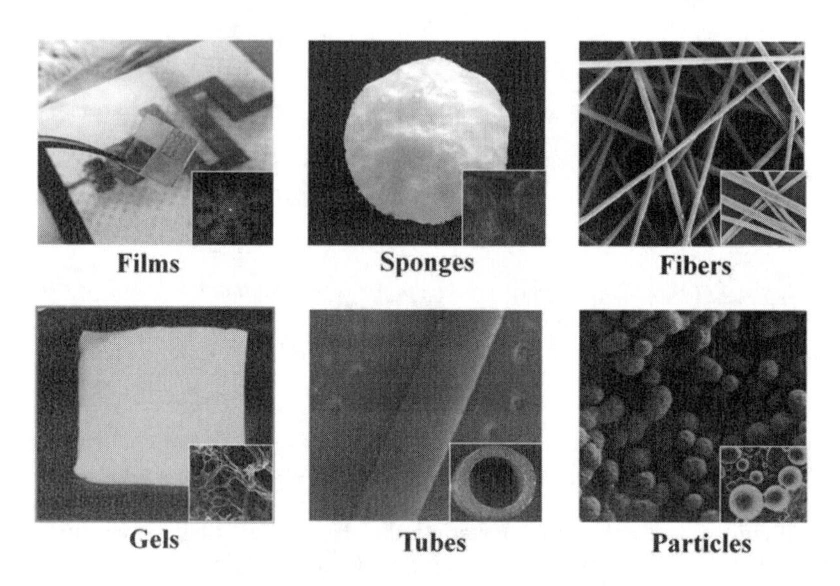

Figure 5. Examples of recently developed protein composite materials. (Partially reproduced with permission from Reference (58). Copyright (2011) Nature Publishing Group).

Table II. Applications of protein-based composites

Material format	*Example*	*Application*	*Reference*
Sponges/ scaffolds	Bone	Tissue engineering	(*5*, *130*)
	Cartilage repair	Tissue engineering	(*81*)
	Bladder tissue	Tissue engineering	(*82*)
	Fibroblast adhesion	Tissue engineering	(*47*)
	Cytocompatible carrier	Tissue engineering	(*59*)
	Chondrocytes	Tissue engineering	(*59*, *103*)
	Mesenchymal stem cells	Tissue engineering	(*47*, *83*)
	Coronary artery smooth muscle cells	Tissue engineering	(*177*)
	Recover leak-point pressure and lumen area	Tissue engineering	(*172*)
	Regenerate urethral sphincter structure	Tissue engineering	(*172*)
Film	Mesenchymal stem cells	Tissue engineering	(*83*, *111*)
	Fibroblast	Tissue engineering	(*84*, *85*)
	Skin	Tissue engineering	(*86*)
	Bone	Tissue engineering	(*127*)
	Peptides	Tissue engineering	(*70*)
	Osteogenesis	Tissue engineering	(*94*)
	Corneal tissue	Tissue engineering	(*17*)
	Orthotopic breast cancer	Tissue engineering	(*174*)
	Vascular tissues	Tissue engineering	(*111*)
Hydrogels/gels	Softer tissue	Tissue engineering	(*128*)
	Enzymatic degradation	Tissue engineering	(*156*)
	Growth factor	Drug delivery	(*173*)
	enzyme entrapment	Drug delivery	(*112*)
Fibers/tubes	Actuating devices	Electronic application	(*95*)
	Biosensing	Electronic application	(*109*, *176*)
	Bio-memristor	Electronic application	(*228*)
	Electrical wire	Electronic application	(*229*)

Continued on next page.

Table II. (Continued). Applications of protein-based composites

Material format	*Example*	*Application*	*Reference*
	Anticancer therapies	Drug delivery	(*109*)
	Electrical current in ultra-low temperature	Electronic application	(*17*)
	Loaded drug	Drug delivery	(*129*)
Microspheres	Drug delivery (i.e vaccines)	Drug delivery	(*140*, *168*, *175*)
	Gene delivery(i.e. embryonic kidne, DNA)	Drug delivery	(*162*, *178*)
Micro-porous	Growth factor delivery	Drug delivery	(*169*, *170*)
	Small molecule (i.e.antibiotics)	Drug delivery	(*139*, *174*, *175*)
Nano-fibers	Drug-eluting stent	Drug delivery	(*138*)
Nano-sheets	Bioactive molecules	Drug delivery	(*140*, *157*)
	Bovine serum albumin	Drug delivery	(*159*)

Table III. Examples of silk composites for materials property improvements

Components	*Morphologies*	*Improved properties*	*References*
Silk - tropoelastin	Sponge	Mechanical elasticity, cell attacment	(*38*, *47*)
Silk - cotton, wool	Film	Strengthen, thermal stability	(*76*, *77*)
Silk - gelatin	Scaffold	Control blend chemistry changes	(*78*)
Silk - chitosan	Scaffold	Mechanical, stabilities, anticoagulant	(*14*, *72*)
Silk - collagen	Film/Hydrogel	Optical, hydrophily	(*91*, *92*)
Silk - carbn nanotubes	Fiber tube, shapeable	High tough, flexible, strain-humidity-sensitive	(*95*)
Silk - PU	Scaffold	High drophilicity, biocompatibility, facilitates cell adhesion	(*100*, *103*)
Silk - PLA	Fibre	Elasticity, ductility, anticoagutation	(*164*)
Silk - PHB	Film	Tensile, elongation, cell adhesion	(*111*)
Silk - CaP	Sponge	Bone cells formation	(*5*, *130*)

Continued on next page.

Table III. (Continued). Examples of silk composites for materials property improvements

Components	*Morphologies*	*Improved properties*	*References*
Silk - graphene	Nano-sheet	Well-ordered hierarchical structure	(*158*)
Silk - Au	Fibre	Strong conductive electricity under ultra-low temperature	(*17*)

Protein-Natural Polymer Composite Materials

As described in Section 2, some of natural protein fibers, such as silkworm silks and spider silks, essentially are protein-protein composites with multiple components, and their thread strengths are even superior to Nylon and Kevlar (*66*). For example, silkworm silk fibers have tensile strength around 0.5~1.3 GPa, and a toughness between 6×10^4~16×10^4 $J\cdot kg^{-1}$ (*1–4*). These features are closely related to the molecular interactions between different protein components in the silk fibers (*4*). Some researchers mimicked the natural spinning process of silk fibers and elucidated their protein interaction mechanisms to control the nano-fibre structures during the spinning (*67*). And they found four key factors governing the self-assembly of silks in aqueous solutions: molecular mobility, charge, interactions between proteins and the solution concentration. Furthermore, self-assembly of silk fibers is a thermodynamic process where kinetics play a key role. Therefore, the physical properties of silk-based materials can be adjusted by these factors, such as through balancing the nanostructure and composition in different protein domains during the fabrication process (*67*).

Hu. *et al.* (*47*) produced a new class of biomedical materials without macrophase separation by blending semi-crystalline silk protein with tropoelastin at different mixing ratios. Both protein polymers used in the study are biocompatible and biodegradable. Moreover, the results revealed that the hydrophobic and hydrophilic interactions between silk and tropoelastin protein chains are the key to control their bulk material properties. Furthermore, the porous composites could support attachment and proliferation of human mesenchymal stem cells, with sufficient mechanical elasticity in micro-/nano-scales. On the other hand, some researchers (*68*) found that domesticated silk fibroin lacked specific cell attachment sites such as arginine-glycine-aspartic acid (RGD) sequence, whereas it was found in other wild-type silkworm silks. Therefore, wild silk based composite materials are designed for *in-vivo* controlling cell differentiations (*69*). Recently, some studies (*70–72*) also demonstrated that conjugating fibroin with starch or cellulose is an alternative approach to customize the cell adhesion and proliferation properties, as well as their material biodegradability and mechanical performance. For example, silk fibroin and oxidized starch proteins were co-cross-linked into a chitosan matrix by using a non-toxic reductive alkylation procedure (*73*, *74*). In this case, silk can be assembled with chitosan materials to form a layer-by-layer structure due to the hydrogen bonds and electrostatic interactions between them (*73–75*).

Besides, tri-component bio-films could be made by blending silk with raw cotton and wool fibers together using ionic liquids, which strengthen the molecular hydrogen bonds in the composites and increase thermal stability of the coagulated films (*76*, *77*). Moreover, adjusting the mixture ratio of silk and gelatin blends (*78*) is a way to control cell proliferation rate and maintain chondrocyte morphology in the silk-gelatin composite materials. Enhancing certain gene expression, *i.e.*, Sox-9, can significantly affect the aggregation process in the microperiodic silk-gelatin scaffolds (*79*). Tomita *et al.* (*80–82*) also demonstrated that a new composite material, silk protein with human collagen sequences, can be produced by transgenic silkworms, which would be a novel tool in the future to produce genetically designed protein composites by silkworms.

In addition, protein-polysaccharide blends have also been investigated as coating materials for porous membranes, which promoted fibroblast adhesion and proliferation, and increased the scaffold strength and hydrophilicity (*47*). Meanwhile, the experiments of cell cultivation (*e.g.*, mesenchymal stem cells (*47*, *83*), fibroblasts (*68*, *84*, *85*)) on various protein-polysaccharide blend substrates demonstrated that these composite materials are suitable for regeneration of corneal tissues, skins and bones. For example, cellulose is the most widely distributed polysaccharides in nature. And it is inexpensive, biodegradable and moisture absorbent. Fibers composed of 70 wt% cellulose and 30 wt% *Bombxy mori* silk fibroin can improve the elastic moduli of the blend materials significantly (*86*). Adding cell adhesive peptides into the composite materials could stimulate growth of bone tissues *in vivo* (*17*, *68*, *70*). Chitin is another polysaccharide material refined from seafood and shellfish. After adding 6 wt % silk fibroin into the wet-spun chitin fibers, the relative elasticity of chitin-silk composite was improved effectively and could be used for sock manufactures (*5*). However, a study on silk fibroin and hyaluronic acid composites (*87*) revealed that hyaluronic acid (a natural polysaccharide found in animals) did not promote the silk fibroin to form strong beta-sheet secondary structures. Therefore, such composite systems can be used for soft tissue regenerations in the future.

In order to improve their mechanical stability and anticoagulant property, silk matrices were also blended with chitosan (*14*, *72*). Porous scaffolds of silk fibroin-chitosan complex can be fabricated for supporting *in vitro* chondrogenesis of mesenchymal stem cells during cartilage repair (*81*). Sionkowska *et al.* (*88*) successfully crosslinked silk fibroin with chitosan and generated a group of three-dimensional (3D) silk fibroin-chitosan composite sponges, with the pore sizes of 20 ~ 150 μm. The sponges had sufficient mechanical integrity for implantation, and their physical properties can be modified by changing the ratio of silk and chitosan components (*89*, *90*).

Moreover, different chemical and physical parameters during processing the composites were modified to better control the phase transitions, morphologies and properties of final blend materials. By freeze drying, silk fibroin scaffolds with collagen and GAG content can not only serve as a cytocompatible carrier for chondrocytes, but also prolong the duration of cultivated cells growth and proliferation on the composite materials (*59*). Composite films of *Bombxy mori* fibroins with rat-tail collagens were also prepared by casting from acetic acid solutions, then air dried and treated with methanol in order to induce beta-sheet configuration in silk proteins. The composite was optically sheer, smooth and hydrophilic (*91*, *92*). Gelatin is a protein produced via hydrolysis of collagen. When silk and gelatin are mixed together in the methanol aqueous solutions, it was able to induce silk-gelatin hydrogel and beta-sheet crystals in the matrix (*5*, *93*). Keratin is another fibrous protein with protective functions in connective tissues grouped by disulfide bonds. Immersion of the keratin-silk composite fibers in methanol also induced beta-sheet structures in the composite fibers, which further improved their mechanical stability (*93*).

A recent study (*94*) reported that silk and collagen protein materials were sent to the International Space Station with nearly 18 months flying in the low-Earth orbit. The study found that individual protein materials were chemically cross-linked by the space radiations, and their ability to crystallize in methanol was reduced. These findings urge researchers to design special composite materials in the future to protect these protein materials in the extreme space environments, especially since some proteins such as collagens are the major component of tissues and organs in our body.

Protein-Synthetic Polymer Composite Materials

A large number of studies have illustrated that blending protein with synthetic polymer is an effective way to overcome the drawbacks of protein materials (*11*, *15*, *59*, *90*). For example, proteins are brittle in dry state and become extremely soft in water (*5*, *13*), blending synthetic polymers into proteins could improve improving the entire material properties. In addition, many synthetic polymers have unique characteristics, *e.g.*, light weight, strong corrosion resistance, high dielectric properties, good thermal insulation performance and controllable degradability in organic solvents (*91*). The difference in their mechanical, thermal, electrical properties and structures may be due to the significant differences between protein and synthetic polymer backbones (*95*). Therefore, new composite materials based on proteins and synthetic polymers are designable and have recently brought increasing interests in material sciences (*96–99*).

For example, water-insoluble polyurethane is a new organic polymer material which has many advantages, *i.e.* low density, good softness, long storage life, excellent wear resistance and remarkable shock absorption *etc.* (*100*) BMIMCl (1-butyl-3-methylimidazolium chloride) is environment-friendly ionic liquid and has been used to dissolve biomacromolecules like wool, cellulose and chitin together with polyurethane to form composite materials. Composites from silk fibroin and polyurethane were also prepared from BMIMCl solution (*100–102*), which provided high hydrophilicity, good biocompatibility, and facilitated cell adhesion and water insolubility. The soft segment of polyurethane can easily interact with silk fibroin chains when they were combined together (*103*).

Poly(D,L-lactic acid) (PDLLA) is a fully biodegradable synthetic polymer material, which can be obtained from grains. The composite sponges of silks with

PDLLA could be prepared by a solvent casting and leaching technique (*104*), which provided a perfect 3D porous structure to support extracellular matrix production and mineralization. Under both non-inflammatory and inflammatory conditions, higher levels of cartilage matrix gene expression were observed in silk and collagen scaffolds than in the pure PDLLA or poly(lactic acid) (PLA) scaffolds (*105*). Therefore, a silk-PLA composite was designed by combining lower content of silk fibers with PLA through injection and extrusion technique. Silk fibers can be well bonded with the PLA matrix, and the composite's elasticity, ductility (*18*) and glass transition temperature (*18*, *19*) were superior to those of pure polymeric materials, while their coefficient of linear thermal expansions was reduced due to the increase of crystallinity in the composite materials (*106*, *107*).

Poly(methyl methacrylate) (PMMA) is another transparent synthetic material with superior texture (*108*). An protein-based inverse opal was fabricated by inserting colloidal PMMA crystals, as active layers, into the fibroin network, which can be used as resorbable biosensors to systematically target and destroy unwanted tissues in anticancer therapies, as well as for various environmental sensing applications (*109*). In addition, composite scaffolds made by filling silk fibroin into the pores of a poly(ε-caprolactone) foam were tested. Initial tests showed that these silk-PCL materials can support cell adhesion and proliferation of human fibroblast cells *in vitro*, better than foams composed solely of poly(ε-caprolactone) (*110*).

Poly[(R)-3-hydroxybutyric acid] (PHB) is a natural polyester, which is involved in the body's microbial carbon and energy storage process. It has a good biocompatibility and can be completely degraded into water and carbon dioxide, and absorbed in an ecological environment. The composite films of PHB hybridized with the silk protein were prepared by a recombinant *escherichia coli* system (*111*). The study revealed that silk-PHB composite films are amorphous and its properties such as the tensile strength and the elongation ratio were improved. In addition, the cross-linked hydrogels from poly(γ-glutamic acid) - gelatin composites are environment-friendly, biocompatible and biodegradable, which can be used to entrap the easily-deactivated enzymes and promote thermostability of the materials (*112*). Proteins can also be assembled with poly(methacrylic), poly(N-vinylcaprolactam) and tannic acids to form stable multilayer films through hydrogen bonds (*75*, *113*). Using vinyl monomer (acryl amide) and Ce^{IV} initiator system, the coplymers of silk fibroin and acryl amide could improve some properties of the materials such as water retention capacity, water staining and tensile strengths (*114*).

Besides, protein materials have been blended with nylon66 (*5*, *108*), poly(acrylamide) (*5*, *115*), poly(acrylonitrile) (*5*, *116*), poly(allylamine) (*5*, *116*), poly(epoxides) (*5*, *117*), poly(ethylene oxide) (*5*, *117*, *118*), poly(pyrrole) (*5*, *119*), poly(styrene) (*5*, *120*), poly(vinyl alcohol) (*5*, *121–123*) and poly(aspartic acid) (*5*, *124–126*) to improve material's water resistance, cell-differentiation and adherence ability, mechanical strength, electrical conductivities and other physical and chemical properties.

Protein-Inorganic Composite Materials

Inorganic materials such as CdTe (*15*), magnetite [16] and gold (*16*, *17*) nanoparticles can be used to functionalize protein composites for fluorescent, magnetic and electronic applications due to their intrinsic properties. Protein-inorganic materials can be used as biomimetic muscle with a superior density (*11*), or as contact (*13*) or shadow (*14*) masks in different environments. In addition, it has been proved that gold-functionalized protein fibers can maintain electrical conductivity at ultra-low temperatures (*17*) Nano-silica is an important ultra-fine inorganic material due to its strong adsorption, high chemical purity and heat resistance, as well as its superior stability and reinforcement performance. Therefore, silica particles can be incorporated into protein films, such as silk-silica and collagen-silica composite films, and these materials can be utilized in osteogenesis for bone repair in the future (*94*, *127*).

Granular sodium chloride (NaCl) in various particle sizes can be added into protein solution to produce 3D protein-salt scaffolds. For example, salt-leached silk porous scaffolds with particle sizes 350~450 μm have been produced by silk-bound water-salt interactions, which reduced the β-sheet crystals in silk materials and resulted in low elastic modulus scaffolds. Thus, these silk scaffolds can be used in softer tissue regeneration since it overcomes the excessive stiffness of the scaffolds (*128*). Lammel *et al.* also used an all-aqueous salting out process with potassium phosphate (> 0.75 M) to control the particle sizes of silk fibroin (500 nm ~ 2 mm). By varying the pHs of silk-potassium phosphate solution, and the loaded drugs can be successfully absorbed into the silk particle matrix by charge-charge interactions and diffusion motions (*129*).

Calcium phosphate (CaP) is one of the main components of bones and tooth enamels which tends to form biomineral composites with proteins in nature. The uniform CaP-silk composite scaffolds can be fabricated by incorporating the CaP hybrid powders into silk solution and then freeze-drying the composite solution, which can be used for improving stem cell differentiation and bone formation in the body (*130*). The hydroxyapatite nano-particles can be also deposited onto the surfaces of silk scaffolds in simulated body fluids (*131–133*). Alternatively, through a spinning coating process, hydroxyapatite nano-particles could be embedded within the protein fibers, which improved the bone formation on these protein materials (*5*).

Carbon nano-tubes (CNTs) are one-dimensional nanomaterials with many unusual mechanical, electrical and chemical properties. CNTs have perfect hexagonal structures with tens of coaxial layers. The layers hold a fixed distance of approximately 0.34nm, with a diameter around 2~20 nm (*96*, *97*). Steven *et al.* (*95*) functionalized carbon nano-tubes into silk fibres in water by a mechanical shearing method. In the structure of those CNT-silk composites, it is found that the carbon nano-tubes are polar with positive charges at the amine sites, whereas silk fibres are neutral or negatively charged due to the amino acid side chain groups (*98*). Therefore, the carrier can transport charges easily inside the composites by inter-tube charge hopping (*98*, *99*). This silk-CNT composite material has outstanding properties such as high toughness, flexible shape and good strain-humidity sensitivity, which could be used for self-monitoring actuating devices or other electronic applications in the future.

Protein-Drug Composite Materials

In order to achieve sustainable drug delivery and release during clinical therapy, colloidal micro- or nano-particle carriers have been developed by various natural protein polymers (*14*, *134*). These systems were able to reduce the toxic side effects of drugs and limit the burst release of drugs in a certain period, while improving associated cellular bioactivity at the same time (*14*, *135*). As an effective drug delivery system, biodegradable polymer has to meet many requirements, including preventing denaturation or degradation of drugs, controlling the drug release rates, and generating non-toxic, low cost and easily processed carriers (*14*).

Proteins are promising candidates for drug delivery due to their non-cytotoxicity, customizable properties, as well as their capabilities to load multiple drugs and be fabricated into various morphologies (*136*, *137*). Germershaus *et al.* (*136*) found that silk fibroin can be fabricated into excellent drug delivery carriers due to their strong supermolecular associations with the loaded drugs. Various antibiotics (*i.e.* penicillin, tetracycline) and vaccines (*i.e.* measles, mumps and rubella) can be also inserted into the silk protein matrices and maintain their bioactivities and stabilities at 25 °C, 37 °C, or 45 °C for more than 6 months (*138*). Qin *et al.* (*139*) reported that aspirin loaded nano-fibers from silk fibroin-PLA blends decrease their average diameters (from 210 nm to 80 nm) with the increase of blending ratios of silk from 0 to 1.1%. Spider silk protein eADF4(C16) particles (sequence from European garden spider) are colloidally stable in solution, which are reliable to encapsulate different drug compounds (*140*). Hofer *et al.* (*141*) prepared eADF4(C16) particles as a drug delivery system by a micromixing method. They reported that macromolecular drugs as small proteins could be loaded into eADF4(C16) particles with an average size of 521 ± 8.3 nm. Moreover, the strong electrostatic interactions between positively charged lysozyme molecules and negatively charged eADF4(C16) particles resulted in an almost 100% loading efficiency in the carriers. At the same time, the effects of pH and ionic strength on drug release have been observed. Through studying the thermodynamic assembly process of eADF4(C16) protein into microparticles, it was found that formation of these solid particles can be improved by adding kosmotropic salts (*142*). For instance, the conversion of monomeric soluble spider silk proteins into solid protein particles with high beta-sheet contents and smooth surfaces can be easily triggered by adding potassium phosphate into the solution (*14*).

Figure 6 displaced a carrier loading and drug release mechanism of silk protein particles. Initially, drug molecules (1) were adsorbed on the surface of a silk particle by electrostatic forces (2). Once the particle surface is saturated, drugs begin to diffuse into the particle matrix (3). Then the drug molecules will interact with the protein matrix by attractive electrostatic forces. Next, the loaded particles would be incubated in the release media, and drug molecules will transport back to the particle surface due to the concentration gradient as well as the swelling and degradation of the particle (4). After a certain incubation period, the drug molecules in the loaded particle will be released from the particle surface to the solvent.

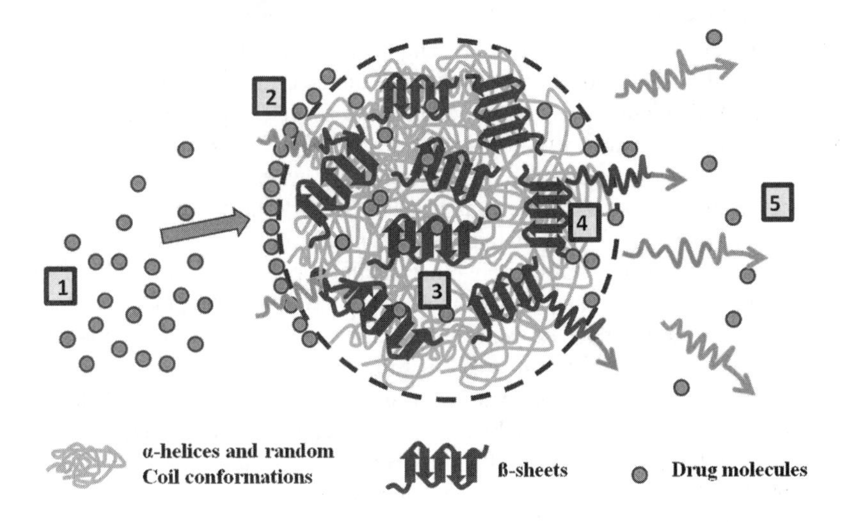

Figure 6. Drug loading and release mechanism of silk particles. The release is driven by electrostatic and hydrophobic interactions as well as concentration gradients, including: (1) attraction, (2) saturation and diffusion, (3) binding, (4) transport to the surface, and (5) release. (Reproduced with permission from Reference (140). Copyright (2011) Elsevier Ltd.).

As aforementioned, protein hydrogels have also been successfully used as biomaterials for several decades due to their low cytotoxicity (*143*), excellent mechanical properties (*144*, *145*) and biocompatibility (*146–150*). The gelation of protein solution can be induced by pH change, ultrasonication, or vortexing (*124*, *151–154*). In addition, bound water can help induce protein interactions with each other (*155*). Moreover, the secondary structures of proteins can be used to regulate the hydrogel degradation rates in enzyme solutions (*156*). Thereby, manipulating the release of bioactive molecules from protein hydrogels is achievable. For example, a dual-drug release system based on silk nano-particles in silk hydrogels was produced previously. Results showed that the release times of incorporated model drugs from silk nanoparticles to hydrogels were fast, while their continuing releases from silk hydrogels to water were relatively slow. Adjusting the physical properties of material matrix, *e.g.*, beta-sheet contents, size of the silk nano-particles, would help control and manipulate the drugs release rates. This silk-based dual-drug delivery system provided a promising polymeric material platform for the delivery of various bioactive molecules in the future (*157*).

In another aspect, many studies demonstrated that the drug release rate and the release time of silk fibroin drug delivery carriers could be controlled by adding other bioactive components. Combining silk with graphene by self-assemble silk fibroin onto graphene nano-sheets displayed tunable hierarchical nanostructures, which has been applied to the drug delivery applications recently (*158*). A 3D drug delivery scaffold system combined silk fibroin protein and calcium alginate beads was also studied recently, by using two different model compounds, bovine serum albumin (66 kDa) and FITC–Inulin (3.9 kDa), during their *in-vitro* release. The drug release time from silk-calcium alginate beads composites was prolonged without initial bursts for 35 days, as compared to the time from pure calcium alginate beads (*159*). Some hydrophilic anti-tumor drugs, such as doxorubicin, can be assembled into the silk-based ionomeric polymers, which can be gradually released out *in vivo* at a certain pH condition (*160*). Microspheres formed by recombinant proteins can be used as drug delivery systems (*161*). For example, silk-elastin protein materials, such as SELP-47K hydrogels, were recently used in controlled drug and gene deliveries (*162*). In addition, wool keratin and silk fibroin can interact with each other in the molecular level and generate keratin-silk composite matrices for the release of active compounds (*163*). The blended keratin-silk films cast from formic acid can maintain a slower biological degradation than these cast from aqueous solutions, since formic acid can induce the crystallization of silk fibroin as well as increase the amount of beta-sheet structures in keratin. The formed blends showed better strength and elasticity than those of pure silk materials (*163*). PLA and regenerated silk fibroin have also been combined uniformly in 70/30 trifluoroacetic acid/ dichloromethane solutions, and used as a drug delivery carrier via their electrospun composite nanofibers (*139*). Results showed that both the drug release rate and the anticoagulation property of PLA-silk blends were better than those of pure PLA films (*139*, *164*). In comparison to the domesticated silkworm silks, tropical tasar silk protein was also engineered into nano-particles to carry anti-cancer drug for specific cancer cells, which showed better bioactivities than the domesticated silks (*165*, *166*). The interaction between silk fibroin and two model proteins, protamine and polylysine, was affected and optimized by the ionic strength and the type of salt deployed. [136] The study of protease concentration and ethylenediamine tetraacetic acid being carried in the silk showed that controlling the carrier degradation rate and manipulating the proteolytic activity, is critical for silk-based drug delivery carriers. Ethylenediamine tetraacetic acid can reduce the degradation rates and inhibit proteolysis, whereas protease concentration can yield an opposite outcome (*167*). To form silk/ PVA blend films at different ratios, Wang *et al.* (*168*) prepared silk micro- and nanospheres with controllable sizes (300 nm to 20 μm) and changeable shapes (film, spheres), using polyvinyl alcohol (PVA) as a continuous phase to separate silk solution into micro- and nanospheres. These findings can be directly applied to silk-based drug delivery systems.

Poly(lactide-co-glycolide) (PLGA) microparticles loaded with growth factors were embedded into a silk fibroin scaffold to become a microporous scaffold complex. The release rates of these growth factors could be controlled by using different types of PLGA microparticles (*169*, *170*). Shi *et al.* (*171*) developed a tissue engineered bulking agent that consisted of adipose-derived stem cells (ADSCs) and silk fibroin microspheres to treat stress urinary incontinence caused by severe intrinsic sphincter deficiency (ISD). The findings showed that silk fibroin microspheres alone could work effectively in the short-term, while tissue engineered bulking agent that combined silk fibroin microspheres with ADSCs exhibited promising long-term efficacy. This study developed a new strategy of tissue engineered bulking agent for future ISD therapy (*172*). Pallotta *et al.* mixed silk gels with platelet gels (a fibrin network containing activated platelets) for growth factors release. The gel composites extended the release of growth factors without inhibiting gel-forming ability. The mechanical properties and the release rates of the resultant gel, which reflect contributions from both gelling components, can also be tuned by manipulating the final silk protein concentration (*173*). Seib *et al.* (*174*) loaded doxorubicin into silk films and directly applied the composite system to tumor therapy. By manipulating silk crystallinity, the doxorubicin release rate could be controlled from immediate release to prolonged release over 4 weeks. This study demonstrated that doxorubicin-silk composite films can provide local *in-vivo* control of human breast cancer. Furthermore, silk fibroin and polyacrylamide can be blended together to form a semi-interpenetrating hydrogel network for drug delivery studies. The studies showed that the maximum pore size of microporous particles was 50 ± 11 nm and the maximum compressive strength of the gels was close to 241.9 ± 5.5 kPa (*175, 176*).

Several studies also showed that silk-like recombinant proteins can be employed as a carrier for drugs such as paclitaxel, clopidogrel or heparin, while providing excellent adhesion and viability on the growth and proliferation of human aortic endothelial cells (HAECs), human coronary artery cells and smooth muscle cells (*177–179*). For example, biodegradable silk-like block copolymers with plasmid DNA can be produced by ionic interactions and further be utilized for gene delivery into human embryonic kidney cells (*178*). Recombinant spider silk matrices can also provide a platform for efficient culturing of neural stem cells with positive effects on cell viability, cell self-renewal differentiation and oligodendrocyte differentiation (*179*).

In conclusion, protein-based composite materials are able to provide a drug delivery platform to immobilize enzymes, growth factors, small proteins or small molecules for various medical applications. Table IV lists a summary of protein-based composites that have been recently used for drug delivery and release, including their immobilized types as well as the specific loaded materials.

Table IV. Drug delivery applications of protein-based composites. (Partially reproduced with permission from Reference (*180*). Copyright (2012) Wiley).

Immobilized Type	*Loaded materials*	*References*
Enzyme	Horseradish peroxidase (HRP)	(*181, 182*)
	Glucose oxidase (GOx)	(*183, 184*)
	Lipase	(*183, 185*)
	Tyrosinase, Invertase	(*51, 186*)
	Cholesterol oxidase (ChOx)	(*187, 188*)
	Ribonuclease	(*189*)

Continued on next page.

Table IV. (Continued). Drug delivery applications of protein-based composites

Immobilized Type	*Loaded materials*	*References*
	L-asparaginase (ASNase)	(*190*)
	Alkaline phosphatase (ALP)	(*191*)
	beta-glucosidase, Uricase	(*192*, *193*)
	Heme proteins (myoglobin, horseradish, peroxidase, hemoglobin, and catalase)	(*194*)
	Phenylalanine ammonia-lyase (PAL)	(*195*)
	Organophosphorus Hydrolase (OPH)	(*196*)
Growth Factor	bone morphogenetic protein 2 (BMP-2)	(*197*, *198*)
	fibroblast growth factor 2 (FGF-2)	(*199*)
	insulin growth factor I (IGF-I)	(*200*)
	parathyroid hormone (PTH)	(*201*, *202*)
	basic fibroblast growth factor (bFGF)	(*203*)
	nerve growth factor (NGF)	(*204*)
	glial cell line-derived neurotrophic factor (GDNF)	(*205*)
	vascular endothelial growth factor (VEGF)	(*206*)
Small Protein	Staphylococcal low molecular weight protein A (LPA)	(*207*)
	Hemoglobin	(*208*)
	Normal murine IgG	(*209*)
	Green fluorescent protein (GFP)	(*210*)
	NeutrAvidin	(*211*)
	Insulin	(*212*)
	Murine anti-TGFb IgG1 monoclonal antibody	(*213*)
Small Molecule	Adenosine, Curcumin	(*214*, *215*)
	Chlorophyll beta-carotene Astaxanthin	(*216*)
	Antioxidants from crude olive leaf extract (oleuropein and rutin),	(*217*, *218*)
	Doxycycline Ciprofloxacin	(*219*)
	Various antibiotics (penicillin, tetracycline, rifampicin, erythromycin),	(*220*)
	Tetracycline, Doxorubicin	(*182*, *221*)
	Phenol-sulfon-phthlein (Phenol Red)	(*208*)
	4-amino-benzoic acid	(*222*)

Conclusions

A variety of natural protein-based composite materials have been creatively designed, produced and used in medical sciences. These protein-based composite materials can be processed into hydrogels, tubes, sponges, composites, fibers, microspheres and thin films blends with a number of applications (*1*). In this chapter, we briefly introduced several important fibrous proteins and their protein-based composite materials for different applications. We envisage that with many new techniques, more novel protein-based composite materials will be discovered in the near future, and such composite materials will not only be used in biomedical sciences and engineering, but also broadly useful in various industrial fields. For example, recent reports also showed that silk fibroin has been used to develop new bio-memristors (*228*) and electrical wires (*95*). In addition, silk, wool, and cashmere guard hair fibers were milled into fine particles, and served as economical alternatives to remove heavy toxic metal ions from pollution water or as antibacterial materials in the factory (*229–235*). All these new developments implicate that protein composite materials will be of great importance due to their good characteristics and highly tunable structures in the future.

Acknowledgments

The authors thank the Rowan University Start-up Grants, Rowan University 2013-2014 Seed Funding Program (10110-60930-7460-12), NSF-MRI Program (DMR- 1338014), and Nanjing Normal University Scholarship for Overseas Studies Foundation of China (2013) for support of this research.

References

1. Hu, X.; Cebe, P.; Weiss, A. S.; Omenetto, F.; Kaplan, D. L. *Mater. Today* **2012**, *15*, 208–215.
2. Almine, J. F.; Bax, D. V.; Mithieux, S. M.; Nivison-Smith, L.; Rnjak, J.; Waterhouse, A.; Wise, S. G.; Weiss, A. S. *Chem. Soc. Rev.* **2010**, *39*, 3371–3379.
3. Wise, S. G.; Weiss, A. S. *Int. J. Biochem. Cell Biol.* **2009**, *41*, 494–497.
4. Cebe, P.; Hu, X.; Kaplan, D. L.; Zhuravlev, E.; Wurm, A.; Arbeiter, D.; Schick, C. *Sci. Rep.* **2013**, *3*, 1130.
5. Hardy, J. G.; Scheibel, T. R. *Prog. Polym. Sci.* **2010**, *35*, 1093–1115.
6. Elvin, C. M.; Vuocolo, T.; Brownlee, A. G.; Sando, L.; Huson, M. G.; Liyou, N. E.; Stockwell, P. R.; Lyons, R. E.; Kim, M.; Edwards, G. A.; Johnson, G.; McFarland, G. A.; Ramshaw, J. A. M.; Werkmeister, J. A. *Biomaterials* **2010**, *31*, 8323–8331.
7. Qin, G. K.; Lapidot, S.; Numata, K.; Hu, X.; Meirovitch, S.; Dekel, M.; Podoler, I.; Shoseyov, O.; Kaplan, D. L. *Biomacromolecules* **2009**, *10*, 3227–3234.
8. Brodsky, B; Baum, J. *Nature* **2008**, *453*, 998–1007.

9. Lee, S. M.; Pippel, E.; Gösele, U.; Dresbach, C.; Qin, Y.; Chandran, C. V.; Bräuniger, T.; Hause, G.; Knez, M. *Science* **2009**, *324*, 488–492.
10. Holland, C.; Vollrath, F.; Ryan, A. J.; Mykhaylyk, O. O. *Adv. Mater.* **2012**, *24*, 105–109.
11. Hagenau, A.; Scheidt, H. A.; Serpell, L.; Huster, D.; Scheibel, T. *Macromol. Biosci.* **2009**, *9*, 162–168.
12. Askarieh, G.; Hedhammar, M.; Nordling, K.; Saenz, A.; Casals, C.; Rising, A.; Johansson, J.; Knight, S. D. *Nature* **2010**, *465*, 236–238.
13. Hardy, J.; Romer, L.; Scheibel, T. *Polymer* **2008**, *49*, 4309–4474.
14. Kristina, S.; Andreas, L.; Thomas, S. *Macromol. Biosci.* **2010**, *10*, 998–1007.
15. Jin, H.; Kaplan, D. *Nature* **2003**, *424*, 1057–1061.
16. Inoue, S.; Tanaka, K.; Arisaka, F.; Kimura, S.; Ohtomo, K.; Mizuno, S. *J. Biol. Chem.* **2000**, *275*, 40517–40528.
17. Sutherland, T. D.; Young, J. H.; Weisman, S.; Hayashi, C. Y.; Merritt, D. J. *Annu. Rev. Entomol.* **2010**, *55*, 171–259.
18. Sponner, A.; Vater, W.; Monajembashi, S.; Unger, R. E.; Grosse, F.; Wiesshart, K. *PLoS One* **2007**, *2*, e998–1008.
19. Mülhaupt, R. *Macromol. Chem. Phys.* **2010**, *211*, 121–126.
20. Koski, K. J.; Akhenblit, P.; McKiernan, K.; Yarger, J. L. *Nat. Mater.* **2013**, *12*, 262–267.
21. Gosline, J. M.; Denny, M. W.; DeMont, M. E. *Nature* **1984**, *309*, 551–552.
22. Blackledge, T. A.; Rigueiro, J. P.; Plaza, G. R.; Perea, B.; Navarro, A.; Guinea, G. V. *Sci. Rep.* **2012**, *2*, 782.
23. Ortega-Jimenez, V. M.; Dudley, R. *Sci. Rep.* **2013**, *3*, 2108.
24. Matsuhira, T.; Yamamoto, K.; Osaki, S. *Polym. J.* **2013**, *45*, 1167–1169.
25. Towie, N. *Nature* **2006**, *444*, 519–529.
26. Kameda, T. *Polym. J.* **2012**, *44*, 876–881.
27. Weisman, S.; Haritos, V. S.; Church, J. S.; Huson, M. G.; Mudie, S. T.; Rodgers, A. J. W.; Dumsday, G. J.; Sutherland, T. D. *Biomaterials* **2010**, *31*, 2695–2700.
28. Smith, M. J.; White, K. L., Jr; Smith, D. C.; Bowlin, G. L. *Biomaterials* **2009**, *30*, 149–159.
29. Sutherland, T. D.; Peng, Y. Y.; Trueman, H. E.; Weisman, S.; Okada, S.; Walker, A. A.; Sriskantha, A.; White, J. F.; Huson, M. G.; Werkmeister, J. A.; Glattauer, V.; Stoichevska, V.; Mudie, S. T.; Haritos, V. S.; Ramshaw, J. A. M. *Sci. Rep.* **2013**, *3*, 2864.
30. Elvin, C. M.; Carr, A. G.; Huson, M. G.; Maxwell, J. M.; Pearson, R. D.; Vuocolo, T.; Liyou, N. E.; Wong, D. C. C.; Merritt, D. J.; Dixon, N. E. *Nature* **2005**, *437*, 999–1002.
31. Menezes, G. M.; Teulé, F.; Lewis, R. V.; Silva, L. P.; Rech, E. L. *Polymer Journal* **2013**, *45*, 997–1006.
32. Yang, Y. J.; Choi, Y. S.; Jung, D.; Park, B. R.; Hwang, W. B.; Kim, H. W.; Cha, H. J. *NPG Asia Mater.* **2013**, *5*, 1–7.
33. Reichl, S.; Borrelli, M.; Geerling, G. *Biomaterials* **2011**, *32*, 3375–3386.
34. Reddy, N.; Yang, Y. Q. *Biotechnol. Bioeng.* **2009**, *103*, 1016–1022.

35. Lee, K. Y.; Kong, S. J.; Park, W. H.; Ha, W. S.; Kwon, I. C. *J. Biomat. Sci-Polym. E.* **1998**, *9*, 905–914.
36. Higgins, J. S.; Lipson, J. E.; White, R. P. *Philos. Trans. R. Soc., A* **2010**, *368*, 1009–1025.
37. Prum, R. O.; Dufresne, E. R.; Quinn, T.; Waters, K. J. R. *Soc. Interface.* **2009**, *6*, S253–S265.
38. Wanker, E. E.; Rovira, C.; Scherzinger, E.; Hasenbank, R.; Walter, S.; Tait, D.; Colicelli, J.; Lehrach, H. *Hum. Mol. Genet.* **1997**, *6*, 487–495.
39. Lee, H. J.; Mochizuki, N.; Masuda, T.; Buckhout, T. J. *J. Exp. Bot.* **2012**, *63*, 695–709.
40. Rich, R. L.; Myszka, D. G. *Curr. Opin. Biotechnol.* **2000**, *11*, 54–61.
41. Huber, W.; Mueller, F. *Curr. Pharm. Des.* **2006**, *12*, 3999–4021.
42. Heim, R.; Tsien, R. Y. *Curr. Biol.* **1996**, *6*, 178–260.
43. Zhang, B.; Park, B. H.; Karpinets, T.; Samatova, N. F. *Bioinformatics* **2008**, *24*, 979–986.
44. Pajula, K.; Taskinen, M.; Lehto, V. P.; Ketolainen, J.; Korhonen, O. *Mol. Pharm.* **2010**, *7*, 795–804.
45. Widmaier, J. M.; Meyer, G. C. *J. Therm. Anal. Calorim.* **1982**, *23*, 193–215.
46. Hu, X.; Kaplan, D.; Cebe, P. *Macromolecules* **2006**, *39*, 6161–6170.
47. Hu, X.; Wang, X. L.; Rnjak, J.; Weiss, A. S.; Kaplan, D. L. *Biomaterials* **2010**, *31*, 8121–8131.
48. Fox, T. G.; Flory, P. J. *J. Appl. Phys.* **1950**, *21*, 581–591.
49. Gordon, M.; Taylor, J. S. *J. Appl. Chem.* **1952**, *2*, 495–499.
50. Kwei, T. K.; Pearce, E. M.; Pennacchia, J. R.; Charton, M. *Macromolecules* **1987**, *20*, 1174–1176.
51. Chen, H.; Hu, X.; Cebe, P. *J. Therm. Anal. Calorim.* **2008**, *93*, 201–206.
52. Meaurio, E.; Zuza, E.; Sarasua, J. R. *Macromolecules* **2005**, *38*, 1207–1215.
53. Fujita, M.; Sawayanagi, T.; Abe, H.; Tanaka, T.; Iwata, T.; Ito, K.; Fujisawa, T.; Maeda, M. *Macromolecules* **2008**, *41*, 2852–2858.
54. Chen, H. P.; Pyda, M.; Cebe, P. *Thermochim. Acta* **2009**, *492*, 61–66.
55. Mohamed, A. A.; Rayas-Duarte, P. *Food. Chem.* **2003**, *81*, 533–545.
56. Zhang, J.; Mungara, P.; Jane, J. *Polymer* **2001**, *42*, 2569–2578.
57. Blamires, S. J.; Wu, C. C.; Wu, C. L.; Sheu, H. S.; Tso, I. M. *Biomacromolecules* **2013**, *14*, 3484–3490.
58. Rockwood, D. N.; Preda, R. C.; Yücel, T.; Wang, X. Q.; Lovett, M. L.; Kaplan, D. L. *Nat. Protoc.* **2011**, *6*, 1612–1631.
59. Talukdar, S.; Nguyen, Q. T.; Chen, A. C.; Sah, R. L.; Kundu, S. C. *Biomaterials* **2011**, *32*, 8927–8937.
60. Lu, Q.; Hu, X.; Wang, X. Q.; Kluge, J. A.; Lu, S. Z.; Cebe, P.; Kaplan, D. L. *Acta Biomater.* **2010**, *6*, 1380–1387.
61. Lu, Q.; Wang, X. Q.; Hu, X.; Cebe, P.; Omenetto, F.; Kaplan, D. L. *Macromol. Biosci* **2010**, *10*, 359–368.
62. Lu, Q.; Zhang, X. H.; Hu, X.; Kaplan, D. L. *Macromol. Biosci.* **2010**, *10*, 289–298.
63. Wang, X. Q.; Yucel, T.; Lu, Q.; Hu, X.; Kaplan, D. L. *Biomaterials* **2010**, *31*, 1025–1035.

64. Lammel, A. S.; Hu, X.; Park, S.; Kaplan, D. L.; Scheibel, T. R. *Biomaterials* **2010**, *31*, 4583–4591.
65. Wittmera, C. R.; Hua, X.; Gauthiera, P. C.; Weismanb, S.; Kaplana, D. L.; Sutherland, T. D. *Acta Biomater.* **2011**, *7*, 3789–3795.
66. Liu, Y.; Shao, Z. Z.; Zhou, P.; Chen, X. *Polymer* **2004**, *45*, 7705–7710.
67. Lu, Q.; Zhu, H.; Zhang, C. C.; Zhang, F.; Zhang, B.; Kaplan, D. L. *Biomacromolecules* **2012**, *13*, 826–832.
68. Wang, Y. Z.; Kim, H. J.; Vunjak-Novakovic, G.; Kaplan, D. L. *Biomaterials* **2006**, *27*, 6064–6082.
69. Morgan, A. W.; Roskov, K. E.; Lin-Gibson, S.; Kaplan, D. L.; Becker, M. L.; Simon, C. G., Jr *Biomaterials* **2008**, *29*, 2556–2563.
70. Baran, E. T.; Tuzlakoğlu, K.; Mano, J. F.; Reis, R. L. *Mater. Sci. Eng., C* **2012**, *32*, 1314–1322.
71. Bessa, P. C.; Balmayor, E. R.; Azevedo, H. S.; Nürnberger, S.; Casal, M.; Griensven, M. V.; Reis, R. L.; Red, H. *J. Tissue Eng. Regener. Med.* **2010**, *4*, 349–355.
72. Zhou, L.; Wang, Q.; Wen, J. C.; Chen, X.; Shao, Z. Z. *Polymer* **2013**, *54*, 5035–5042.
73. Prachayawarakorn, J.; Hwansanoet, W. *Fibers Polym.* **2012**, *13*, 606–612.
74. Prachayawarakorn, J.; Sangnitidej, P.; Boonpasith, P. *Carbohydr. Polym.* **2010**, *81*, 425–433.
75. Kozlovskaya, V.; Baggett, J.; Godin, B.; Liu, X. W.; Kharlampieva, E. *ACS Macro. Lett.* **2012**, *1*, 384–387.
76. Silva, R. D.; Wang, X. G.; Byrne, N. *Cellulose* **2013**, *20*, 2461–2468.
77. Goujon, N.; Rajkhowa, R.; Wang, X.; Byrne, N. *J. Appl. Polym. Sci.* **2012**, *128*, 4411–4416.
78. Das, S.; Pati, F.; Chameettachal, S.; Pahwa, S.; Ray, A. R.; Dhara, S.; Ghosh, S. *Biomacromolecules* **2013**, *14*, 311–321.
79. Numata, K.; Katashima, T.; Sakai, T. *Biomacromolecules* **2011**, *12*, 2137–2144.
80. Tomita, M.; Munetsuna, H.; Sato, T.; Adachi, T.; Hino, R.; Hayashi, M.; Shimizu, K.; Nakamura, N.; Tamura, T.; Yoshizato, K. *Nat. Biotechnol.* **2002**, *21*, 52–58.
81. Bhardwaj, N.; Kundu, S. C. *Biomaterials* **2012**, *33*, 2848–2857.
82. Tu, D. D.; Chung, Y. G.; Gil, E. S.; Seth, A.; Franck, D. *Biomaterials* **2013**, *34*, 8681–8689.
83. Cranford, S. W.; Tarakanova, A.; Pugno, N. M.; Buehler, M. J. *Nature* **2012**, *482*, 72–76.
84. Keten, S.; Xu, Z.; Ihle, B.; Buehler, M. J. *Nat. Mater.* **2010**, *9*, 359–367.
85. Vollrath, F. *Nature* **2010**, *466*, 319–319.
86. Marsano, E.; Canetti, M.; Conio, G.; Corsini, P.; Freddi, G. *J. Appl. Polym. Sci* **2007**, *104*, 2187–2196.
87. Garcia-Fuentes, M.; Giger, E.; Meinel, L.; Merkle, H. P. *Biomaterials* **2008**, *29*, 633–642.
88. Sionkowska, A.; Płanecka, A. *J. Mol. Liq.* **2013**, *178*, 5–14.
89. Miranda, E. S.; Silva, T. H.; Reis, R. L.; Mano, J. F. *Tissue Eng., Part A* **2011**, *17*, 2663–2674.

90. Dash, M.; Chiellini, F.; Ottenbrite, R. M.; Chiellini, E. *Prog. Polym. Sci.* **2011**, *36*, 981–1014.
91. Lv, Q.; Hu, K.; Feng, Q. L.; Cui, F. Z. *J. Appl. Polym. Sci.* **2008**, *109*, 1577–1584.
92. Cirillo, B.; Morra, M.; Catapano, G. *Int. J. Artif. Organs* **2004**, *27*, 60–68.
93. Gil, E. S.; Spontak, R. J.; Hudson, S. M. *Macromol. Biosci.* **2005**, *5*, 702–9.
94. Hu, X.; Raja, W. K.; An, B.; Tokareva, O.; Cebe, P.; Kaplan, D. L. *Sci. Rep.* **2013**, *3*, 3428.
95. Steven, E.; Saleh, W. R.; Lebedev, V.; Acquah, S. F. A.; Laukhin, V.; Alamo, R. G.; Brooks, J. S. *Nat. Commun.* **2013**, *4*, 2435–2440.
96. Liu, Z.; Sun, X. M.; Nakayama-Ratchford, N.; Dai, H. J. *ACS Nano.* **2007**, *1*, 50–56.
97. Behnam, A.; Sangwan, V. K.; Zhong, X. Y.; Lian, F. F.; Estrada, D.; Jariwala, D.; Hoag, A. J.; Lauhon, L. J.; Marks, T. J.; Hersam, M. C.; Pop, E. *ACS Nano.* **2013**, *7*, 482–490.
98. O'Brien, J. P.; Fahnestock, S. R.; Termonia, Y.; Gardner, K. H. *Adv. Mater.* **1998**, *10*, 1185–1195.
99. Beese, A. M.; Sarkar, S.; Nair, A.; Naraghi, M.; An, Z.; Moravsky, A.; Loutfy, R. O.; Buehler, M. J.; Nguyen, S. B. T.; Espinosa, H. D. *ACS Nano.* **2013**, *7*, 3434–3446.
100. Liu, X. Y.; Zhang, C. C.; Xu, W. L.; Liu, H. T.; Ouyang, C. X. *Mater. Lett.* **2011**, *65*, 2489–2491.
101. Neves, S. C.; Moreira Teixeira, L. S.; Moroni, L.; Reis, R. L.; Van Blitterswijk, C. A.; Alves, N. M.; Karperien, M.; Mano, J. F. *Biomaterials* **2011**, *32*, 1068–1079.
102. Shao, H. J.; Lee, Y. T.; Chen, C. S.; Wang, J. H.; Young, T. H. *Biomaterials* **2010**, *31*, 4695–4705.
103. Nakazawa, Y.; Asano, A.; Nakazawa, C. T.; Tsukatani; Asakura, T. *Polym. J.* **2012**, *44*, 802–807.
104. Stoppato, M.; Stevens, H. Y.; Carletti, E.; Migliaresi, C.; Motta, A.; Guldberg, R. E. *Biomaterials* **2013**, *34*, 4573–4581.
105. Kwon, H.; Sun, L.; Cairns, D. M.; Rainbow, R. S.; Preda, R. C.; Kaplan, D. L.; Zeng, L. *Acta Biol.* **2013**, *9*, 6563–6575.
106. Cheung, H. Y.; Lau, K. T.; Tao, X. M.; Hui, D. *Composites, Part B* **2008**, *39*, 1026–1033.
107. Ho, M. P.; Wang, H.; Lau, K. T.; Leng, J. S. *J. Appl. Polym. Sci.* **2013**, *127*, 2389–2396.
108. Ferriol, M.; Gentilhomme, A.; Cocheza, M.; Ogetb, N.; Mieloszynski, J. L. *Polym. Degrad. Stab.* **2003**, *79*, 271–281.
109. MacLeod, J.; Rosei, F. *Nat. Mater.* **2013**, *12*, 98–100.
110. Chen, G.; Zhou, P.; Mei, N.; Chen, X.; Shao, Z. Z.; Pan, L. F.; Wu, C. G. *J. Mater. Sci. Mater. Med.* **2004**, *15*, 671–677.
111. Numata, K.; Doi, Y. *Polym. J.* **2011**, *43*, 642–647.
112. Xie, J. F.; H. Zhang, W.; Li, X.; Shi, Y. L. *Biomacromolecules* **2014**, *15*, 690–697.
113. Perez-Rigueiro, J.; Viney, C.; Llorca, J.; Elices, M. *J. Appl. Polym. Sci.* **2000**, *75*, 1270–1277.

114. Das, A. M.; Chowdhury, P. K.; Saikia, C. N.; Rao, P. G. *Ind. Eng. Chem. Res.* **2009**, *48*, 9338–9345.
115. Freddi, G.; Tsukada, M.; Beretta, S. *J. Appl. Polym. Sci.* **1999**, *71*, 1563–1571.
116. Sun, Y. Y.; Shao, Z. Z.; Hu, P.; Yu, T. Y. *J. Polym. Sci., Part B: Polym. Phys.* **1997**, *35*, 1405–1414.
117. Padma, P. S.; Rai, S. K. *J. Reinf. Plast. Compos.* **2006**, *25*, 565–574.
118. Jin, H. J.; Park, J.; Valluzzi, R.; Cebe, P.; Kaplan, D. L. *Biomacromolecules* **2004**, *5*, 711–717.
119. Guimard, N. K.; Gomez, N.; Schmidt, C. E. *Prog. Polym. Sci.* **2007**, *32*, 876–921.
120. Swinerd, V. M.; Collins, A. M.; Skaer, N. J. V.; Gheysens, T.; Mann, S. *Soft Matter* **2007**, *3*, 1377–1380.
121. Yamaura, K.; Kuranuki, N.; Suzuki, M.; Tanigami, T.; Matsuzawa, S. *J. Appl. Polym. Sci.* **1990**, *41*, 2409–2425.
122. Tsukada, M.; Freddi, G.; Crighton, J. S. *J. Polym. Sci., Part B: Polym. Phys.* **1994**, *32*, 243–248.
123. Nogueira, G. M.; Aimoli, C. G.; Weska, R. F.; Nascimento, L. S.; Beppu, M. M. *Key Eng. Mater.* **2008**, *361-363*, 503–506.
124. Kim, H. J.; Kim, U. J.; Kim, H. S.; Li, C. M.; Wada, M.; Leisk, G. G.; Kaplan, D. L. Bone tissue engineering with premineralized silk scaffolds. *Bone* **2008**, *42*, 1226–1234.
125. Zhao, J.; Zhang, Z. C.; Wang, S.; Sun, X.; Zhang, X.; Chen, J.; Kaplan, D. L.; Jiang, X. Q. *Bone* **2009**, *45*, 517–527.
126. Jiang, X.; Zhao, J.; Wang, S.; Sun, X.; Zhang, X.; Chen, J.; Kaplan, D. L.; Zhang, Z. Y. *Biomaterials* **2009**, *30*, 4522–4532.
127. Weska, R. F.; Nogueira, G. M.; Vieira, W. C.; Beppu, M. M. *Key Eng. Mater.* **2009**, *396-398*, 187–190.
128. Yao, D. Y.; Dong, S.; Lu, Q.; Hu, X.; Kaplan, D. L.; Zhang, B. B.; Zhu, H. S. *Biomacromolecules* **2012**, *13*, 3723–3729.
129. Lammel, A. S.; Hu, X.; Park, S. H.; Kaplan, D. L.; Scheibel, T. R. *Biomaterials* **2010**, *31*, 4583–4591.
130. Zhang, Y. F.; Wu, C. T.; Friis, T.; Xiao, Y. *Biomaterials* **2010**, *31*, 2848–2856.
131. Kong, X. D.; Cui, F. Z.; Wang, X. M.; Zhang, M.; Zhang, W. *J. Cryst. Growth* **2004**, *270*, 197–202.
132. Wang, J.; Zhou, W.; Hu, W.; Zhou, L; Wang, S.; Zhang, S. *J. Biomed. Mater. Res., Part A* **2011**, *99A*, 327–334.
133. Kino, R.; Ikoma, T.; Yunoki, S.; Monkawa, A; Matsuda, A.; Kagata, G; Asakura, T.; Munekata, M.; Tanaka, J. *Key Eng. Mater.* **2006**, *309–311*, 1169–1172.
134. Won, Y. W.; Kim, Y. H. *Macromol. Res.* **2009**, *17*, 464–468.
135. Altikatoglu, M.; Kuzu, H. *Pol. J. Chem. Technol.* **2010**, *12*, 12–16.
136. Germershaus, O.; Werner, V.; Kutscher, M.; Meinel, L. *Biomaterials* **2014**, *35*, 3427–3434.
137. Lawrence, B. D.; Cronin-Golomb, M.; Georgakoudi, I.; Kaplan, D. L.; Omenetto, F. G. *Biomacromolecules* **2008**, *9*, 1214–1220.

138. Zhang, J.; Pritchard, E.; Hu, X.; Valentin, T.; Panilaitis, B.; Omenetto, F. G.; Kaplan, D. L. *Proc. Natl. Acad. Sci. U.S.A.* **2012**, *109*, 11981–11986.
139. Qin, J. W.; Jiang, Y. Y.; Fu, J. J.; Wan, Y. Q.; Yang, R. H.; Gao, W. D.; Wang, H. B. *Iran. Polym. J.* **2013**, *22*, 729–739.
140. Lammel, A.; Schwab, M.; Hofer, M.; Winter, G.; Scheibel, T. *Biomaterials* **2011**, *32*, 2233–2240.
141. Hofer, M.; Winter, G.; Myschik, J. *Biomaterials* **2012**, *33*, 1554–1562.
142. Exler, J. H.; Hummerich, D.; Scheibel, T. *Angew. Chem., Int. Ed. Engl.* **2007**, *46*, 3559–3562.
143. Altman, G. H.; Diaz, F.; Jakuba, C.; Calabro, T.; Horan, R. L.; Chen, J.; Lu, H.; Richmond, J.; Kaplan, D. L. *Biomaterials* **2003**, *24*, 401–416.
144. Numata, K.; Kaplan, D. L. *Adv. Drug Delivery Rev.* **2010**, *62*, 1497–1508.
145. Wang, Y.; Kim, H. J.; Vunjak-Novakovic, G.; Kaplan, D. L. *Biomaterials* **2006**, *27*, 6064–6082.
146. Anumolu, R.; Gustafson, J. A.; Magda, J. J.; Cappello, J.; Ghandehari, H.; Pease, L. F. *ACS Nano.* **2011**, *5*, 5374–5382.
147. Mandal, B. B.; Kundu, S. C. *Nanotechnology* **2009**, *20*, 355101–355115.
148. Zhang, Y. Q.; Shen, W. D.; Xiang, R. L.; Zhuge, L. J.; Gao, W. J.; Wang, W. B. *J. Nanopart. Res.* **2007**, *9*, 885–900.
149. Zhang, Y. Q.; Wang, Y. J.; Wang, H. Y.; Zhu, L.; Zhou, Z. Z. *Soft Matter* **2011**, *7*, 9728–9736.
150. Zhu, L.; Hu, R. P.; Wang, H. Y.; Wang, Y. J.; Zhang, Y. Q. *J. Agric. Food Chem.* **2011**, *59*, 10298–10302.
151. Hu, X. A.; Lu, Q. A.; Sun, L.; Cebe, P.; Wang, X. Q.; Zhang, X. H.; Kaplan, D. L. *Biomacromolecules* **2010**, *11*, 3178–3188.
152. Yucel, T.; Cebe, P.; Kaplan, D. L. *Biophys. J.* **2009**, *97*, 2044–2050.
153. Wang, X. Q.; Kluge, J. A.; Leisk, G. G.; Kaplan, D. L. *Biomaterials* **2008**, *29*, 1054–1064.
154. Matsumoto, A.; Chen, J.; Collette, A. L.; Kim, U. J.; Altman, G. H.; Cebe, P.; Kaplan, D. L. *J. Phys. Chem. B* **2006**, *110*, 21630–21638.
155. Numata, K.; Katashima, T.; Sakai, T. *Biomacromolecules* **2011**, *12*, 2137–2144.
156. Numata, K.; Kaplan, D. L. *Biomacromolecules* **2010**, *11*, 3189–3195.
157. Numata, K.; Yamazaki, S.; Naga, N. *Biomacromolecules* **2012**, *13*, 1383–1389.
158. Ling, S. J.; Li, C. X.; Adamcik, J.; Wang, S. H.; Shao, Z. Z.; Chen, X.; Mezzenga, R. *ACS Macro. Lett.* **2014**, *3*, 146–152.
159. Mandal, B. B.; Kundu, S. C. *Biomaterials* **2009**, *30*, 5170–5177.
160. Serban, M. A.; Kaplan, D. L. *Biomacromolecules* **2010**, *11*, 3406–3412.
161. Breslauer, D. N.; Muller, S. J.; Lee, L. P. *Biomacromolecules* **2010**, *11*, 643–647.
162. Dandu, R.; Megeed, Z.; Haider, M.; Cappello, J.; Ghandehari, H.; Svenson, S. Silk-Elastinlike Hydrogels: Thermal Characterization and Gene Delivery. In *Polymeric Drug Delivery II: Polymeric Matrices and Drug Particle Engineering*; Svenson, S., Ed.; ACS Symposium Series 924; American Chemical Society: New York, 2006; Chapter 10, pp 150−168.

163. Vasconcelos, A.; Freddi, G.; Cavaco-Paulo, A. *Biomacromolecules* **2008**, *9*, 1299–1305.
164. Almeria, B.; Deng, W. W.; Fahmy, T. M.; Gomez, A. *J. Colloid Interface Sci.* **2010**, *343*, 125–133.
165. Cheema, S. K.; Gobin, A. S.; Rhea, R.; Lopez-Berestein, G.; Newman, R. A.; Mathur, A. B. *Int. J. Pharmaceut.* **2007**, *341*, 221–229.
166. Kundu, B.; Saha, P.; Datta, K.; Kundu, S. C. *Biomaterials* **2013**, *34*, 9462–9474.
167. Pritchard, E. M.; Valentin, T.; Boison, D.; Kaplan, D. L. *Biomaterials* **2011**, *32*, 909–918.
168. Wang, X. Q.; Yucel, T.; Lu, Q.; Hu, X.; Kaplan, D. L. *Biomaterials* **2010**, *31*, 1025–1035.
169. Wanga, Y. Z.; Kima, H. J.; Vunjak-Novakovicb, G.; Kaplan, D. L. *Biomaterials* **2006**, *27*, 6064–6082.
170. Wenk, E.; Meinel, A. J.; Wildy, S.; Merkle, H. P.; Meinel, L. *Biomaterials* **2009**, *30*, 2571–2581.
171. Wang, Y. Z.; Rudym, D. D.; Walsh, A.; Abrahamsen, L.; Kim, H. J.; Kim, H. S.; Kirker-Head, C.; Kaplan, D. L. *Biomaterials* **2008**, *29*, 3415–3428.
172. Shi, L. B.; Cai, H. X.; Chen, L. K.; Wu, Y.; Zhu, S. A.; Gong, X. N.; Xia, Y. X.; Ouyang, H. W.; Zou, X. H. *Biomaterials* **2014**, *35*, 1519–1530.
173. Pallotta, I.; Kluge, J. A.; Moreau, J.; Calabrese, R.; Kaplan, D. L.; Balduini, A. *Biomaterials* **2014**, *35*, 3678–3687.
174. Seib, F. P.; Kaplan, D. L. *Biomaterials* **2012**, *33*, 8442–8450.
175. Mandal, B. B.; Kapoor, S.; Kundu, S. C. *Biomaterials* **2009**, *30*, 2826–2836.
176. Benfenati, V.; Toffanin, S.; Capelli, R.; Camassa, L. M. A.; Ferroni, S.; Kaplan, D. L.; Omenetto, F. G.; Muccini, M.; Zamboni, R. *Biomaterials* **2010**, *31*, 7883–7891.
177. Wang, X. Y.; Zhang, X. H.; Castellot, J.; Herman, I.; Iafratid, M.; Kaplan, D. L. *Biomaterials* **2008**, *29*, 894–903.
178. Numata, K. J.; Subramanian, B.; Currie, H. A.; Kaplan, D. L. *Biomaterials* **2009**, *30*, 5775–5784.
179. Lewicka, M.; Hermanson, O.; Rising, A. U. *Biomaterials* **2012**, *33*, 7712–7717.
180. Pritchard, E. M.; Dennis, P. B.; Omenetto, F.; Naik, R. R.; Kaplan, D. L. *Biopolymers* **2012**, *97*, 479–498.
181. Lu, Q.; Wang, X.; Hu, X.; Cebe, P.; Omenetto, F.; Kaplan, D. L. *Macromol. Biosci.* **2010**, *10*, 359–368.
182. Tsioris, K.; Raja, W. K.; Pritchard, E. M.; Panilaitis, B.; Kaplan, D. L.; Omenetto, F. G. *Adv. Mater.* **2012**, *22*, 330–335.
183. Lu, S. Z.; Wang, X.; Uppal, N.; Kaplan, D. L.; Li, M. Z. *Front. Mater. Sci. China* **2009**, *3*, 367–373.
184. Demura, M.; Asakura, T. *J. Membr. Sci.* **1991**, *59*, 39–52.
185. Cameotra, S. S.; Mahanta, P.; Goswami, P. *Appl. Biochem. Biotechnol.* **2009**, *157*, 593–600.
186. Acharya, C.; Kumar, V.; Sen, R.; Kundu, S. C. *Biotechnol. J.* **2008**, *3*, 226–233.
187. Yoshimizu, H.; Asakura, T. *J. Appl. Polym. Sci.* **1990**, *40*, 127–134.

188. Saxena, U.; Goswami, P. *Appl. Biochem. Biotechnol.* **2010**, *162*, 1122–1131.
189. Cordier, D.; Couturier, R.; Grasset, L.; Ville, A. *Enz. Microb. Technol.* **1982**, *4*, 249–255.
190. Zhang, Y. Q.; Zhou, W. L.; Shen, W. D.; Chen, Y. H.; Zha, X. M.; Shirai, K.; Kiguchi, K. *J Biotechnol.* **2005**, *120*, 315–326.
191. Grasset, L.; Cordier, D.; Couturier, R.; Ville, A. *Biotechnol. Bioeng.* **1983**, *25*, 1423–1434.
192. Miyairi, S.; Sugiura, M.; Fukui, S. *Agric. Biol. Chem.* **1978**, *42*, 1661–1667.
193. Zhang, Y. Q. *Biotechnol. Adv.* **1998**, *16*, 961–971.
194. Wu, Y.; Shen, Q.; Hu, S. *Anal. Chim. Acta* **2006**, *558*, 179–186.
195. Inoue, S.; Matsunaga, Y.; Iwane, H.; Sotomura, M.; Nose, Y. *Biochem. Biophys. Res. Commun.* **1986**, *141*, 165–170.
196. Dennis, P. B.; Walker, A. Y.; Dickerson, M. B.; Kaplan, D. L.; Naik, R. R. *Biomacromolecules* **2012**, *13*, 2037–2045.
197. Karageorgiou, V.; Tomkins, M.; Fajardo, R.; Meinel, L.; Snyder, B.; Wade, K.; Chen, J.; Vunjak-Novakovic, G.; Kaplan, D. L. *J. Biomed. Mater. Res., Part A* **2006**, *78*, 324–334.
198. Bessa, P. C.; Balmayor, E. R.; Hartinger, J.; Zanoni, G.; Dopler, D.; Meinl, A.; Banerjee, A.; Casal, M.; Redl, H.; Reis, R. L.; Van- Griensven, M. *Tissue Eng., Part C* **2010**, *16*, 937–945.
199. Wenk, E.; Murphy, A. R.; Kaplan, D. L.; Meinel, L.; Merkle, H. P.; Uebersax, L. *Biomaterials* **2010**, *31*, 1403–1413.
200. Uebersax, L.; Merkle, H. P.; Meinel, L. *J. Controlled Release* **2008**, *127*, 12–21.
201. Wenk, E.; Wandrey, A. J.; Merkle, H. P.; Meinel, L. *J. Controlled Release* **2008**, *132*, 26–34.
202. Sofia, S.; McCarthy, M. B.; Gronowicz, G.; Kaplan, D. L. *J. Biomed. Mater. Res.* **2001**, *54*, 139–148.
203. Wongpanit, P.; Ueda, H.; Tabata, Y.; Rujiravanit, R. *J. Biomater. Sci.* **2010**, *21*, 1403–1419.
204. Uebersax, L.; Mattotti, M.; Papaloizos, M.; Merkle, H. P.; Gander, B.; Meinel, L. *Biomaterials* **2007**, *28*, 4449–4460.
205. Madduri, S.; Papalo¨zos, M.; Gander, B. *Biomaterials* **2010**, *31*, 2323–2334.
206. Zhang, W.; Wang, X.; Wang, S.; Zhao, J.; Xu, L.; Zhu, C.; Zeng, D.; Chen, J.; Zhang, Z.; Kaplan, D. L.; Jiang, X. *Biomaterials* **2011**, *32*, 9415–9424.
207. Kikuchi, J.; Mitsui, Y.; Asakura, T.; Hasuda, K.; Araki, H.; Owaku, K. *Biomaterials* **1999**, *20*, 647–654.
208. Lawrence, B. D.; Cronin-Golomb, M.; Georgakoudi, I.; Kaplan, D. L.; Omenetto, F. G. *Biomacromolecules* **2008**, *9*, 1214–1220.
209. Lu, Q.; Wang, X.; Zhu, H.; Kaplan, D. L. *Acta Biomater.* **2011**, *7*, 2782–2786.
210. Putthanarat, S.; Eby, R. K.; Naik, R. R.; Juhl, S. B.; Walker, M. A.; Peterman, E.; Ristich, S.; Magoshi, J.; Tanaka, T.; Stone, M. O.; Farmer, B. L.; Brewer, C.; Ott, D. *Polymer* **2004**, *45*, 8451–8457.
211. Wang, X.; Kaplan, D. L. *Macromol. Biosci.* **2011**, *11*, 100–110.
212. Zhang, Y. Q.; Ma, Y.; Xia, Y. Y.; Shen, W. D.; Mao, J. P.; Zha, X. M.; Shirai, K.; Kiguchi, K. *J. Biomed. Mater. Res., Part B* **2006**, *79*, 275–283.

213. Guziewicz, N.; Best, A.; Perez-Ramirez, B.; Kaplan, D. L. *Biomaterials* **2011**, *32*, 2642–2650.
214. Wilz, A.; Pritchard, E. M.; Li, T.; Lan, J. Q.; Kaplan, D. L.; Boison, D. *Biomaterials* **2008**, *29*, 3609–3616.
215. Szybala, C.; Pritchard, E. M.; Lusardi, T. A.; Li, T.; Wilz, A.; Kaplan, D. L.; Boison, D. *Exp. Neurol.* **2009**, *219*, 126–135.
216. Ishii, A.; Furukawa, M.; Matsushima, A.; Kodera, Y.; Yamada, A.; Kanai, H.; Inada, Y. *Dyes Pigm.* **1995**, *27*, 211–217.
217. Baycin, D.; Altiok, E.; Ülkü, S.; Bayraktar, O. *J. Agric. Food. Chem.* **2007**, *55*, 1227–1236.
218. Choi, H. M.; Bide, M.; Phaneuf, M.; Quist, W.; Logerfo, F. *Text. Res. J.* **2004**, *74*, 333–342.
219. Gupta, V.; Aseh, A.; Ríos, C. N.; Aggarwal, B. B.; Mathur, A. B. *Int. J. Nanomed.* **2009**, *4*, 115–122.
220. Pritchard, E. M.; Valentin, T.; Panilaitis, B.; Omenetto, F.; Kaplan, D. L. *Adv. Funct. Mater.* **2013**, *23*, 854–861.
221. Tao, H.; Siebert, S. M.; Pritchard, E. M.; Sassaroli, A.; Panilaitis, B. J. B.; Brenckle, M. A.; Amsden, J. J.; Fantini, S.; Kaplan, D. L.; Omenetto, F. G. *Proc. Natl. Acad. Sci.* **2012**, *109*, 19584–19589.
222. Tsioris, K.; Tilburey, G. E.; Murphy, A. R.; Domachuk, P.; Kaplan, D. L.; Omenetto, F. G. *Adv. Funct. Mater.* **2010**, *20*, 1083–1089.
223. Gregory, H.; Altman, F. D.; Jakuba, C. *Biomaterials* **2003**, *24*, 401–416.
224. Cunniff, P. M.; Fossey, S. A.; Auerbach, M. A.; Song, J. W.; Kaplan, D. L.; Adams, W. W.; Eby, R. K.; Mahoney, D.; Vezie, D. L. *Polym. Adv. Technol.* **1994**, *5*, 401–410.
225. Pins, G. D.; Christiansen, D. L.; Patel, R.; Silver, F. H. *Biophys. J.* **1997**, *73*, 2164–2172.
226. Engelberg, I.; Kohn, J. *Biomaterials* **1991**, *12*, 292–304.
227. Gosline, J. M.; Guerette, P. A.; Ortlepp, C. S.; Savage, K. N. *J. Exp. Biol.* **1999**, *202*, 3295–3303.
228. Mukherjee, C.; Hota1, M. K.; Naskar, D.; Kundu, S. C.; Maiti1, C. K. *Phys. Status Solidi A* **2013**, *210*, 1797–1805.
229. Wang, L. J.; Singh, A.; Wang, X. G. *Fibers Polym.* **2008**, *9*, 509–514.
230. Patil, K.; Rajkhowa, R.; Dai, J. X.; Tsuzuki, T.; Lin, T.; Wang, X. G. *Powder Technol.* **2012**, *219*, 179–185.
231. Patil, K.; Smith, S. V.; Rajkhowa, R.; Tsuzuki, T.; Wang, X. G.; Lin, T. *Powder Technol.* **2012**, *218*, 162–168.
232. Patil, K.; Wang, X. G.; Lin, T. *Powder Technol.* **2013**, *245*, 40–47.
233. Tang, B.; Wang, J. F.; Xu, S. P.; Afrin, T.; Xu, W. Q.; Lu, S.; Wang, X. G. *J. Colloid Interface Sci.* **2011**, *356*, 513–518.
234. Rajkhowa, R.; Gil, E. S.; Kluge, J.; Numata, K.; Wang, L. J.; Wang, X. G.; Kaplan, D. L. *Macromol. Biosci.* **2010**, *10*, 599–611.
235. Rockwood, D. N.; Gil, E. S.; Park, S. H.; Kluge, J. A.; Grayson, W.; Bhumiratana, S.; Rajkhowa, R.; Wang, X. G.; Kim, S. J.; Vunjak-Novakovic, G.; Kaplan, D. L. *Acta Biomater.* **2011**, *7*, 144–151.

Chapter 12

Lightweight Composites Reinforced by Agricultural Byproducts

Xin Yu*[,1] and Helan Xu[2]

[1]International School, Zhejiang Fashion Institute of Technology, No. 495 Fenghua Road, Jiangbei District, Ningbo, Zhejiang, China, 315211
[2]Department of Textiles, Merchandising and Fashion Design, 234, HECO Building, University of Nebraska-Lincoln, Lincoln, Nebraska 68583-0802, United States
***E-mail: sisi_yu_xin@hotmail.com**

Reinforcing lightweight composites with different agricultural byproducts is cost effective and environmentally friendly due to their abundant availability, low cost, renewability and biodegradability. The agricultural byproducts reinforced composites are high value-added products with a variety of applications, such as panels, boards, concrete and starch-based films in automobile, building and packing industries. Wheat and rice straw, rice husk, rice husk ash, bagasse, coir, corn stalk, banana fiber and pineapple leaf fiber are lignocellulosic biomass that have been successfully utilized to reinforce polymeric matrices. The mechanical properties of neat matrix could be improved by incorporation of the agro-fibers, while the density could be decreased or retained. However, incompatibility between hydrophilic agro-based reinforcements and hydrophobic matrices decreases mechanical performance, dimensional and thermal stability of composites. Chemical or physical modifications are essential to enhance the interfacial compatibility of fiber and matrix, resulting in improved mechanical properties, dimensional and thermal stability.

Introduction

Every year, agricultural industry produces billions of tons of lignocellulosic residues (*1*, *2*). Agricultural byproducts from wheat, rice, sugarcane, coconut, pineapple, corn and banana include rice straw (*3*), wheat straw (*4*, *5*), rice husk (*6*), rice husk ash (*7*), sugarcane bagasse (*8*, *9*), coir (*10*), pineapple leaf (*11*), corn stalk (*12*) and banana leaf (*13*). These lignocellulosic byproducts are abundant, cheap and annually renewable (*2*). Lignocellulosics are referred to dry plant matter, which mainly consist of cellulose, hemicellulose and lignin as shown in Table I. The proportion of the constituents of the lignocellulosics depends on their age, source and extraction method. Cellulose is the main structure component and acts as the backbone providing strength and stability. The properties and applications of lignocellulosic fibers are to a large extent determined by the amount of cellulose in the fiber. Hemicellulose is the filler between cellulose and lignin but contributes little to the stiffness and strength of fibers. Lignin acts as glue to provide structural support and compressive strength to plant tissue and individual fibers. Lignin protects carbohydrates in fibers from chemical and physical damage. The content of lignin in lignocellulosics influences the structure properties, morphology and flexibility of the fibers (*2*).

The lignocellulosic biomass could be valuable resources for various industrial applications. At present, most of byproduct biomass, such as wheat and rice straws are left on the ground to decompose or are burnt in fields (*2*). However, the byproducts could be potential resources to be used as pulp (*2*), biofuel (*1*, *2*), cellulose fibers (*2*) and composties (*14*, *15*) in the fields of paper, energy, packing, automobile and construction. Currently, the materials used in these areas, such as gasoline, wood, bast fibers are not environmentally friendly or with high carbon emission, because they are either derived from petroleum resources, or need abundant land, water and fertilizer to grow. Moreover, with the decline of the global farmlands the availability of bast fibers will be significantly decreasing (*16*). Introduction of agricultural byproducts as potential alternatives to natural fibers will ease the environmental burdens, lower the product prices and increase farmers' profits.

Composites are a class of materials that comprise two or more distinct components with different physical and/or chemical properties. Usually, composites consist of a strong constituent, the reinforcement, and a ductile or tough phase, the matrix. The reinforcement is embedded in the matrix and the combination of them into composite materials can achieve property improvements (*17*). Lightweight composites have densities lower than the combined densities of the materials built the composites due to the presence of voids inside the composites (*18*, *19*). Both agricultural byproducts and natural fibers are processed into lightweight composites if the matrix before processing is also in fibrous forms.

Natural fiber reinforced composites have drawn a great deal of attention. The natural fibers such as flax, kenaf, hemp, jute and sisal are cheap, light weight, biodegradable, and carbon neutral (*17*, *20*). More and more natural fiber composites have been utilized in automotives to replace glass fiber composites for door panels, seat backs and the interior (*21*), and as building materials such as fiberboards and panels (*22*). With lower densities than glass fibers', natural fiber composite products are lighter than those from glass fibers with the same proportion in composites. Using lightweight composites in vehicles for instance has the potential to save fuel consumption and it has practical significance due to the diminishing fuel resources (*18*, *19*). Additionally, natural fibers are biodegradable whereas glass fibers are not. The tendency of using natural fibers replacing glass fibers in automobile interior parts is also driven by an European Directive that for all end-of life vehicles, the reuse and recovery shall be increased to a minimum of 95% by an average weight per vehicle by 2015 (*23*).

The fibers extracted from agricultural byproducts have similar properties to natural fibers, which makes agro-based fibers suitable as composite reinforcements. Moreover, agro-fibers are cheap, consistently and abundantly available. This makes them potential and feasible as alternatives to natural fibers (*16*). As a result, a large number of researches have been done recently on the lightweight composites reinforced with agro-based lignocellulosic materials, including straw (*4*, *24*), rice husk (*25*, *26*), rice husk ash (*27*), bagasse (*8*), coir fiber (*10*), banana fiber (*13*), corn stalk (*12*) and pineapple leaf fiber (*28*), etc. Their physical, mechanical and thermal properties have been investigated and the potential applications have been discussed. Normally, with the presence of agro-fibers in the matrix, the mechanical properties of materials would be improved compared to neat matrix. However, a number of factors including fiber parameters, fiber loading level, type of polymer matrix, chemical treatments, surface modifications and adhesives would have effect on mechanical, thermal and dimensional properties of composites.

Agricultural byproducts are usually prepared in forms of fibers or particles to polymers. The agro-materials are good sources for nanofibers and cellulose in some nanocomposite applications as well (*29–32*). The extraction of nanofibers and cellulose is usually achieved by chemical processing. Considering the lignocellulosic nature of agro-fibers, the general chemical processing of lignocellulosics is firstly to deal with the separation of its three main components, cellulose, hemicellulose and lignin. The cellulose is the useful material for composite products. Therefore, the aim of extraction is to remove the hemicellulose and lignin from agro-fibers. Native lignin, due to its three dimensionally cross-linked structures, is insoluble in any solvent. So, in order to extract lignin from lignocellulosics, lignin has to be partially degraded to lower molecular weight fragments that are soluble in the solvent (*33*). The pulping processing is largely applied as the first step of agro-fiber extraction. During this processing, alkaline solvent could dissolve lignin and the following acid treatment can hydrolyze hemicellulose (*30*). Bleaching is followed for the removal of residual lignin. Acetic acid is a popular choice combined with hydrogen peroxide, chlorine, hypochlorite and/or chlorine dioxide (*33*, *34*). Refining is the final step to separate pulp and isolate micro- and nano-scale fibers by mechanical techniques such as using a high-shear ultrafine friction grinder (*31*, *32*).

Table I. Estimated potential production and fiber composition of agricultural residues

Agricultural Product	*Agricultural Byproduct*	*Residue Production MMT/year*	*Cellulose%*	*Hemicellulose %*	*Lignin %*	*Ash %*	*References*
Rice	Rice straw	457	28-36	23-28	12-14	14-20	(*1*, *2*)
Rice	Rice husk	222	25-35	18-21	26-31	15-17	(*1*, *42*)
Wheat	Wheat straw	475	33-38	26-32	17-19	6-8	(*1*, *2*)
Sugarcane	Bagasse	505	32-48	19-24	23-32	1.5-5	(*1*, *2*)
Coconut	Coir	–	43	0.25	45	2	(*43*)
Banana	Banana leaves	184	60-65	6-8	5-10	4.7	(*1*, *2*)
Corn	Corn stalk	1267	38-40	28	7-21	3.6-7.0	(*1*, *2*)
Pineapple	Pineapple leaves	10	70-82	18	5-12	0.7-0.9	(*1*, *2*)

MMT: Million metric tons.

One limitation of agro-fiber reinforced composites is the poor interfacial adhesion between fibers and the matrix due to the hydrophilic nature of fibers incompatible with hydrophobic polymers (*2*, *16*). Thus, the introduction of suitable methods or adhesives to fibers or composites is necessary to improve the interfacial adhesion resulting in better properties of composites. According to a number of literatures on interfacial improvements, normally there are two methods, fiber modifications and the adhesive incorporation. Alkali and acid treatments are common solvents to treat fibers. Fibers such as wheat straws (*38*), cornhusks (*16*) and bagasse (*39*) would have better adhesion to the matrix due to the removal of noncellulosic materials in fiber after treatments. Enzyme treatments also improve interfacial adhesion for wheat straws by producing finer fibers and increasing the fiber aspect ratio (*40*, *41*). Some physical methods, like steam cooking which enhances the wettability of wheat straw, result in better interfacial interaction (*24*). Introduction of adhesives such as saline coupling agents (*36*) and maleic anhydride (*37*) to agro-based composites has been proved useful and efficient. Treatment of polypropylene (PP) with maleic anhydride successfully increases the interaction between wheat straws and PP. Maleic anhydride acts as bridges between fibers and the matrix, which contributes to better encapsulation of wheat straw particles by the plastic. (*37*).

The aim of this work is to introduce the current development of cost-effective, environmentally friendly and abundantly available agro-fiber reinforced lightweight composites. The agro-based reinforcements refer to wheat and rice straw, rice husk, rice husk ash, bagasse, coir, banana leaf fibers, corn stalks and pineapple leaf fibers.

Wheat and Rice Straw

Among the agriculture plants in the world, wheat occupies the largest planting area with the highest annual production, and rice is the primary food for more than 40 % of the world's population, with about 596 million tons of rice and 570 million tons of rice straw produced annually in the world (*5*, *6*). Both rice and wheat straw are lignocellulosic agricultural byproducts and have similar chemical composition consisting of cellulose, hemicellulose and lignin mainly. Straw fibers are suitable sources as reinforcement to lightweight composites for building materials due to their low density, high toughness, reduced dermal and respiratory irritation and good biodegradability properties. Straw fibers may be considered as alternatives to natural wood for particleboards (*44*). A wide range of research papers have been reported the application of rice and wheat straws as reinforcements.

To obtain desirable performance, the mechanical properties, dimensional stability and thermal properties of straw reinforced composites have been widely studied. Some examples of composite materials are summarized in Table II.

Table II. Some examples of composites reinforced by wheat/rice straw

Fiber loading wt%	*Matrix*	*Modification method*	*Additives*	*Procedure*	*Application*	*References*
40	Polyester	n/a	n/a	Mould press and post cured	Lightweight building material	(*4*)
25	PP	n/a	MA-PP	Injection molded	n/a	(*37*)
10,20,25, 30	PP	n/a	MA-g-PP	Injection molded	n/a	(*46*)
88	UF	Ethanol-benzene treatment	Silane coupling agent	Hot pressing	Particleboard	(*36*)
90	Phenolic resin	Acetylation treatment	n/a	Hot pressing	MDF	(*35*)
85	UF	Steam cooking	n/a	Platen pressing	MDF	(*24*)
21,27 (A mixture of straw and bagasse)	Gypsum	n/a	n/a	Cold pressing	Gypsum-bonded particleboard	(*48*)
5,10,15	Thermoplastic starch	Chemical extraction; ultrasonic technique	Glycerol	Film casting	Bio-nanocomposite film	(*29*)

n/a: Not applicable; PP:Polypropylene; UF: Urea-formaldehyde; MA-PP: Maleic anhydride modified polypropylene; MA-g-PP: Maleic anhydride grafted polypropylene.

In the study of White and Ansell, mechanically crushed wheat straw stems were incorporated in a polyester resin matrix. The straw fibers considerably improved the stiffness, strength and toughness of the resin, and reduced the density. The specific flexural stiffness was about half of softwoods, thus it is envisaged that alternative methods for processing the fibers and introduction of other types of resin would improve the composite properties further (*4*). Hornsby applied hammer-milled wheat straw fibers at 25 wt% loading level to reinforce thermoplastic polypropylene (PP) matrix. It resulted in a significant increase in tensile modulus (2.63 GPa) compared to unfilled polymer (1.18 GPa) (*37*, *45*). In another case of wheat straw-PP composites, it utilized ground wheat straw combining with 3 wt% coupling agent maleic anhydride grafted polypropylene (PP-g-MA). It was observed that with the increase of fiber content, the tensile strength, tensile modulus and storage modulus were gradually increased while impact strength and thermal stability declined. The coupling agent PP-g-MA helped improve the tensile properties and impact strength of the composites. It enhances the interface adhesion between straw particles and PP and brings better encapsulation of straw particles by the plastic. The decreased impact strength may be attributed to points of stress concentrations for crack initiation with the presence of wheat straw in PP matrix and some thermal degradation during compounding. The less thermal stability of the composites is because wheat straw fiber has a lower thermal stability than PP (*46*). Panthapulakkal evaluated the suitability of utilizing wheat straw as an alternative to wood flour. The wheat straw was pretreated by a 3-week fungi treatment before compounding. Incorporation of wheat straw in PP resin did not change tensile strength of the composites; however, addition of 5 % of coupling agent (maleated PP) increased the tensile strength. The effect of compounding process (a high shear and a low shear mixing) on mechanical properties was studied. The high shear compounding process caused extensive break down of wheat straw fibers and made the aspect ratio of the straw fibers and the milled straw become close. As a result, the reinforcing effect of the fibers decreased and showed a similar effect to that of ground wheat straw. Fungal treatment showed slight improvement in strength; however the treatment considerably improved the modulus of the composites. The results of tensile and flexural properties showed that wheat straw could be used as an alternative to wood fiber filled composites (*44*).

Cellulose and nanofibers extracted from rice straw reinforce thermoplastic starch for bio-nanocomposite films by film casting technique. There are cellular bundles in raw rice straw covered by an outer epidermis that has a concentrated layer of silica on the surface (Figure 1). Chemical treatments aim at degradation of silica layer and removal of lignin and hemicelluloses to remain cellulose fibers. Rice straw was firstly soaked in NaOH solution and subsequently hydrolyzed by dilute HCl solution. And then, the hydrolyzed pulp was treated by alkali and followed by bleaching. Finally, the dried pulp was separated and the nanofibers were isolated by mechanical refining. After the purification, rice straw microbiers with a dimension of around 5-20 μm were obtained (Figure 2). The effect of the fiber content on the composites indicated the mechanical and water resistant properties have been improved with increasing content of cellulose fibers (*29*).

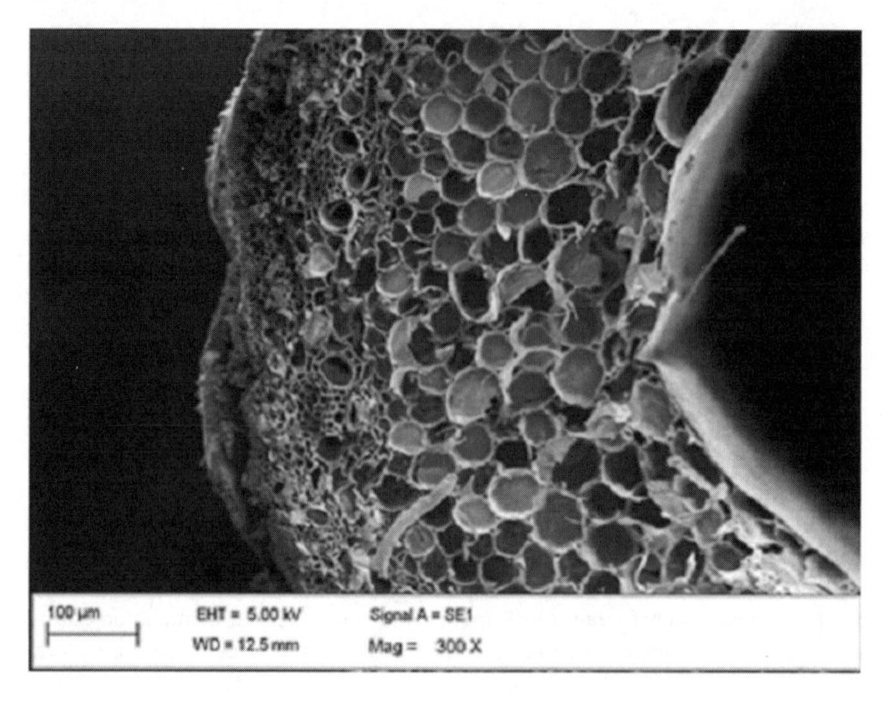

Figure 1. SEM image of rice straw cross section. (Reproduced with permission from reference (29). Copyright 2014 Springer Press.)

Apart from polymeric matrices, wheat straw fibers were also utilized as reinforcement in cement-bonded and gypsum-bonded particleboards (*47*, *48*). A study has proved the feasibility of using wheat straw particles for the production of gypsum-bonded particleboards (*48*). The effect of rice straw content on concrete composites for hollow blocks was analyzed. A certain loading of 10 wt% rice straw in combination of early strength agent (Al_2 $(SO4)_3$, $CaCl_2$) in concrete would impart optimal properties to the composite materials (*49*).

The properties of the composites largely depend on the interfacial interaction between the reinforcement and the matrix. In order to improve the interfacial adhesion between agro-straws and matrices resulting in enhanced properties of composites. Adhesives and fiber modification methods were properly studied and introduced.

Agro-straws have a problem of bonding with urea-based resins (*50*). The waxy coating on the epidermis of straw stem causes problems in bonding this material with conventional urea formaldehyde (UF) resins (*51*). The UF-bonded agro-straw boards were reported to have undesirable properties, however high quality boards can be produced using isocyanate resin. MDI is a complex mixture of the isomers of di-, tri-isocyanates and higher polymeric aromatic species derived from side reactions and generally sold as PMDI (polymeric MDI) and EMDI, an emulsion of PMDI in water (*52*). Two combinations of UF: EMDI and UF: PMDI resin systems have been evaluated the effect on the properties of composites in comparison with those of pure isocyanate composites. It has been showed that UF: EMDI formulations can be successfully used as an alternative to pure isocyanate resin. With appropriate combination of UF: PMDI, straw made particleboards can compete with wood made ones as they satisfy the requirements of related European standards (*51*, *52*).

Isocyanate resin is more expensive than UF resin, whereas silane coupling agents (SCA) are generally considered to be cost effective. The effect of SCA on the properties of UF bonded reed and wheat straw particleboards was investigated. Introduction of SCA to the composites upgraded the performance of UF-bonded boards, while the boards without SCA were relatively lower than those of commercial particleboards (*5*). It was reported that the internal bonding strength (IB) and thickness swelling (TS) of wheat straw-UF particleboard were significantly improved by addition of amino SCA. The dimensional stability could be improved by increasing the content of SCA (*53*). SCA were also applied to rice straw composites followed by electron beam (EB) irradiation. The property improvement of the composites was confirmed by the presence of SCA, and the further enhanced performance was observed due to the EB irradiation in which produces more free radicals to form chemical bonding between rice straw fibers and polymers (*54*). It is the case that the EB irradiation treatment also improved the physical and mechanical properties of rice straw-polystyrene and rice straw reinforced poly (vinyl alcohol) (PVA) composites (*3*). A comparison of the effect of SCA and ethanol-benzene treatment on wheat straw-UF particleboards indicated that ethanol-benzene treatment is more effective than SCA (*53*).

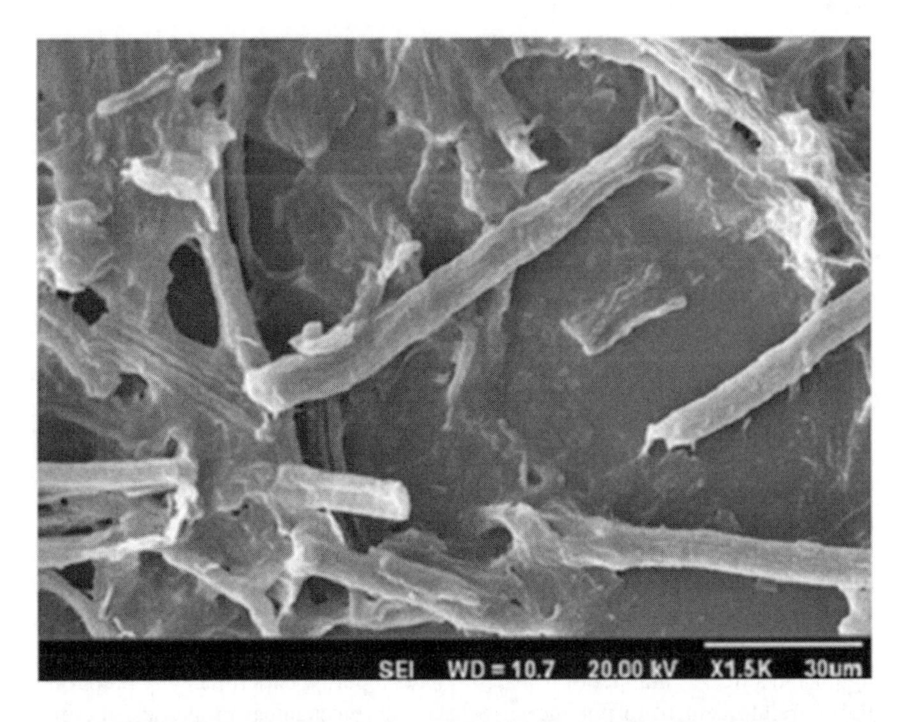

Figure 2. SEM image of rice straw microfibers after chemical purification. (Reproduced with permission from reference (29). Copyright 2014 Springer Press.)

To manufacture low-formaldehyde emission particleboards from wheat straw and UF resins, the straw particles were pretreated by spraying urea on them and subsequently compounded with UF. The urea pretreatment of wheat straw helped urea to react with free formaldehyde in the UF in formation of methylol groups, which contributed to the reduction of free formaldehyde and the improvement of physical and mechanical properties by means of modulus of rupture (MOR), modulus of elastic (MOE), IB and impact strength (*36*).

Enhanced wettability of straw fibers could improve the interfacial adhesion with polymers. The wettability of straw fibers could be increased by NaOH treatment. It was considered that the hemicellulose and lignin were partially decomposed and the lipophilic substance composition of wax-like layer was effectively degraded from the observation of SEM (*38*). The high temperature steam treatment can upgrade the wettability of straw fibers as well. A high performance UF-bonded wheat straw medium density fiberboard (MDF) was prepared by steam cooking wheat straws before compounding. The wettability of the straw was improved after cooking treatment. As a result, all the properties of the straw MDF, except the TS property, could meet the requirements of JIS fiberboard standard (*24*). The steam treatment was also utilized in a case of wheat straw and poly (3-hydroxybutyrate-co-hydroxyvalerate) (PHBV) composites. This treatment removed a significant part of the non-fibrous components (lignin and hemicellulose) and also changed fiber morphology which can favor interfacial adhesion with plastic matrix (*55*).

Besides the physical modifications, enzymes such as cellulases and lipases were introduced as fiber modification as well. It was observed by electron spin resonance (ESR) that free radical content of wheat straw clearly increased after the cellulose treatment, along with the reduction of surface wax due to the lipase treatment. The surface wax is one of the main adhesion inhibitors to polymers (*40, 41*).

A crosslinked rice straw fiber (CRSF) combined with glycidyl methacrylate grafted polylactide (PLA-g-GMA) were compounded into PLA polymer. The evaluated properties indicated that PLA-g-GMA-CRSF had noticeably superior mechanical properties because of greater compatibility between the polymer and the CRSF (*56*).

Rice Husk

Rice husk (RH) is one of rice residues after rice harvest and most of RH is burnt after harvest. Burning RH and other agriculture residues in wide areas not only results in serious environment issues, but also wastes precious resources. Development of value added materials from RH can reduce the environmental issues (*6*). The lignocellulosic RH mainly consists of cellulose, hemicellulose, lignin and silica. A number of researches have reported experimentally preparing RH as reinforcement to polymeric resins. The mechanical properties, thermal stability, water resistance, biodegradability of the composites have been evaluated. Some cases of RH reinforced composites are showed in Table III.

Table III. Some examples of composites reinforced by rice husk/rice husk ash

Fiber loading wt%	*Matrix*	*Modification method*	*Additives*	*Procedure*	*Application*	*References*
50 (RH)	Polyester	Polyester resin dissolved in styrene	Catalyst; Initiator	Compression pressing	n/a	(*6*)
50 (RH)	PP	n/a	Coupling agent	Hot pressing	n/a	(*6*)
10,20,30 (RH)	HDPE/ Natural rubber	n/a	Liquid natural rubber	Hot pressing	n/a	(*25*)
10,20,30,40 (RHF)	PBS	n/a	n/a	Melt blending, extrusion and pelletizing.	Biocomposites	(*60*)
5,10,15,20 (RHA)	Portland cement	n/a	Fine aggregate; Coarse aggregate; Superplasticizers	Mould casting	Concrete	(*15*)
70,80,90,95,97.5,99 (RHA)	Aluminium hydroxide	n/a	Boric acid	Heat curing	Mortar	(*69*)
10,20 (RHA)	Epoxy	White RHA and black RHA of different compositions	n/a	Mixing	Paint	(*7*)

RH: Rice husk; RHF: Rice husk flour; RHA: Rice husk ash; PP: Polypropylene; HDPE: High density polyethylene: PBS: polybutylene succinate; n/a: Not applicable.

Pure cellulose derived from RH was employed to reinforce poly (lactic acid) (PLA). The composites induced a significant improvement of the mechanical properties compared to neat PLA. The milled RH was firstly treated with H_2SO_4 and subsequently with KOH in order to hydrolyze hemicellulose and remove impurities and extract silica. During the final step, $NaClO_2$ was used to remove amorphous cellulose and lignin remaining the pure cellulose. Figure 3 shows a schematic representation of the extraction procedure of cellulose from RH (*30*). In the case of PLA-RH composites, the flexural modulus of pure PLA was increased by the presence of RH, while the flexural and impact strength declined. The thermal stability of the virgin PLA was decreased and the biodegradability was slightly improved by addition of RH (*57*). In a study of composites of poly (vinyl chloride) (PVC) reinforced by bagasse, rice straw, RH and pine, PVC-RH composites showed the best dimensional stability in water (*58*). El-Saied et al. studied RH and rice straw as fillers in polyester-based thermosetting and thermoplastic PP matrices. The results indicated that the strength properties and water resistance of PP-RH composites with coupling agents were higher than those of RH reinforced polyester composites together with catalyst (copper tween) and initiator (methyl ethyl ketone peroxide, MEKP). RH is adequate as reinforcement to PP matrix while polyester matrix prefers to use in blend with rice straw (*6*).

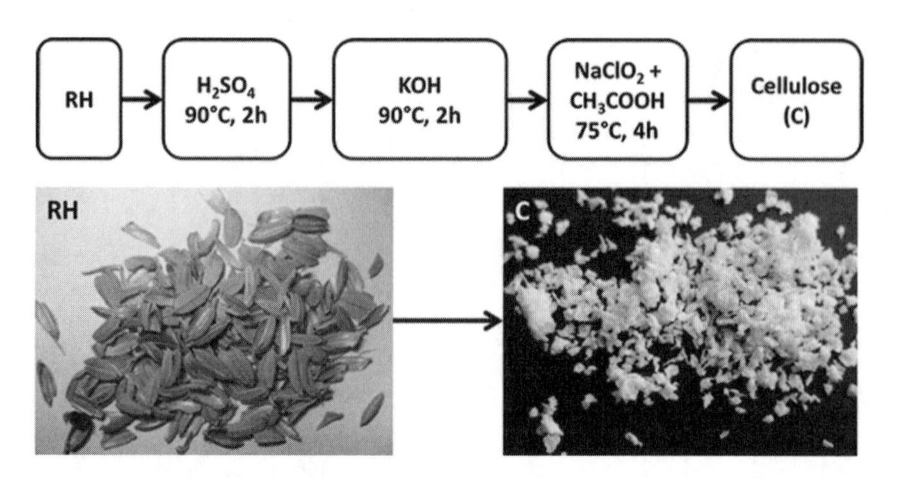

Figure 3. Schematic representation of the cellulose extraction from RH. (Reproduced with permission from reference (30). Copyright 2014 Springer Press.)

The incorporation of reinforcement into a matrix is carried out with the aim of enhancing the specific mechanical properties of composites. The main factors of reinforcement governing reinforcing a matrix, apart from fiber loading, are specific surface area, dispersion in a matrix and interaction to a matrix (*25, 29*). A composite panel, made from a mixture of RH and wattle (acacia minosa) tannin based resin, had a drastic improvements in stiffness through a slight physical modification by hammer-milling RH particles and removal of very fine particles.

The milling increases the specific surface area of particles and this contributes to enhance the interfacial bond strength. The removal of very fine particles helped a better dispersion of RH in matrix (*59*). RH reinforced thermoplastic natural rubber (TPNR) composites had much better mechanical properties by addition of liquid natural rubber as the compatibilizer. The compartibilizer increased the interaction between filler and matrix and improved the homogeneity of RH distribution in the matrix (*25*).

The thermal stability of composites has an important effect on the manufacturing system (*60*). By increasing lignocellulosic fillers, rice husk flour (RHF), the thermal stability of PP, high-density polyethylene (HDPE), low-density polyethylene (LDPE) and polybutylene succinate (PBS)-RHF composites decreased (*60–62*). It is a logical consequence of the lower thermal stability of lignocellulosics than the matrices. However RHF is a suitable material for preventing the thermal expansion of the composites caused by atmospheric changes (*62*). The thermal stability of the PP-RHF composites is largely dependent on the compatibility and interfacial adhesion between the lignocellulosics and the polymers. Incorporation of a coupling agent maleated polypropylene (MAPP) can improve matrix-filler compatibility and interfacial adhesion as well as the dynamic-mechanical properties of the PP-RHF composites (*62*, *63*).

Rice Husk Ash

Rice husk contains about 75% organic matter and around 25% is converted into ash during the burning process. The burning residue is known as rice husk ash (RHA). RHA contains around 85%-90% amorphous silica which is highly reactive in nature and can be produced by combustion of RH under controlled conditions, at relatively low temperature (500-600 °C) and low rate of combustion (*64*). The change from amorphous to crystalline silica starts at 800 °C and completes at 900 °C (*65*). Fixing the temperature and the time of heat treatment while changing the amount of treated RH, RHA of different compositions can be obtained. Black RHA contains less silica (more carbon), and white RHA with more silica (less carbon, Figure 4) (*7*). Introduction of low carbon content RHA as a supplementary raw material for bamboo pulp-cement composites contributed to lower permeability, higher interfacial adhesion and better durability performance than did high carbon content RHA. (*66*). Milling the RHA to get higher surface area can increase the silica reactivity. There is a growing demand for fine amorphous silica for the production of special cement, concrete mixtures, high performance and low permeability concrete and paints, etc (*7*, *15*, *27*, *64*, *66*). Table III demonstrates some RHA based composites.

RHA is suitable as supplementary cementitious material. The ordinary Portland cement of cellulose-cement composites could be partially replaced by RHA up to 30 wt% without impairing mechanical behavior. The strength loss of cellulose based cement composites is mainly due to interfacial debonding and alkali erosion, which are governed by moisture movement through the porous network of composites. One approach for the development of composites with improved durability is partial replacement of ordinary Portland cement by low carbon content RHA (*66*). By addition of RHA to Portland cement paste, the resistance to acid had a notable improvement (*15*, *65*, *67*). Saturated water absorption (WA) of RHA concrete diminished by addition of super plasticizer (Sulphonated Naphthalene) and the porosity decreased from 4.70% to 3.45% when the replacement level increased from 5% to 20% (*15*). The effects of silica forms on concrete mechanical and durability performance have been evaluated in comparison of three types of silica, amorphous RHA, partial crystalline and crystalline RHA. Amorphous RHA performed higher pozzolanic reactivity and showed excellent properties compared to other types of silica (*68*).

Figure 4. SEM of white ash of RH. (Reproduced with permission from reference (64). Copyright 2003 Springer Press.)

RHA also has positive effect on mechanical improvement of epoxy paint protecting steel against different corrosive environments. Black and white RHAs of were prepared and compounded with epoxy. The presence of RHA in epoxy paint enhanced the wear resistance, scratch resistance and elongation. The white ash had better improvement effect on the wear resistance due to the presence of more silica (*7*). Being rich in silica, RHA can be a source of silica in the alumino silicate composites (ASC). Aluminium hydroxide ($Al(OH)_3$) powder was used as the aluminium source. Stable ASC mortars with high RHA : $Al(OH)_3$ mass ratios of 90:10 to 97.5:2.5 were prepared with boric acid. The ASC mortars showed good resistance to H_2SO_4 solution with a slight strength loss after 90-day immersion (*69*).

Sugarcane Bagasse

Sugarcane bagasse is a fibrous residue of the sugarcane milling process. Nearly 30% of sugarcane will turn into bagasse and it produces around 505 million metric tons of bagasse annually all over the world (*1*, *2*). The large volume of bagasse wastes may have harmful effect upon the environment if they are not suitably treated (*70*). Due to the high production of sugar from sugarcane, the product residue bagasse normally needs dispose (*71*). Using bagasse fibers to reinforce polymers serves as a useful solution for its abundant disposal. With increasing wood cost, cheap, environmentally friendly alternative sources for plastic composites are highly needed (*58*, *72*). The feasibility of bagasse as reinforcement for lightweight composites has been largely investigated. Various types of composites have been prepared by incorporation of bagasse, such as raw bagasse fibers, pretreated bagasse fibers, bagasse residues and nanofibers extracted from bagasse. Treatments and adhesives were also introduced to study their effect on the composite properties. Some of bagasse based composites were studied and listed in Table IV.

In a case of bagasse reinforced UF composites for the surface layer of three-layer particleboards with targeted density of 0.7 g/cm^3, the surface layer had a bagasse-wood ratio of 4:6 bonded with UF resin. The bending strength performed better compared to commercial wood particleboards, which indicates that bagasse has a positive effect on the bending strength of boards (*73*). The effect of bagasse source (pith and rind) and bagasse content on the properties of bagasse-PVC composites showed that increasing fiber content (up to 40 wt%) had a positive influence on tensile strength, MOE and storage modulus. This may be ascribed to the stress transferred to strong fibers especially in the direction of elongation; in addition, the rind bagasse-PVC composites offered superior elastic response compared to pith/PVC composites; however, the later had better thermal stability and interfacial bonding. The presence of bagasse decreased the impact strength and it may be due to the poor compatibility between the fiber and the matrix (*58*, *70*). The geometry, morphology and slenderness ratio (L-D) of bagasse fibers were investigated to study the effect on the properties of bagasse panels. With the slenderness radio increasing from 3 to 26, the panel properties were positively enhanced by means of MOR, MOE, IB and TS (*9*). The effect of fiber loading level on the thermal stability of polyurethane-bagasse composites was evaluated. Increasing the bagasse content resulted in impaired thermal stability by thermogravimetry (TG-DTG). It is due to low degradation temperatures of lignocellulosic fibers incompatible with higher degradation temperatures of polymers (*74*).

Table IV. Some examples of composites reinforced by bagasse

Fiber loading wt%	*Matrix*	*Modification method*	*Additives*	*Procedure*	*Application*	*References*
10, 20, 30, 40	PVC	n/a	n/a	Hot pressing	n/a	(*70*)
30	PVC	n/a	Impact modifier (SEBS)	Compression molded	n/a	(*58*)
10	Epoxy	Alkali Treatment with NaOH	Calcium carbonatepowder	Hardener adding and Cold molded	n/a	(*39*)
0, 5, 10, 20, 30	EVA	Acid-alkali treatment	TiO_2	Melt mix intercalation combined Extruded shape	n/a	(*72*)
21,27 (A mixture of straw and bagasse)	Gypsum	n/a	n/a	Cold pressing	Gypsum-bonded particleboard	(*48*)
15	Potato starch	Pulping combined refining	Glycerol	Film casting	Biodegradable packaging and Biocomposite medical application	(*31*)

n/a: Not applicable; EVA: Ethylene co-vinyl acetate; PVC: Polyvinyl chloride; SEBS: Styrene-ethylene-butylene-styrene; TiO_2: Titanium dioxide.

Bagasse fibers, similar to other natural fibers, have the problem of poor adhesion with polymers. Suitable treatments on bagasse fibers prior to compounding and introduction of interfacial adhesives are necessary to enhance the composite properties. In a comparison of the untreated and NaOH chemically treated bagasse reinforced epoxy composites, the chemical treatment imparted the composites better mechanical properties in terms of flexural strength, ultimate tensile strength and compression strength (*39*). Sulphuric acid followed by NaOH pretreated bagasse fibers were applied to ethylene vinyl acetate (EVA) copolymer in combination of titanium dioxide (TiO_2) nanoparticles. There was a decrease in tensile strength by increaseing fiber content but an increase after the addition of TiO_2 at lower bagasse loading. At lower fiber loading, there is stress transfer from the matrix to the filler as the TiO_2 is able to fill in the air spaces created by the sugarcane bagasse fibers (*72*).

Using sugarcane bagasse to produce biofuel (e.g.ethanol) has been extensively studied recently due to its wide availability and low cost. bioethanol can be achieved by the fermentation process of bagasse. The process leaves bagasse residues, hard-to-hydrolyzed fibers and un-reacted lignin components. Some studies have empolyed the bagasse residues to reinforce biodegradable matrices such as PLA and PVA (*71*, *75*). The residues showed a less improvement on physical properties compared with raw and mild acid pretreated bagasse fibers (Figure 5 shows the morphologies of three types of bagasse fibers). It is due to a significant increase of lignin content and the reduction of the aspect ratio of residues after fermentation (*75*). The additional treatments on the residues, disk refining and ultrasonication were carried out to gain reasonable mechanical properties. Consequently, the tensile modulus was significantly improved while the tensile strength slightly increased. The mechanical treatments were proven to effectively increase the surface area of bagasse residues thus improve the interaction between the residues and the matrix (*71*).

Bagasse based particleboards bonded with inorganic binders, gypsum, cement, have been demonstrated in various researches (*8*, *49*, *76*). In the case of gypsum-bonded particleboards (GBPB) reinforced by various mixtures of bagasse and wheat straw, it examined the feasibility of these agricultural residues to use in commercial GBPB manufacturing. TS, WA, MOR, MOE and IB strength of the boards were evaluated. The study found that a decrease of the lignocellulosic material-gypsum ratio resulted in a decrease of MOR and MOE, due to the presence of high brittleness and low MOE of the gypsum. However, panels bonded with the lower amount of gypsum had higher WA and TS and lower IB strength (*49*). In a study of Poland cement bonded bagasse particleboards, ground bagasse particles and sodium silicate (Na_2SiO_3) as accelerator were used. As the particle size decreased while the content of additive increased, the MOR increased steadily. However, the IB and TS of the boards did not surpass the minimum requirements of BISON type HZ and CEN EN (2007) (*76*).

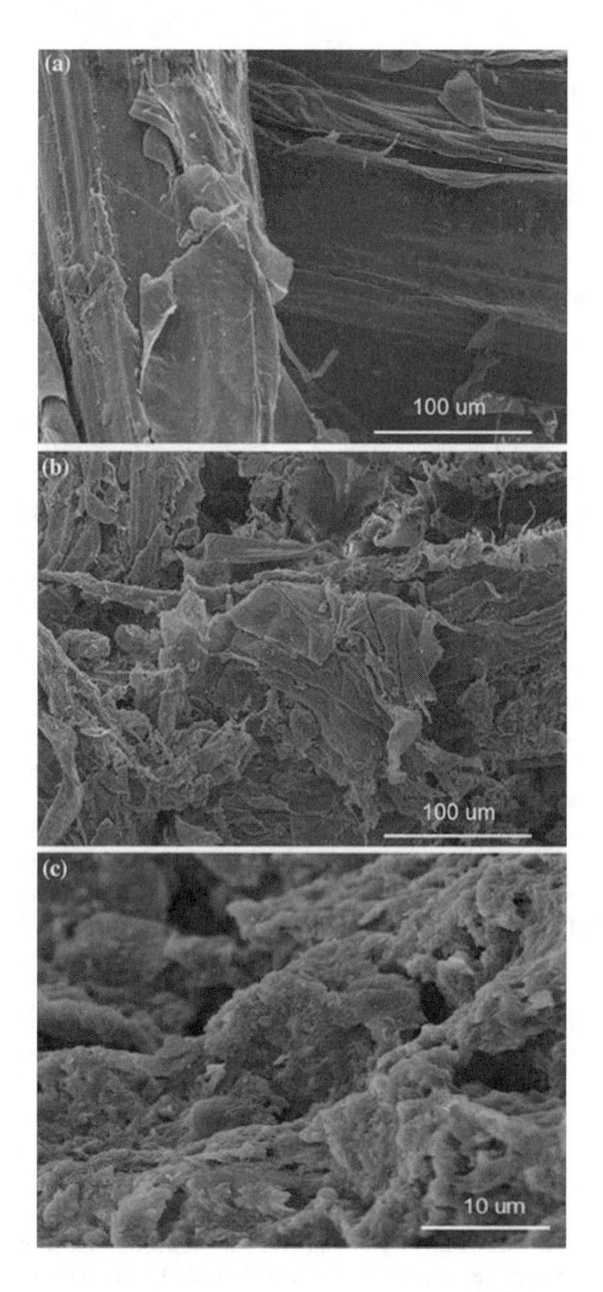

Figure 5. SEM images of three types of bagasse residues: a. raw bagasse residues; b. pretreated bagasse residues; c. fermentation bagasse residues. (Reproduced with permission from reference (75). Copyright 2013 Springer Press.)

Microfibrillated cellulose or nanofibers extracted from bagasse have potential applications in nanocomposites (*31*, *32*, *34*). Crude cellulose can be extracted from sugarcane bagasse after pretreatments with sulfuric acid and followed by NaOH. The further removal of residual lignin with acetic acid and sodium chloride, it obtained bleached cellulose (*34*). It is observed that nanofibers (average diameter 26.5 nm; aspect ratio 247) derived from unbleached bagasse pulp could be strongly attached to the starch matrix by scanning electron microscopic (SEM). There are strong interactions between fibers and the matrix at low fiber content (not above 10 wt%). The composites doubled tensile strength and tripled Young's modulus compared to neat starch. It may be due to the formation of a continuous network at low nanofiber loadings. The nanocomposite film has a potential application in biodegradable packing and medical science (*31*). Bleached cellulose demonstrated better film-forming ability with hydrous niobium phosphate for the solvent system used was only effective for the dissolution of bleached cellulose. It reveals that the presence of cellulose impurities (eliminated during the bleaching procedure) can be particularly disturbing to combination of cellulose and inorganic matrix (*34*). The NaOH and xylanase enzyme pretreated bagasse pulp enabled the chitosan-bagasse cellulose nanocomposites to have higher dry and wet tensile strength than those fabricated from untreated bagasee pulp (*32*).

Coir

Coconut husk is available in large quantities as residues from coconut production in many countries. Coconut coir fiber derived from coconut husk has the lowest thermal conductivity and bulk density compared to other natural fibers. Coir fiber is used as a wide variety of floor furnishing materials, yarns and ropes, etc (*77*). Development of new coir based products of higher added values finds the interest in utilization of coir as a composite reinforcement (*78*). The addition of coir reduces the thermal conductivity of the composites and yields a lightweight product for building as an alternative to wood composite materials (*77*).

A great deal of studies on coir fiber composites has been carried out and some examples are demonstrated in Table V. Biswas investigated the influence of coir fiber length on coir-epoxy composites. With the fiber length increasing, it helped improve composites resistance to impact as well as compatibility between fibers and matrix (*77*). However, the fiber loading level, compared to fiber length, plays a major role on the improvement of tensile, flexural and impact strength of coir-polyester composites (*78*). In the application of automotive interior panel, coir-PP composites were prepared with different coir loading levels combining with 3 wt% MA-g-PP as coupling agents. The water resistance and IB strength were negatively influenced while the flexural and tensile strength, the hardness and the flame resistance were improved by increasing the coir content (*10*).

Table V. Some examples of composites reinforced by coir

Fiber loading wt%	*Matrix*	*Modification method*	*Additives*	*Procedure*	*Application*	*References*
40,50,60,70	PP	n/a	Coupling agent (MAPP)	Hot pressing	Panel	(*10*)
5	Epoxy	Bromination and Stannous chloride treatment	Hardener (Amine)	Cold molded	Fire retardant material	(*85*)
0,5,10,15,20	Aluminium; Silicon carbide; Graphite; Aluminium oxide; Zirconia oxide	n/a	Zirconia oxide; Paper ash	Powder metallurgy technique	Brake pad	(*79*)
10,15,25	Polyester	n/a	Low-melting point polyester fiber	Needle-punching technique and thermal bonding	Insulation board	(*84*)
30 (vl %)	Unsaturated polyester derived from PET waste	Silane on alkalized coconut	Initiator (cobalt octoate); Accelerator (MEKPO)	Curing	Recycled composites	(*83*)
10,15,20,25	PP	Oxidation and Coupling reaction	n/a	Injection molded	n/a	(*82*)

n/a: Not applicable; PP: Polypropylene; MAPP: Maleic anhydride modified polypropylene; PET: Polyethylene Terephthalate; MEKPO: Methyl ethyl ketone peroxide.

Due to the good wear resistance, coir was applied to the development of asbestos-free brake pad materials in vehicles. Coir powders were utilized as reinforcement in aluminum matrix along with silicon carbide (abrasive material), graphite (solid lubricant), zirconium oxide (friction modifier) and the resin (binder). The composite brake pad containing 5 wt% of coir powder had the promising physical and mechanical properties in terms of porosity, hardness, compressive strength, wear weight loss and thickness loss. Thus, coir fiber is a potential candidate material for the mass-scale fabrication of asbestos-free brake pads (*79*).

Coir fiber is a good source to cellulose nanowhiskers for the potential nanocomposite applications. The cellulose nanowhiskers were prepared by tree pretreatments eliminating non-cellulosic components, hemicellulose and lignin in combination with acid hydrolysis. The route of benzene-ethanol extraction followed by sodium chlorite and KOH pretreatments was the best to obtain cellulose nanowhiskers with the highest crystallinity index, degree of polymerization, and thermal stability (*80*).

Common to other natural fibers, the use of coir fibers as reinforcement in resin matrix composites is limited by the poor interfacial adhesion between fibers and matrices. A two-step chemical treatment was prepared. Firstly, fibers were soaked in sodium sulfite ($NaSO_3$) removing lignin from the surface. Secondly, fibers were treated in a mixture of acetic anhydride and sulfuric acid to reduce the number of free hydroxyl groups of cellulose. It is observed by SEM and FTIR that the reduction of polarity on coir surface and the removal of the lignin rich outer surface layer after the treatments (*81*). Another treatment containing oxidization by $NaIO_4$ and coupling reaction with p-aminophenol was prepared to enhance interfacial adhesion between coir fibers and PP matrix. The hydrophilic nature of coir is significantly reduced upon chemical treatments. Therefore, fiber-matrix polarity gap has been largely minimized leading to improvement in interfacial adhesion and mechanical properties compared with untreated coir reinforced PP composites (*82*). A biological treatment utilizing two organisms, pseudomonas putida and phanerochaete chrysosporium was introduced to treat unretted coir. The treatment partially degraded lignin and helped produce whiter and softer coir fiber having better tensile strength and elongation properties than chemically treated ones (*43*). Application of saline treatment on alkalized coconut fiber improved the tensile and impact strength and lowered the water uptake of coir reinforced unsaturated polyester composites. The unsaturated polyester resin was prepared by recycling polyethylene terephthalate (PET) waste through glycolysis and polyesterification reactions (*83*).

Coir has the lowest thermal conductivity as well as good acoustic resistance due to its porous structures compared to other nature fibers. Thus it has the potential applications in thermal and acoustic insulation. In the research of Huang et al., coir fiber was laminated with 2D-PETF and 12D-PETF respectively by needle punching techniques to prepare two types of coir fiber-PET composite boards. Both composite boards possess excellent thermal and acoustic insulation as well as fire resistant properties with the increasing content of coir fiber (*84*). The treatments of de-waxed coir fibers were carried out in saturated bromine water and stannous chloride solution to impart coir fibers fire resistant properties as bromine and chorine compounds are the most regular halogen-containing fire retardants. The treated coir fibers were then ground to nano-size and compounded with epoxy resin. The obtained composites presented significantly improved fire resistant properties (*85*).

The mechanical and machinable characteristics of coir-polyester and coir woven fabric reinforced polyester composites were investigated through the analysis of nonlinear mathematical equations correlating thrust force, torque and tool wear parameters. This can be used to find the optimum values of machining parameters for drilling to reduce tool wear and prediction of machinability characteristics in the field of natural fiber reinforced composites (*86*, *87*).

With the addition of coir fiber into concrete system as reinforcement, it contributed to enhanced compressive strength, tensile and flexural performance as well as better resistance to sulphate attack while the rate of increments was lower than conventional concrete specimen at later curing ages (*88*).

Banana Fiber

Banana fiber is a lignocellulosic bast fiber with relatively good mechanical properties. The banana consists of 43.46% cellulose, 38.54% hemicellulose, 9% lignin, and 9% others (*13*). The fibers extracted from various varieties of banana plants present different physical-chemical properties (*89*). There have been a number of reports about the use of banana fibers as reinforcing components in polymer matrices.

In the case of banana fiber reinforced polyester composites, at certain levels of fiber loading (less than 19 wt %), the composite properties were inferior to that of the pure polyester matrix. However, when the composites with long fiber strand and at sufficiently high content of 30 wt%, the flexural strength reached to 97 MPa and the flexural elastic modulus upgraded to 6.5 GPa as well as 1.6 times increase in fracture toughness over the value of the neat resin (*90*). Zaman et al. compared physico-mechanical properties of banana fiber reinforced PP composites with various treatments. The composites containing 40 wt% of banana fiber without treatments showed improved mechanical properties compared to neat PP sheet in terms of tensile strength, tensile modulus and impact strength. Furthermore, UV radiation was carried out to irradiate the fibers and the matrix, which resulted in further improvement on mechanical properties. Subsequently, the optimized banana fibers (irradiated by 75 UV passes) were soaked in 2-hydroxyethyl methacrylate (HEMA) monomer solutions along with methanol and benzyl peroxide. Significant improvement on mechanical properties was observed after the treatment (*13*). In a study of MDF boards, bananas stems and the mid rib of banana leaves were compounded with UF. Compared with the market MDF boards, both banana stem and the mid rib reinforced MDF boards performed better in dimensional stability (WA, TS) and mechanical properties (MOR, MOE), and the former was more efficient in property improvement (*91*).

Banana fibers were also successfully introduced to biodegradable composites (*92*, *93*). Various banana fibers extracted from different varieties of banana trees, were prepared to reinforce tamarind seed gum from the endosperm of roasted seeds of the tamarind tree. The Red banana fiber composites possess the highest tensile strength while Poovan fiber composites show the lowest. In addition, the investigation on fire retardant property of tamarind seed gum composites revealed that untreated and varnish coated banana fibers contributed to good fire retardant characteristics (*92*). In the case of PLA based biocomposites as shown in Table VI, the Bis-(3-triethoxy silyl -propyl) tetrasulfane (Si69) surface treated banana fibers were incorporated with PLA at different fiber loading levels. Biocomposites having 30 wt% of fiber content showed better mechanical properties by means of tensile strength and modulus. The surface treated composites possessed superior mechanical properties and better thermal stability due to improved interfacial addition between fibers and matrix (*93*).

Corn Stalk

Corn stalk, one of agricultural residues available in large quantity all over the world, can be made into composite boards with reasonable properties for a variety of applications in ceiling panels, bulletin boards and core materials. Bavan and Kumar concluded that maize stalk fibers are appreciable as reinforcement to composites due to their good morphological features with favorable thermal degradation properties that can withstand the polymer environment (*94*). Thamae et al. developed a composite material consisting of corn stalk outer ring and wasted linear low-density polyethylene (LLDPE). The results indicated that the mechanical properties of composites either reinforced with whole corn stalk or with outer rings are quite similar in terms of flexural properties (*12*). In the case of PP-corn stalk composites as shown in Table VI, the composites containing 40 wt% of corn stalk possess the best tensile and flexural properties with 2.5 wt% Eastman G-3003 compatibilizing agents (*95*). Babatunde studied cement-bonded particleboards of 6 mm thickness using maize stalk particles at three levels of board density and additive concentrations respectively. With increasing board density and additive concentrations, the MOR, MOE increased while the TS and WA decreased. It is concluded that maize stalk particles are suitable materials for cement-bonded particleboards. (*14*).

Pineapple Leaf Fiber

Pineapple leaf fiber (PALF) is the waste from pineapple plant cultivation. PALF is very hygroscopic and relatively inexpensive, an important natural fiber that exhibits high specific strength and stiffness. The potential of PALF as reinforcing fibers in both thermosetting and thermoplastic resins has attracted interest for their excellent specific properties.

Wan Nadirah et al. carried out a study on morphological characteristics, thermal and crystalline properties of PALF and it concluded that PALF has potentials to be as a raw material in composites fabrication (*28*). In the case of PALF reinforced poly (hydroxybutyrate-co-valerate) (PHBV) green composites, the fibers were compounded with PHBV without additional treatments as shown in Table VI. Although both tensile and flexural properties of the composites were better than neat PHBV, the SEM photomicrographs showed an adhesive failure of the interface. Furthermore, the photomicrographs in tensile mode showed partial fibers pull-out indicating weak bonding between the fiber and the matrix. This suggests that fiber surface ought to be treated to improve interfacial adhesion (*11*).

Munawar et al. studied the effects of alkali, mild steam and chitosan treatments on the mechanical properties and the morphological characteristics of the PALF based composites. The properties are largely influenced by treatments and the mild steam treatment resulted in the best improvement of properties compared to other treatments (*96*). George et al. studied PALF reinforced low density polyethylene (LDPE) matrix. It was found that chemical treatments imparted composites better properties than untreated system due to good interfacial adhesion between the matrix and fibers (*97*). Shih et al. prepared sol-gel-modified PALF and PLA composites with coupling agents. The alkali and 3-trieth-oxysilyl propyl isocyanate modified fibers were subsequently treated by tetraethyloxysilan (TEOS) and 3-glycidyloxypropyl trimethoxysilane coupling agents. The composites containing 5 wt% modified PALF presented the best interfacial adhesion, heat resistance and mechanical properties (*98*).

Table VI. Some examples of composites reinforced by banana fiber/corn stalk/pineapple leaf fiber

Fiber loading wt%	*Matrix*	*Modification method*	*Additives*	*Procedure*	*Application*	*References*
80 bananas stem; 80 mid rib of banana leaf	UF	NaOH treatment and refining	n/a	Hot pressing	MDF board	(*91*)
10,20,30,40 (Banana fiber)	PLA	NaOH treatment and silanes	n/a	Sheet hot pressing	Biocomposites	(*93*)
30,40,50 (Corn stalk Flour)	PP	n/a	MAPP	Injection molded	n/a	(*95*)
25 (surface layer); 50 (middle layer) (PALF)	PHBV	n/a	n/a	Hot pressing	Biocomposites	(*11*)
5 (PALF)	PLA	Alkali and silane coupling treatment; Sol-gel-modification by TEOS	n/a	Compression molding	Biocomposites	(*99*)

MDF: Medium density fiber; UF: Urea formaldehyde; n/a: Not applicable; PLA: Poly lactic acid; PP: Polypropylene; MAPP: Maleic anhydride grafted polypropylene; PHBV: poly (hydroxybutyrate-co-valerate); PALF: Pineapple leaf fiber; TEOS: tetraethyloxysila.

Conclusion

Development of lightweight composites reinforced by agricultural byproducts is cost effective and environmentally friendly from the economical and environmental points of view. The agricultural byproducts, such as wheat and rice straw, rice husk, rice husk ash, bagasse, coir fiber, banana fiber, corn stalk, and pineapple leaf fiber are abundant in nature, biodegradable, annually renewable and cheap. Their lignocellulosic and ligntweight properties make them appreciable reinforcement materials for lightweight composites. In various applications, agro-byproducts can be made into fibers, particles, ash and extracted nanofibers and then compounded with thermoplastic, thermosetting and biodegradable polymers. Application of agro-byproducts improves mechanical properties such as tensile strength and flexural strength. Particleboard is one area in which agro-byproducts could be of interest. Agro-straw, sugarcane bagasse, coir, banana leaf fiber and corn stalk have been successfully introduced as raw materials to fabricate high performance boards with conventional UF resin. Besides particleboards, some agro-fibers such as coir with the lowest thermal conductivity as well as good acoustic resistance provide the possibility to manufacture a wide range of coir-based composites for fire retardant, thermal and acoustic insulation. Automotive industry also finds the interest in utilizing agro-based composites for the interior due to their superior mechanical and acoustic performance. Products derived from agro-byproducts such as rice husk ash are suitable as supplementary materials to concrete composites by partially replacing ordinary Portland cement resulting in enhanced resistance to chemical erosion and improved durability. The rich content of cellulose makes agro-byproducts a potential source of cellulosic nanofibers as reinforcement to nanocomposites for packing industry and medical applications.

Although many promising achievements have been made in recent years, there are several challenges that need to be addressed in order to produce agro-based composites on industrial scales. A major issue is the incompatibility between the hydrophilic agro-fibers and hydrophobic polymer matrix. The poor interfacial adhesion leads to undesirable mechanical properties. Thus, fiber modifications and the addition of adhesives are essential to achieve improved mechanical properties. Another obstacle is the low thermal stability of agro-fibers compared with polymer matrices that are used in the composites. This limits the type of polymers that can be used with agro-fibers and the processing temperature. Consequently, more research is needed on fiber modification, introduction of novel polymers to agro-fibers and production of economical adhesives if agro-fiber based lightweight composites are to reach their full potential.

References

1. Santana-Méridas, O.; González-Coloma, A.; Sánchez-Vioque, R. *Phytochem. Rev.* **2012**, *11*, 447–466.
2. Reddy, N.; Yang, Y. *Trends Biotechnol.* **2005**, *23*, 22–7.
3. Ismail, M. R.; Yassen, A. A. M.; Afify, M. S. *Fibers Polym.* **2011**, *12*, 648–656.

4. White, N. M.; Ansell, M. P. *J. Mater. Sci.* **1983**, *18*, 1549–1556.
5. Han, G.; Zhang, C.; Zhang, D.; Umemura, K.; Kawai, S. *J. Wood Sci.* **1998**, *44*, 282–286.
6. El-Saied, H.; Basta, A. H.; Hassanen, M. E.; Korte, H.; Helal, A. *J. Polym. Environ.* **2012**, *20*, 838–847.
7. Azadi, M.; Bahrololoom, M. E.; Heidari, F. *J. Coat. Technol. Res.* **2010**, *8*, 117–123.
8. Carvajal, O.; Valdés, J. L.; Puig, J. *Holz Roh- Werkst.* **1996**, *54*, 61–63.
9. Lee, S.; Shupe, T. F.; Hse, C. Y. *Holz Roh- Werkst.* **2005**, *64*, 74–79.
10. Ayrilmis, N.; Jarusombuti, S.; Fueangvivat, V.; Bauchongkol, P.; White, R. H. *Fibers Polym.* **2011**, *12*, 919–926.
11. Luo, S.; Netravali, A. N. *J. Mater. Sci.* **1999**, *34*, 3709–3719.
12. Thamae, T.; Marien, R.; Chong, L.; Wu, C.; Baillie, C. *J. Mater. Sci.* **2008**, *43*, 4057–4068.
13. Zaman, H. U.; Khan, M. A.; Khan, R. A. *Fibers Polym.* **2013**, *14*, 121–126.
14. Babatunde, A. *J. Forst. Res.* **2011**, *22*, 111–115.
15. Ramasamy, V. *KSCE J. Civ. Eng.* **2011**, *16*, 93–102.
16. Huda, S.; Yang, Y. *Macromol. Mater. Eng.* **2008**, *293*, 235–243.
17. Miao, C.; Hamad, W. Y. *Cellulose* **2013**, *20*, 2221–2262.
18. Zou, Y.; Huda, S.; Yang, Y. *Bioresour. Technol* **2010**, *101*, 2026–2033.
19. Zou, Y.; Reddy, N.; Yang, Y. *J. Appl. Polym. Sci.* **2010**, *116*, 2366–2373.
20. Li, X.; Tabil, L. G.; Panigrahi, S. *J. Polym. Environ.* **2007**, *15*, 25–33.
21. Holbery, J.; Houston, D. *JOM* **2006**, *58*, 80–86.
22. Ndazi, B.; Tesha, J. V.; Bisanda, E. T. N. *J. Mater. Sci.* **2006**, *41*, 6984–6990.
23. Robson, S.; Goodhead, T. C. *J. Mater. Process. Technol.* **2003**, *139*, 327–331.
24. Han, G.; Kawai, S.; Umemura, K.; Zhang, M.; Honda, T. *J. Wood Sci.* **2001**, *47*, 350–355.
25. Jamil, M. S.; Ahmad, I.; Abdullah, I. *J. Polym. Res.* **2006**, *13*, 315–321.
26. Mishra, P.; Chakraverty, A.; Banerjee, H. D. *J. Mater. Sci.* **1986**, *21*, 2129–2132.
27. Rajamani, D.; Surender, R.; Mahendran, A.; Muthusubramanian, S.; Vijayakumar, C. T. *J. Therm. Anal. Calorim.* **2013**, *114*, 883–893.
28. Wan Nadirah, W. O.; Jawaid, M.; Al Masri, A. A.; Abdul Khalil, H. P. S.; Suhaily, S. S.; Mohamed, A. R. *J. Polym. Environ.* **2011**, *20*, 404–411.
29. Nasri-Nasrabadi, B.; Behzad, T.; Bagheri, R. *Fibers Polym.* **2014**, *15*, 347–354.
30. Battegazzore, D.; Bocchini, S.; Alongi, J.; Frache, A.; Marino, F. *Cellulose* **2014**, *21*, 1813–1821.
31. Gilfillan, W. N.; Moghaddam, L.; Doherty, W. O. S. *Cellulose* **2014**, *21*, 2695–2712.
32. Hassan, M. L.; Hassan, E. A.; Oksman, K. N. *J. Mater. Sci.* **2010**, *46*, 1732–1740.
33. Nimz, H. H.; Casten, R. *Holz Roh- Werkst.* **1986**, *44*, 207–212.
34. Pereira, P. H. F.; Voorwald, H. J. C.; Cioffi, M. O. H.; Silva, M. L. C. P.; Rego, A. M. B.; Ferraria, A. M.; Pinho, M. N. *Cellulose* **2013**, *21*, 641–652.

35. Gomez-Bueso, J.; Westin, M.; Torgilsson, R.; Olesen, P. O.; Simonson, R. *Holz Roh- Werkst.* **1999**, *57*, 433–438.
36. Hematabadi, H.; Behrooz, R. *J. Forst. Res.* **2012**, *23*, 497–502.
37. Hornsby, P. R.; Hinrichsen, E.; Tarverdi, K. *J. Mater. Sci.* **1997**, *32*, 1009–1015.
38. Shen, J.-h.; Liu, Z.-m.; Li, J.; Niu, J. *J. Forst. Res.* **2011**, *22*, 107–110.
39. Verma, D.; Gope, P. C.; Shandilya, A. *J. Inst. Eng. India, Ser. D* **2014**, *95*, 27–34.
40. Zhang, Y.; Lu, X.; Pizzi, A.; Delmotte, L. *Holz Roh- Werkst.* **2003**, *61*, 49–54.
41. Hosseini, S. M.; Aziz, H. A.; Syafalni; Kiamahalleh, M. V. *KSCE J. Civ. Eng.* **2013**, *17*, 921–928.
42. Ludueña, L. N.; Vecchio, A.; Stefani, P. M.; Alvarez, V. A. *Fibers Polym.* **2013**, *14*, 1118–1127.
43. Rajan, A.; Senan, R. C.; Pavithran, C.; Abraham, T. E. *Bioprocess Biosyst. Eng.* **2005**, *28*, 165–73.
44. Panthapulakkal, S.; Sain, M. *J. Polym. Environ.* **2006**, *14*, 265–272.
45. Hornsby, P. R.; Hinrichsen, E.; Tarverdi, K. *J. Mater. Sci.* **1997**, *32*, 443–449.
46. Farsi, M. *Fibers Polym.* **2012**, *13*, 515–521.
47. Nazerian, M.; Sadeghiipanah, V. *J. Forst. Res.* **2013**, *24*, 381–390.
48. Nazerian, M.; Kamyab, M. *For. Sci. Pract.* **2013**, *15*, 325–331.
49. Liu, J.; Zhou, H.; Ouyang, P. *J. Wuhan Univ. Technol.* **2013**, *28*, 508–513.
50. Boquillon, N.; Elbez, G. r.; Schönfeld, U. *J. Wood Sci.* **2004**, *50*, 230–235.
51. Grigoriou, A. H. *Wood Sci. Technol.* **2000**, *34*, 355–365.
52. Papadopoulos, A. N. *J. Indian Acad. Wood Sci.* **2011**, *7*, 54–57.
53. Han, G.; Umemura, K.; Wong, E.; Zhang, M.; Kawai, S. *J. Wood Sci.* **2001**, *47*, 18–23.
54. Ismail, M. R.; Yassene, A. A. M.; Abd El Bary, H. M. H. *Appl. Compos. Mater.* **2011**, *19*, 409–425.
55. Avella, M.; Rota, G. L.; Martuscelli, E.; Raimo, M.; Sadocco, P.; Elegir, G.; Riva, R. *J. Mater. Sci.* **2000**, *35*, 829–836.
56. Wu, C.-S.; Liao, H.-T.; Jhang, J.-J.; Yeh, J.-T.; Huang, C.-Y.; Wang, S.-L. *Polym. Bull.* **2013**, *70*, 3221–3239.
57. Yussuf, A. A.; Massoumi, I.; Hassan, A. *J. Polym. Environ.* **2010**, *18*, 422–429.
58. Xu, Y.; Wu, Q.; Lei, Y.; Yao, F.; Zhang, Q. *J. Polym. Environ.* **2008**, *16*, 250–257.
59. Ndazi, B.; Tesha, J. V.; Karlsson, S.; Bisanda, E. T. N. *J. Mater. Sci.* **2006**, *41*, 6978–6983.
60. Kim, H. S.; Yang, H. S.; Kim, H. J.; Lee, B. J.; Hwang, T. S. *J. Therm. Anal. Calorim.* **2005**, *81*, 299–306.
61. Kim, H. S.; Yang, H. S.; Kim, H. J.; Park, H. J. *J. Therm. Anal. Calorim.* **2004**, *76*, 395–404.
62. Yang, H. S.; Wolcott, M. P.; Kim, H. S.; Kim, H. J. *J. Therm. Anal. Calorim.* **2005**, *82*, 157–160.
63. Rosa, S. L.; Nachtigall, S. B.; Ferreira, C. *Macromol. Res.* **2009**, *17*, 8–13.
64. Chandrasekhar, S.; Satyanarayana, K. G.; Pramada, P. N.; Raghavan, P.; Gupta, T. N. *J. Mater. Sci.* **2003**, *38*, 3159–3168.

65. Abalaka, A. E. *Int. J. Concr. Struct. Mater.* **2013**, *7*, 287–293.
66. Souza Rodrigues, C.; Ghavami, K.; Stroeven, P. *Waste Biomass Valorization* **2010**, *1*, 241–249.
67. Hashem, F. S.; Amin, M. S.; El-Gamal, S. M. A. *J. Therm. Anal. Calorim.* **2012**, *111*, 1391–1398.
68. Bui, L. A.-t.; Chen, C.-t.; Hwang, C.-l.; Wu, W.-s. *Int. J. Miner., Metall. Mater.* **2012**, *19*, 252–258.
69. Rattanasak, U.; Chindaprasirt, P.; Suwanvitaya, P. *Int. J. Miner., Metall. Mater.* **2010**, *17*, 654–659.
70. Wirawan, R.; Sapuan, S. M.; Robiah, Y.; Khalina, A. *J. Therm. Anal. Calorim.* **2010**, *103*, 1047–1053.
71. Cheng, Q.; Tong, Z.; Dempere, L.; Ingram, L.; Wang, L.; Zhu, J. Y. *J. Polym. Environ.* **2012**, *21*, 648–657.
72. Vilakati, G. D.; Mishra, A. K.; Mishra, S. B.; Mamba, B. B.; Thwala, J. M. *J. Inorg. Organomet. Polym Mater.* **2010**, *20*, 802–808.
73. Dahmardeh Ghalehno, M.; Nazerian, M.; Bayatkashkooli, A. *Eur. J. Wood Wood Prod.* **2010**, *69*, 533–535.
74. Mothé, C. G.; de Araujo, C. R.; de Oliveira, M. A.; Yoshida, M. I. *J. Therm. Anal. Calorim.* **2002**, *67*, 305–312.
75. Wang, L.; Tong, Z.; Ingram, L. O.; Cheng, Q.; Matthews, S. *J. Polym. Environ.* **2013**, *21*, 780–788.
76. Nazerian, M.; Eghbal, S. H. *J. Indian Acad. Wood Sci.* **2013**, *10*, 86–94.
77. Biswas, S.; Kindo, S.; Patnaik, A. *Fibers Polym.* **2011**, *12*, 73–78.
78. Jayabal, S.; Natarajan, U. *Int. J. Adv. Des. Manuf. Technol.* **2010**, *54*, 639–648.
79. Maleque, M. A.; Atiqah, A. *Arabian J. Sci. Eng.* **2013**, *38*, 3191–3199.
80. Fahma, F.; Iwamoto, S.; Hori, N.; Iwata, T.; Takemura, A. *Cellulose* **2010**, *18*, 443–450.
81. Calado, V.; Barreto, D. W.; D'Almeida, J. R. M. *J. Mater. Sci. Lett.* **2000**, *19*, 2151–2153.
82. Haque, M. M.; Islam, M. S.; Islam, M. N. *J. Polym. Res.* **2012**, *19*.
83. Abdullah, N. M.; Ahmad, I. *Fibers Polym.* **2013**, *14*, 584–590.
84. Huang, C.-H.; Lin, J.-H.; Lou, C.-W.; Tsai, Y.-T. *Fibers Polym.* **2013**, *14*, 1378–1385.
85. Sen, A. K.; Kumar, S. *J. Therm. Anal. Calorim.* **2010**, *101*, 265–271.
86. Jayabal, S.; Natarajan, U. *Bull. Mater. Sci.* **2011**, *34*, 1563–1567.
87. Jayabal, S.; Velumani, S.; Navaneethakrishnan, P.; Palanikumar, K. *Fibers Polym.* **2013**, *14*, 1505–1514.
88. Sivaraja, M.; Kandasamy, ; Velmani, N.; Pillai, M. *Bull. Mater. Sci.* **2010**, *33*, 719–729.
89. Kiruthika, A. V.; Veluraja, K. *Fibers Polym.* **2009**, *10*, 193–199.
90. Zhu, W. H.; Tobias, B. C.; Coutts, R. S. P. *J. Mater. Sci. Lett.* **1995**, *14*, 508–510.
91. Rashid, M. M.; Das, A. K.; Shams, M. I.; Biswas, S. K. *J. Indian Acad. Wood Sci.* **2014**, *11*, 1–4.
92. Kiruthika, A. V.; Priyadarzini, T. R. K.; Veluraja, K. *Fibers Polym.* **2012**, *13*, 51–56.

93. Jandas, P. J.; Mohanty, S.; Nayak, S. K. *J. Polym. Environ.* **2012**, *20*, 583–595.
94. Saravana Bavan, D.; Mohan Kumar, G. C. *Fibers Polym.* **2012**, *13*, 887–893.
95. Nourbakhsh, A.; Hosseinzadeh, A.; Basiji, F. *J. Polym. Environ.* **2011**, *19*, 908–911.
96. Munawar, S. S.; Umemura, K.; Tanaka, F.; Kawai, S. *J. Wood Sci.* **2007**, *54*, 28–35.
97. George, J.; Bhagawan, S. S.; Thomas, S. *J. Therm. Anal.* **1996**, *47*, 1121–1140.
98. Shih, Y.-F.; Huang, R.-H.; Yu, Y.-H. *J. Sol-Gel Sci. Technol.* **2014**, *70*, 491–499.
99. Kim, K.-W.; Lee, B.-H.; Kim, H.-J.; Sriroth, K.; Dorgan, J. R. *J. Therm. Anal. Calorim.* **2011**, *108*, 1131–1139.

Chapter 13

Biopolymer-Based Lightweight Materials for Packaging Applications

Bin Hu*

Thayer School of Engineering, Dartmouth College, 14 Engineering Dr., Hanover, New Hampshire 03755, United States
***E-mail: bin.hu@dartmouth.edu**

This chapter offers an overview of biopolymer-based lightweight materials with respect to their potential for packaging applications. Biopolymer-based materials have garnered increasing attention from packaging markets due to concerns in recent years from both environmental and economic perspectives of traditional petroleum-based polymers. The extensive use of these traditional synthetic polymers has already resulted in serious ecological problems. Recently, biopolymers have been increasingly found in applications such as food, pharmaceutical, and consumer goods packaging, since they are from or partially from renewable resources and potential to be biodegradable and/or compostable. In this review, the classification of biopolymers is introduced according to their production methods or origin resources of materials. Then, the advantages and challenges of these biopolymers over conventional synthetic polymers are compared. Several important biopolymers such as polylactic acid (PLA), starch, and cellulose are discussed in details. The chemical structures, mechanical properties, thermal properties, and recycling of these biopolymers are summarized. In the second part of this review, biofibers, and biocomposites are also discussed since they are important biomaterials for packaging applications.

1. Introduction

Almost all the consumer goods come with packaging. The packaging provides the functions of containing, protecting, preserving, transporting and informing for products. A wide range of materials can be used for packaging applications such as metal, glass, wood, paper, cotton, or polymer based materials. Among them, petroleum-based polymers such as polyethylene terephthalate (PET), polyvinylchloride (PVC), polyethylene (PE), polypropylene (PP), polystyrene (PS), and polyamide (PA) have been used widely as packaging materials because of their light weight, low cost, good mechanical performance, good barrier properties, heat stability, easy-to-process etc.. These properties are coming from their unique microstructures with many repeated units chained together. The final forms of the polymer based packing are bottles, bags, containers, films, foams, coating, industrial wrapping, etc..

However, these polymers are the least recycled packaging materials compared to metals, papers, and glass. Many of these polymers will end up in landfill sites, where they will remain for centuries before full degradation. The wide use of these polymers has already resulted in serious ecological problems. These packaging materials are often contaminated with foodstuffs and biological substances, and therefore recycling is uneconomical and impracticable. The growing environmental concerns impose to packaging materials with ability to be biodegradable and compostable. Moreover, the price of the conventional synthesis polymers relies on the price of petroleum, which has been increasing recently. Therefore, these concerns from both environmental and economic perspectives bring the opportunity of biopolymer-based materials using for packaging applications (*1–6*).

Recently, several biopolymers-based materials have been introduced as the substitutes of petroleum based polymers, such as polylactic acid (PLA), poly-3-hydroxybutyrate (PHB), starch, and cellulosic, etc. Most of them are designed for the packaging applications. These biopolymer-based materials are expected to reduce the environmental impact and lower the dependence on non-renewable resources (*3*).

Most biopolymers-based materials made from renewable resources are biodegradable and especially compostable. They are used increasingly in applications such as food, pharmaceutical, and consumer goods packaging. Composting, which allows disposal of the packages in the soil, is becoming one of the prevailing methods for disposal of these biopolymers-based packaging waste. The packaging waste produced only water, carbon dioxide, and inorganic compounds during biological degradation.

So far, significant technological development has been achieved to produce biopolymer and biocomposites for packaging applications with comparable properties and functionalities compared to those of traditional petroleum-based packaging. Although, the current production cost is still high, many of them have found increasing commercial applications in packaging field.

Lightweight materials are materials with reduced weight. They are important for the packaging industry since they can reduce the cost of materials and transport, and therefore reduce the waste and energy used. Lightweight can be achieved by using low density materials, and by designing novel thin film or foamed structures.

2. Overview of Biopolymer-Based Materials

Biopolymers are polymers produced by living organisms or derived from biomass. They contain monomers which are covalently bonded to form macromolecules. Cellulose, starch and chitin, proteins and polylactic acid (PLA) are all examples of biopolymers in which the monomer units are sugars, amino acids, and lactic acid, respectively (*4*, *7*).

Biopolymers derived from renewable resources are classified into three main categories according to the production methods. Figure 1 shows a schematic overview of the classification of biopolymers based on their origin (*8*, *9*).

1. Biopolymer extracted from natural materials. Examples are polysaccharides such as starch, cellulose, and proteins etc.
2. Biopolymer produced through classical chemical synthesis method from renewable monomers. One example is PLA, which is polymerized from lactic acid obtained by fermentation from starch.
3. The biodegradable polymers obtained via bacterial biosynthesis of natural materials (polyesters of polysaccharides). The best-known biopolymer produced by this method is polyhydroxyalkanoates (PHAs) and polyhydroxybutyrate (PHB) (*8*, *10*, *11*).

In recent years, a number of companies have introduced different types of biopolymers. Table 1 lists their manufactures, brands and main packaging applications (*10*, *12*). According to the materials they are manufactured from, these recent developed biopolymers can also be classified as starch polymers (e.g., Mater-Bi), cellulosic (e.g., Cellophane), aliphatic polyesters (e.g., PLA), bio-based Polyethylene (Bio-PE), and microbial synthesized polyhydroxyalkanoates (e.g., poly-3-hydroxybutyrate (PHB)). The examples given are designed for packaging applications, or with the potential to be used in packaging applications in the near future (*10*, *12*).

As discussed above, biopolymers can be divided into several categories depending on their sources or materials. Among these biopolymers, PLA has the largest impact on industries as a packaging material. However, a considerable amount of research has also been focused on potential packaging application of starch, cellulosic, and PHA, etc. (*11*, *13*).

From a practical point of view, biopolymers have two main advantages over conventional synthetic polymers for packing applications: 1) biopolymers are biodegradable and/or compostable. 2) Biopolymers are available from renewable sources (*13*). After their useful life, it is desirable for the packaging materials to biodegrade in a reasonable time period without causing environmental problems. In this sense, biopolymer-based packaging materials have some beneficial properties over traditional synthetic packaging materials. Biopolymers also reduce the further dependency of packaging materials on petroleum reserves, which have uncertainty of future supply (*3*, *14*, *15*).

There are also challenges using biopolymers in packaging applications. These challenges are related to either processability or the final properties of packaging materials. Biopolymers have disadvantages of brittleness and poor thermal stability. One of the drawbacks of processing PLA in the molten state is that it tends to undergo thermal degradation. Another major challenge is that it is difficult to achieve mechanical and barrier properties as durable as traditional synthetic polymers to match the shelf-life of products while maintaining biodegradability (*8*). These challenges are expected to be overcome by blending biopolymer with other polymers, making nanocomposites, coating with high barrier materials, and/or polymer modifications (*16*). Apart from the technical considerations such as processability and physical properties, biopolymers also have a relatively high cost and the recycling of them also presents some challenges because of sorting and cleaning requirements. So far, the cost still limited the wide adoption of biopolymer-based packaging materials such as PHA and cellulose acetate etc. However, this limitation will fall down as the production capacity increases (*10*, *11*).

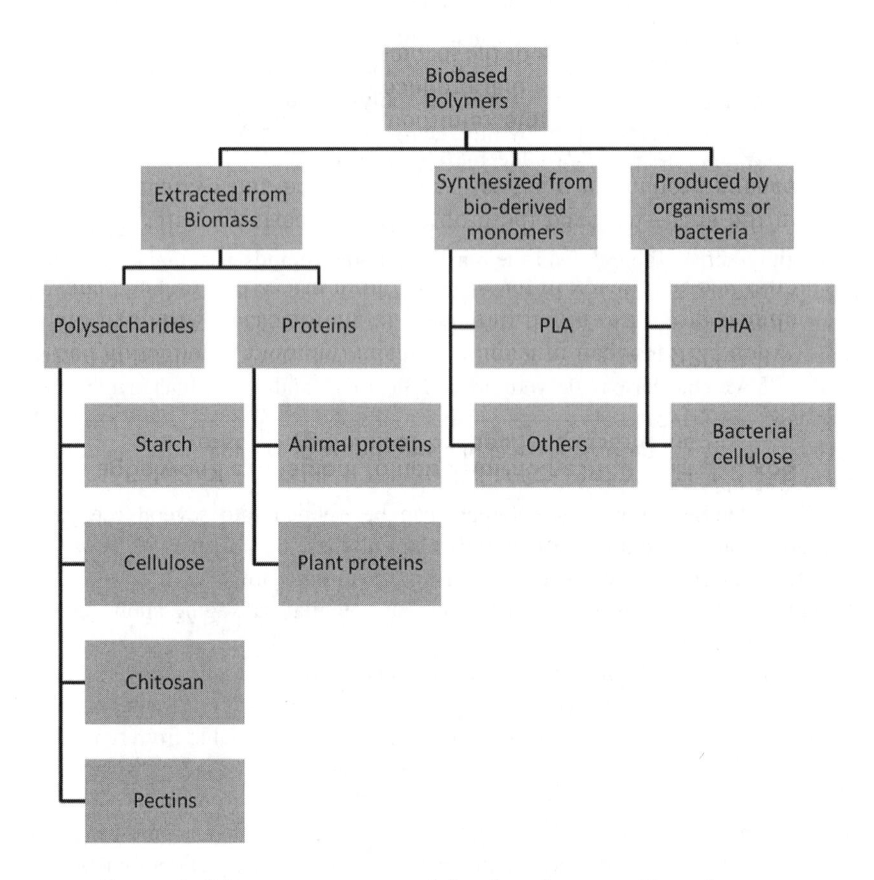

Figure 1. Schematic overview of the classification of biopolymers.

Table 1. Biopolymers and their manufactures

Type of polymers	*Manufactures*	*Brand*	*Applications*
Starch	Dupont	Biomax	Loose fill, bags, films, trays, wrap film
	Biotec	Bioplast	
	Novamount	Mater-Bi	
Cellulosics	Innovia films	Nature Flex	Flexible film
	Eastman Chemical	Tenite	
	FKuR	Biograde	
	Sateri	Sateri	
PLA	BASF	Ecovio	Rigid containers, films, barrier coating
	NatureWorks	Ingeo	
	Cargill Dow	EcoPLA	
	Synbra	Biofoam	
Bio-based PET	Dupont	Biomax	Bottles, trays, films
Bio-based PE	Braskem	Bio-PE	Rigid containers, film wrap, barrier coating
PHA/PHB	Monsano	Biopol	Films, barrier coating, trays
	Biomer	Biomer	

2.1. Important Biopolymers

Polylactic Acid (PLA)

Among numerous kinds of biopolymers, polylactic acid (PLA), sometimes called polylactide, an aliphatic polyester and biocompatible thermoplastic, is currently the most promising and popular material (*16*, *17*).

The monomer unit of PLA, lactic acid, can be manufactured through carbohydrate fermentation or chemical synthesis. The majority of lactic acid is made by bacterial fermentation of carbohydrates. The homofermentative method is used due to high yield of lactic acid and small amount of byproducts in industry (*18*). The sources used for fermentation are simple sugars such as glucose and maltose from corn or potato starch, sucrose from cane/ beet sugar, and lactose from cheese (*6*).

The recent expansion of PLA is due to high molecular weight PLA can be produced in economical methods. PLA can be produced using three methods: 1) direct condensation polymerization, 2) azeotropic dehydrative condensation, 3) polymerization through lactide formation and ring opening polymerization as shown in Figure 2 (*6, 19, 20*). The last method is used for producing commercially available high molecular weight PLA. High molecular weight PLA can be produced based on this method since there is no water generated during the polymerization process. The fist method generates water during each condensation step and results in low molecular weight material due to undesired chain transfer (*6, 21*).

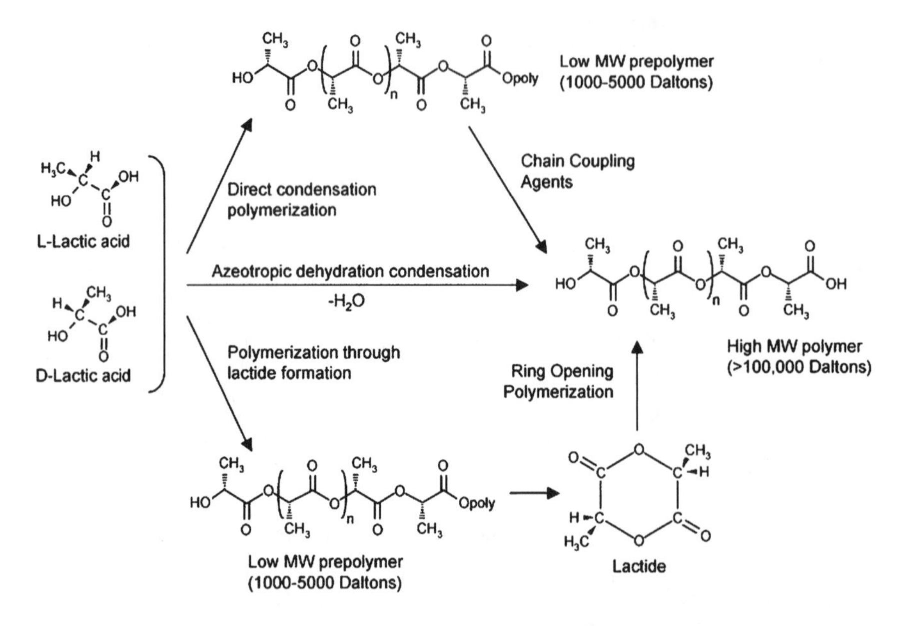

Figure 2. Synthesis of high molecular weight PLA from L- and D-lactic acids. (Reproduced with permission from reference (19), copyright (2008) Elsevier).

PLA has numerous advantages over traditional petroleum based polymers: 1) it can be obtained from a renewable agricultural source (e.g.: corn or potato); 2) it provides significant energy savings; 3) it is recyclable and compostable; 4) it is helpful for improving agricultural economies and 5) their physical and mechanical properties can be tailored through design the different polymer architectures or processing methods (*6*).

In the past, the use of PLA has been limited to biomedical applications due to its bioresorbable characteristics. With the discovery of a new polymerization processes in the past decade, high molecular weight PLA has been economically produced. This expands the use of PLA for packaging applications (*19*). PLA have packaging applications for a broader range of products such as films, foams, food containers, paper coating etc. (*22*)

PLA films have better mechanical performances than PS, but are comparable to those of PET (*6*). PLA films seal well at a temperature lower than melting temperature and shrink at a temperature near their melting temperature. The amount of lactic acid migrating from PLA packaging containers to food is much lower than the amount of lactic acid in food ingredients (*23*). Therefore, PLA is a good material for food packing application (*6, 17, 19, 24, 25*).

Starch

Starch-based materials are the most widespread and economic biomaterials. Starch is composed of two types of molecules: the amylose and the amylopectin. The amylose is linear and helical, while amylopectin are branched. The relative amounts of amylose and amylopectin depend upon the plant sources (wheat, rice, corn, and potato, etc.). The ratio of the two molecules gives starch-based polymers with very different properties. High ratio of amylopectin results in increase of solubility of starch due to the highly branched polymer. In contrast, amlylose is insoluble and hydrolyzed much more slowly (*26*, *27*).

Starch-based polymers, such as Mater-Bi and modified starch, have received great attention in the food packaging applications due to its wide availability, biodegradability, and the low cost (*26*). It can be used as edible and bioresorbable films and coating for packaging applications. However, the starch also presents some drawbacks, such as the strong hydrophilic behavior (poor moisture barrier) and poor mechanical properties when compared to the conventional non-biodegradable polymer based films used in the food packaging industries (*8*).

The morphology and the properties of starch-based polymers can be manipulated easily and efficiently by blending or mixing with synthetic polymers. This approach was successfully adopted by Novamont (*5*). Novamont is one of the leading companies in processing starch-based products under the trademark of Mater-Bi. The blends contain thermoplastic starch that is grafted with polycaprolactone (PCL) to enhance flexibility and moisture resistance. These materials are used for packaging films and sheets. The blends with more than 85 % starch are used for foams and the foams can be used as loose fill to replace PS (*5*).

Cellulose

Cellulose is the most abundant biopolymer on earth and the predominant constituent in cell walls of all plants (*28*). Cellulose is a complex polysaccharide consisting of a linear chain of D-glucose building blocks. Figure 3 shows the molecular structure of cellulose (*28*).

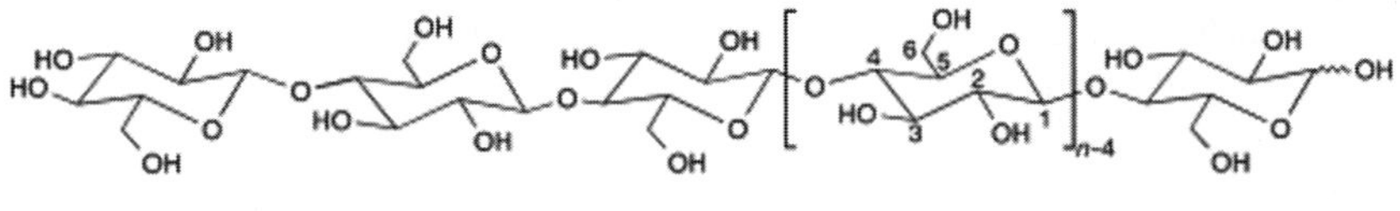

Figure 3. Molecular structure of cellulose (n is the degree of polymerization). (Reproduced with permission from reference (28), copyright (2005) John Wiley and Sons).

The most important raw material sources for the production of cellulosic polymers are cotton fibers and wood. Wood pulp remains the most important raw material for the producing of cellulose. Most of cellulose is used for the production of paper and cardboard. Others are used for the production of cellulose fibers or films and synthesis cellulose esters and ethers. These cellulose derivatives are used for coating laminates, optical films, etc. on industrial scales (*28*).

Cellulose is a polymer and has much more crystalline compared to starch. It is insoluble in aqueous solutions due to tightly packed crystalline structure. This results in a packaging material that is brittle and demonstrates poor flexibility. It is difficult to use cellulose directly in packaging due to its hydrophilic nature. In order to overcome these problems, modification, plasticizing, blending, and coating with other polymers are used. The final mechanical and thermal properties are different depends on the modification methods (*10*, *29*).

Research is focused on the development of cellulose derivatives for its use in packaging applications. Cellulose esters (e. g., cellulose acetate) are one of the cellulose derivatives considered as potentially useful for packaging. Cellulose acetate is synthesized through the reaction of acetic anhydride with cotton linters or wood pulp (*12*). Bipolymer-based cellulose acetate is manufactured by Mazzucchelli (Italy) and Planet polymer (USA) under trade name of BIOCETA and EnviroPlastic, respectively. Both of them are targeted for the manufactures of biodegradable packaging films, containers, and tubes (*12*).

Cellulose can also be converted into a thin transparent film called cellophane. Cellophane can be used for food packaging because of its low permeability to air, oils, greases, bacteria and water (*8*).

Cellulose derivatives are blended with other polymers such as methylcellulose, PCL to lower the water vapor permeability. Adding lipids and mineral or cellulosic fillers may also enhance the water vapor barrier property (*30*).

2.2. Mechanical Properties of Biopolymers

Different packaging applications require different mechanical properties of biopolymers. For the flexible packaging applications such as films, flexibility is the more important property. For the applications of rigid packaging such as containers, the Young's modulus is more critical property.

PLA normally shows higher modulus and lower elongation while starch based biopolymers have higher elongation and lower modulus. Table 2 summarized the mechanical properties of different biopolymers and compared them to petrochemical based polymers (*5*).

Table 2. Mechanical properties of different biopolymers

Polymers	*PLA*	*Mater-Bi*	*PHB*	*Cellophane*	*HDPE*	*LDPE*	*PP*	*PVC*	*PS*
Modulus (MPa)	3500	350~2500	3250	3000	750	100	1500	1500	3150
Elongation (%)	5	20~500	20	22	500	650	600	50	10

2.3. Thermal Stability of Biopolymers

During the conventional polymer melting processing, biopolymers are subject to process at elevated temperatures. Biopolymers must possess adequate thermal stability to prevent degradation and maintain molecular weight and properties. The thermal stability needs to be taken in to consideration.

PLA has similar mechanical properties to conventional polymers. But, the thermal property is not good due to the low T_g of ~60 °C. The melting temperatures of PLA are ranging from 190 to 250 °C. Typically, the processing temperatures are 20~100 °C higher than the melt temperatures. One problem of processing PLA is its tendency to undergo thermal degradation. This problem can be solved by blending PLA with other polymers and adjusting the stereochemistry of the polymers (*16*, *17*).

PHB has a T_g near to the regular ambient temperatures. Biopol (PHB copolymer with 90 % PHB) shows a T_g of 18~22 °C (*31*). PHB shows poor thermal stability at temperatures above the melting point (around 175 °C). Some polymeric additives can increase the thermal stability of PHB (*32*).

Starch and cellulose will thermally decompose at temperature of ~300 °C, which is higher than the decomposition temperature of PLA and PHB. Their thermal decomposition and stability depend on the microstructures, modification and processing conditions (*33*, *34*).

2.4. Recycling and Composting of Biopolymers

Composting is a biological process by which organic material is decomposed by microorganisms into soil-like substances. PLA is one of compostable packaging material, which could loss 45% of its weight in 180 days in thermophilic conditions (*35*).

Biopolymers are different from petroleum-based polymers and they are generally considered unsuitable for conventional recycling, but more suitable for composting. Biopolymers are generally inappropriate for recycling because the biodegradation has been triggered during service life or in the waste stream. They generally have lower thermo-mechanical and chemical resistance than conventional plastics (*3*). Although recycling of many biopolymers could be energetically more favorable than composting, it may not be practical due to sorting and cleaning requirements. Compositing allows biological degradation of the packages and produces water, carbon dioxide and inorganic compounds (*11*).

PLA can be recycled or composted. It can be hydrolyzed with steam or boiling water to lactic acid, which can be recycled back to monomer. This could lead to molecular recycling and would allow the recycling of packaging materials (*6*). Mater-Bi is also recyclable and biodegradable as pure cellulose. All Mater-Bi materials are certified compostable according to international standards such as ASTM D-6400.

3. Overview of Biofibers

Natural fibers play an important role in our daily life in the forms of cloths (cotton fibers) and papers (wood fibers). They are also the basics for different packaging materials. The term biofibers are various kinds of fibers that produced by plants, animals. The biofibers used for packaging application normally refers to plant fibers which are extracted from different kinds of plants (*13*). These biofibers include flax, hemp, jute, sisal, and kenaf etc. Cotton and jute are important natural fibers and have large volume production compared with other cellulosic natural fibers.

Biofibers also includes renewable fibers which are produced from renewable sources like viscose, acetate fibers, and nanocellulose fibers. The packaging application of these renewable fibers is growing. Several obstacles need to be addressed to increase the production. 1) The raw materials for the renewable fibers need to be sustainable. 2) The process to form fiber should be environmentally friendly. 3) The innovation of performance improvement with a reasonable overall cost by using nanofillers and chemical modifications (*13*).

The major chemical compositions of biofibers are cellulose, hemicelluloses and lignin etc. The various constituents of a specific natural fiber depend on the species, age, and extraction process. Table 3 represents the typical chemical composition and structural parameters of different natural fibers. Cellulose is a glucan polymer consisting of a linear chain of several hundred to over ten thousand 1,4-β-linked D-glucose units (*36*, *37*). These units contain hydroxyl groups which form hydrogen bonds inside the cellulose macromolecules, among macromolecules, and between hydroxyl groups in air. As a result, all the natural fibers are hydrophilic in nature and the moisture content can reach to 8-12.6 % (*12*, *38*).

Although the chemical structure of cellulose in biofibers is the same, the degrees of polymerization are different. The mechanical performances of these biofibers are significantly influenced by their degree of polymerization. The properties such as density, tensile strength, and modulus are related to the chemical composition and internal structures of the fibers. A comparison of the physical properties of biofibers to conventional manmade fibers is shown in Table 4 (*12*).

In recent years, biofibers attracted much attention of researchers and scientists due to their advantages over traditional fibers such as glass and carbon fibers (*39*). The advantages of biofibers over man-made fibers are low cost, low density, comparable tensile properties, renewability and biodegradability (*40*).

The main drawback of biofibers is their hydrophilic nature which reduces the compatibility with hydrophobic polymeric matrix during fabrications. The other disadvantage is the relatively low processing temperature required due to the possibility of fiber degradation and/or the possibility of volatile emissions that could affect packaging properties. The processing temperatures for most of the biofibers are thus limited to about 200 °C, although it is possible to use higher temperatures for short periods (*12*).

Table 3. Chemical composition and structural parameters of biofibers. (Reproduced with permission from reference (*12*), copyright (2000) John Wiley and Sons).

Type of fiber	*Cellulose wt%*	*Lignin wt%*	*Hemicellulose wt%*	*Pectin wt%*	*Wax wt%*	*Moisture content wt%*
Jute	61-71.5	12-13	13.6-20.4	0.2	0.5	12.6
Flax	71	2.2	18.6-20.6	2.3	1.7	10
Hemp	70.2-74.4	2.7-5.7	17.9-22.4	0.9	0.8	10.8
Ramie	68.6-76.2	0.6-0.7	13.1-16.7	1.9	0.3	8
Kenaf	31-39	15-19	21.5	-	-	-
Sisal	67-78	8-11	10-14.2	10	2	11
PALF	70-82	5-12	-	-	-	11.8
Henequen	77.6	13.1	4-8	-	-	-
Cotton	82.7	-	5.7	-	0.6	-
Coir	36-43	41-45	0.15-0.25	3-4	-	8

Table 4. Mechanical properties of biofibers. (Reproduced with permission from reference (*12*), copyright (2000) John Wiley and Sons).

Fiber	*Density* g/cm^3	*Diameter* μm	*Tensile strength MPa*	*Young's modulus GPa*	*Elongation at break %*
Jute	1.3-1.45	25-200	393-773	13-26.5	1.16-1.5
Flax	1.5	-	345-1100	27.6	2.7-3.2
Hemp	-	-	690	-	1.6
Ramie	1.5	-	400-938	61.4-128	1.2-3.8
Sisal	1.45	50-200	468-640	9.4-22	3-7
PALF	-	20-80	413-1627	34.5-82.51	1.6
Cotton	1.5-1.6	-	287-800	5.5-12.6	7-8
Coir	1.15	100-450	131-175	4-6	15-40
E-glass	2.5	-	2000-3500	70	2.5
S-glass	2.5	-	4570	86	2.8
Aramid	1.4	-	3000-3150	63-67	3.3-3.7
Carbon	1.7	-	4000	230-240	1.4-1.8

4. Overview of Biocomposites

Biocomposites are composite materials including one or more phases derived from a biological origin. The reinforcements include plant fibers such as cotton, flax, hemp, wood fibers, or regenerate cellulose fibers (viscose). Matrices are polymers derived from renewable resources such as starches, or vegetable oils (*41*, *42*).

Over the years, many researchers have investigated the biocomposites with natural fiber as the reinforcements. The using of this biocomposites in packaging application is increasing due to the reducing cost, light weight, environmental friendly and competitive mechanical proprieties (*12*, *14*, *43*, *44*).

The production of 100% biocomposites is still a challenge since the limitation of bio-resins. A solution to this would be combining a small amount of petroleum based resin with bio-resin to meet the performance for packaging application. Thermoplastic starch obtained from biomass is used for biocomposites are reported in (*45*). However, it is shows drawbacks of hydrophilic character and variation of the physical properties. A large number of biopolymers are now commercially available. They show a variety of properties and can compete with petroleum based polymers for composite manufacturing.

Xia and Larock explored the vegetable oils from soybean, peanut, walnut, and sunflowers for this application (*46*, *47*). Dried soy proteins are also used for bio-resin and show higher modulus than currently used epoxy resin. It can be used for rigid packaging with a proper moisture barrier (*1*). Relatively water-resistant biodegradable soy-protein composites made from bioabsorbable polyphosphate fillers and soy protein isolate are under investigated in (*48*, *49*). Ganjyal and Yang used Starch acetate blended with corn stalk fibers at different concentrations for the production of biodegradable extruded foams (*50*). Biocomposites made from biopolymer with nanoscale fillers demonstrate improved mechanical properties, water vapor barrier properties and thermal stability without sacrificing biodegradability due to the nano size dispersion (*26*, *42*, *51*).

The main disadvantages of natural fibers for biocomposites are the poor compatibility between fiber and matrix, the relatively high water absorption. The solution for this is modifying the fiber surface to improve the adhesion between fiber and matrix. The modification methods are following in two categories. One is physical method including thermal, mechanical, and plasma treatments, the other is chemical method including silane, alkaline, maleated coupling, and enzyme treatments (*52*). Extensive research is carried out to study the interface inside of biocomposites. Many reports show the improvement of the physical and mechanical properties of biocomposites by modifying the fiber surface. Among these treatments, maleated and silane treatments are becoming better choices due to the beneficial results (*52–54*). The enzyme technology to modify the natural fiber surface is increasing substantially due to environment friendly and cost effective (*55*).

5. Conclusion

Currently, the packaging industry is looking for lightweight packaging materials for reducing of raw materials use, and reducing of waste and transportation costs. For a long time, petroleum-based polymers have been the most common packaging materials because of their desired features such as softness, lightness and transparency. However, the heavy use of these polymers has resulted in serious ecological problems due to non-biodegradability of these polymers.

In recent years, the development of biodegradable packaging materials from biopolymers (e.g., renewable natural resources) has received increasing attentions. Most of these biopolymers are biodegradable, compostable, and available from renewable sources. Because of these advantages, biopolymer and their based materials are gradually used for packaging field such as films packaging for food products, loose film used for transport packaging, service packaging like carry bags, cups, plates and cutlery, biowaste bags, bags and compostable articles in agricultural fields.

Historically, biopolymer and their based materials have important applications in the medical field, where function is more important than cost. They are unlikely to replace all the petroleum-based polymers for packaging applications, where cost is more important than environmental issues. Biopolymers fulfill the environmental concerns but they also show some limitations in terms of physical performances such as thermal resistance, and barrier and mechanical properties that are associated with the costs. More research, including nanotechnology, and the introduction of smart and intelligent molecules, needs to be done to meet the market requirement for packaging application with satisfied quality, shelf-life, and microbiological safety.

Acknowledgments

The author thanks Dr. Helan Xu for useful comments and suggestions.

References

1. Plackett, D. V. *Biopolymers: New Materials for Sustainable Films and Coatings*; Wiley Online Library: London, 2011.
2. Rhim, J.-W.; Ng, P. K. *Crit. Rev. Food Sci. Nutr.* **2007**, *47*, 411–433.
3. Davis, G.; Song, J. H. *Ind. Crops Prod.* **2006**, *23*, 147–161.
4. Mohanty, A. K.; Misra, M.; Drzal, L. T. *Natural Fibers, Biopolymers, and Biocomposites*; Taylor & Francis: Boca Raton, FL, 2005.
5. Gross, R. A.; Kalra, B. *Science* **2002**, *297*, 803–807.
6. Auras, R.; Harte, B.; Selke, S. *Macromol. Biosci.* **2004**, *4*, 835–864.
7. Chandra, R.; Rustgi, R. *Prog. Polym. Sci.* **1998**, *23*, 1273–1335.
8. Petersen, K.; Væggemose Nielsen, P.; Bertelsen, G.; Lawther, M.; Olsen, M. B.; Nilsson, N. H.; Mortensen, G. *Trends Food Sci. Technol.* **1999**, *10*, 52–68.

9. Johansson, C.; Bras, J.; Mondragon, I.; Nechita, P.; Plackett, D.; Simon, P.; Svetec, D. G.; Virtanen, S.; Baschetti, M. G.; Breen, C. *BioResources* **2012**, *7*, 2506–2552.
10. Babu, R. P.; O'Connor, K.; Seeram, R. *Progress in Biomaterials* **2013**, *2*, 1–16.
11. Siracusa, V.; Rocculi, P.; Romani, S.; Rosa, M. D. *Trends Food Sci. Technol.* **2008**, *19*, 634–643.
12. Mohanty, A. K.; Misra, M.; Hinrichsen, G. *Macromol. Mater. Eng.* **2000**, *276-277*, 1–24.
13. Johansson, C.; Bras, M.; Mondragon, I.; Nechita, P.; Plackett, D.; Šimon, P.; Svetec, D. G.; Virtanen, S.; Baschetti, M. G.; Breen, C. *BioResources* **2012**, *7*.
14. Mohanty, A. K.; Misra, M.; Drzal, L. T. *J. Polym. Environ.* **2002**, *10*, 19–26.
15. Tharanathan, R. N. *Trends Food Sci. Technol.* **2003**, *14*, 71–78.
16. Madhavan Nampoothiri, K.; Nair, N. R.; John, R. P. *Bioresour. Technol.* **2010**, *101*, 8493–8501.
17. Garlotta, D. *J. Polym. Environ.* **2001**, *9*, 63–84.
18. Mobley, D. P. *Plastics from Microbes: Microbial Synthesis of Polymers and Polymer Precursors*; Hanser Publishers: Munich, 1994.
19. Lim, L. T.; Auras, R.; Rubino, M. *Prog. Polym. Sci.* **2008**, *33*, 820–852.
20. Hartmann, M. H. In *Biopolymers from Renewable Resources*; Kaplan, D., Ed.; Springer: Berlin, 1998; pp 367–411.
21. Benson, R. D.; Borchardt, R. L.; Gruber, P. R.; Hall, E. S.; Iwen, M. L.; Kolstad, J. J. Google Patents US 5258488 A, 1992.
22. Datta, R.; Henry, M. *J. Chem. Technol. Biotechnol.* **2006**, *81*, 1119–1129.
23. Conn, R. E.; Kolstad, J. J.; Borzelleca, J. F.; Dixler, D. S.; Filer, L. J., Jr; Ladu, B. N., Jr; Pariza, M. W. *Food Chem. Toxicol.* **1995**, *33*, 273–283.
24. Steinbüchel, A.; Doi, Y. *Biopolymers, Polyesters III – Applications and Commercial Products*; Wiley: Weinheim, 2002.
25. Weber, C.; Haugaard, V.; Festersen, R.; Bertelsen, G. *Food Addit. Contam.* **2002**, *19*, 172–177.
26. Avella, M.; De Vlieger, J. J.; Errico, M. E.; Fischer, S.; Vacca, P.; Volpe, M. G. *Food Chem.* **2005**, *93*, 467–474.
27. Liu, H.; Xie, F.; Yu, L.; Chen, L.; Li, L. *Prog. Polym. Sci.* **2009**, *34*, 1348–1368.
28. Klemm, D.; Heublein, B.; Fink, H.-P.; Bohn, A. *Angew. Chem.* **2005**, *44*, 3358–3393.
29. Eichhorn, S. J.; Baillie, C. A.; Zafeiropoulos, N.; Mwaikambo, L. Y.; Ansell, M. P.; Dufresne, A.; Entwistle, K. M.; Herrera-Franco, P. J.; Escamilla, G. C.; Groom, L.; Hughes, M.; Hill, C.; Rials, T. G.; Wild, P. M. *J. Mater. Sci.* **2001**, *36*, 2107–2131.
30. Paunonen, S. *BioResources* **2013**, *8*, 3098–3121.
31. Van de Velde, K.; Kiekens, P. *Polym. Test.* **2002**, *21*, 433–442.
32. Hong, S.-G.; Gau, T.-K.; Huang, S.-C. *J. Therm. Anal. Calorim.* **2011**, *103*, 967–975.
33. Shafizadeh, F.; Bradbury, A. G. W. *J. Appl. Polym. Sci.* **1979**, *23*, 1431–1442.

34. Liu, X.; Wang, Y.; Yu, L.; Tong, Z.; Chen, L.; Liu, H.; Li, X. *Starch* **2013**, *65*, 48–60.
35. Kale, G.; Kijchavengkul, T.; Auras, R.; Rubino, M.; Selke, S. E.; Singh, S. P. *Macromol. Biosci.* **2007**, *7*, 255–277.
36. Hua, L.; Zadorecki, P.; Flodin, P. *Polym. Compos.* **1987**, *8*, 199–202.
37. Crawford, R. L. *Lignin Biodegradation and Transformation*; Wiley: New York, 1981.
38. Bledzki, A. K.; Reihmane, S.; Gassan, J. *J. Appl. Polym. Sci.* **1996**, *59*, 1329–1336.
39. Saheb, D. N.; Jog, J. *Adv. Polym. Technol.* **1999**, *18*, 351–363.
40. Li, X.; Tabil, L. G.; Panigrahi, S. *J. Polym. Environ.* **2007**, *15*, 25–33.
41. Fowler, P. A.; Hughes, J. M.; Elias, R. M. *J. Sci. Food Agric.* **2006**, *86*, 1781–1789.
42. Zhao, R.; Torley, P.; Halley, P. J. *J. Mater. Sci.* **2008**, *43*, 3058–3071.
43. Pilla, S. *Handbook of Bioplastics and Biocomposites Engineering Applications*; John Wiley & Sons: Hoboken, NJ, 2011; Vol. 81.
44. Coles, R.; Kirwan, M. J. *Food and Beverage Packaging Technology*; 2 ed.; John Wiley & Sons: London, 2011.
45. Belgacem, M. N.; Gandini, A. *Monomers, Polymers and Composites from Renewable Resources*; Elsevier: Oxford, U.K., 2011.
46. Xia, Y.; Larock, R. C. *Green Chem.* **2010**, *12*, 1893–1909.
47. Xia, Y.; Quirino, R. L.; Larock, R. C. *J. Renewable Mater.* **2013**, *1*, 3–27.
48. Otaigbe, J. U.; Adams, D. O. *J. Environ. Polym. Degrad.* **1997**, *5*, 199–208.
49. Jane, J.; Lim, S.; Paetau, I.; Spence, K.; Wang, S. Biodegradable Plastics Made from Agricultural Biopolymers. In *Polymers from Agricultural Coproducts*; Fishman, M. L., Friedman, R. B., Huang, S. J., Eds.; ACS Symposium Series 575; American Chemical Society: Washington, DC, 1994; pp 92–100.
50. Ganjyal, G.; Reddy, N.; Yang, Y.; Hanna, M. *J. Appl. Polym. Sci.* **2004**, *93*, 2627–2633.
51. Sorrentino, A.; Gorrasi, G.; Vittoria, V. *Trends Food Sci. Technol.* **2007**, *18*, 84–95.
52. Faruk, O.; Bledzki, A. K.; Fink, H.-P.; Sain, M. *Prog. Polym. Sci.* **2012**, *37*, 1552–1596.
53. Mishra, S.; Naik, J.; Patil, Y. *Compos. Sci. Technol.* **2000**, *60*, 1729–1735.
54. Cantero, G.; Arbelaiz, A.; Llano-Ponte, R.; Mondragon, I. *Compos. Sci. Technol.* **2003**, *63*, 1247–1254.
55. Bledzki, A. K.; Mamun, A. A.; Jaszkiewicz, A.; Erdmann, K. *Compos. Sci. Technol.* **2010**, *70*, 854–860.

Chapter 14

Hollow Micro-/Nano-Particles from Biopolymers: Fabrication and Applications

Jian Qian*

Department of Pharmaceutical Chemistry, University of Kansas, Multidisciplinary Research Building, 2030 Becker Drive, Lawrence, Kansas 66047, United States
***E-mail: jianqian@ku.edu**

Hollow micro-/nano-particles from biopolymers are very promising for various applications due to their unique physical and chemical properties. This chapter is devoted to the progress made in recent years in the fabrication and applications of hollow micro-/nano-particles from biopolymers. This chapter will first highlight the various fabrication methods, including template method, emulsion method and self-assembly method. Fundamental aspects of these formation processes and their effects on the particle properties are also discussed in this chapter. Then various applications in drug delivery, tissue engineering and wastewater treatment will be introduced.

1. Introduction

Polymeric micro-/nano-particles containing interior cavities, or so called 'hollow particles', have gained intensive attentions due to their characteristics of low density, high specific surface area, and large useful inner spaces for guest molecules encapsulation, and have shown promising potential of applications or already been used in the area of cosmetics, catalyst carrier, industrial coatings, microencapsulation and drug delivery (*1–9*). For example, the large surface area and the large fraction of void space inside the particles have been successfully used to adsorb, encapsulate and sustainably release various materials such as drugs, proteins, and DNA. The void within the hollow particles provides not only useful spacious compartments but also unique light scattering properties. The low density also makes them used as light-weight fillers (*4*).

In order to meet the different requirements of multiple applications, various types of hollow micro-/nano-particles with different physical properties and chemical functionalities were prepared by employing different functional polymeric materials, including petroleum-derived polymers and biopolymers (*2*, *3*, *6*, *7*, *9–18*). Biopolymers are becoming more and more attractive than petroleum-derived polymers, not only because of the increasing oil price and the environment issues from petroleum industry, but also the advantages of biopolymer themselves. Biopolymers, such as carbohydrates, proteins and other biosynthetic polymers, are produced by living organisms, including plants, animals and bacterias. They are always sustainable and renewable, and could become less costly as rapid technological progress. In addition, biopolymers are biocompatible and biodegradable, could be fully degraded by microorganisms and produce limited burden on environment. Moreover, the biocompatibility and biodegradability offer safer applications in biomedical fields, particularly in the fields of tissue engineering and drug delivery (*19–21*). Consequently, hollow micro-/nano-particles from biopolymers have all the properties of synthetic counterparts as well as being intrinsically biodegradable, abundant in nature, renewable, nontoxic, and relatively cheap. In addition, they possess a high content of functional groups including hydroxyl, amino, and carboxylic acid groups for further chemical modification and functionalization.

In the past decades, various of techniques, such as template method (*6–8*, *14*, *22–35*), emulsion method (*36–46*), self- assembly method (*9–11*, *18*, *47–53*) and other methods (*3*, *4*, *54–57*), have been successfully developed to fabricate hollow micro-/nano-particles. These techniques could not only create interior hollow structures of the particles, but also control the particle size and shape, volume fraction, shell thickness, shell permeability and surface functionality (*6–8*, *16*, *58*). In this chapter, the recent fabrication methods of hollow micro-/nano-particles are summarized, and the fundamental formation mechanism of each method is discussed. Several hollow particles fabricated by these methods from several types of biopolymers, including dextran (*52*, *59–62*), chitosan (*63–68*), collagen (*22*, *23*, *25*), hyaluronic acid (*69*, *70*), polylactic acid (*41*, *43*, *71–75*), zein (*76*, *77*) or the combination of the polymers (*33*, *78–84*) (as shown in Table 1.), will be introduced. We will not cover all the applications of hollow micro-/nano-particles from biopolymers, but focus on the recent and hot research areas such as drug delivery, tissue engineering and waste water treatment.

Table 1. Some hollow micro-/nano-particles from biopolymers

Polymers	*Fabrication method*	*Size*	*Encapsulated component*	*Application*	*Reference*
Dextran sulfate/poly-L-arginine	LbL	2 μm	protein	Drug delivery	Ref. (*115*)
Chitosan/alginate	LbL	3-5 μm	Doxorubicin	Drug delivery	Ref. (*78*)
Collagen	Template	4 μm	Nerve growth factor	Drug delivery	Ref. (*22*)
Collagen	Template	0.1, 1, and 10 μm	pDNA polyplexes	Gene delivery	Ref. (*25*)
Poly(c-glutamic acid)/ chitosan	Self-assembly & LbL	~150 nm	Doxorubicin	Drug delivery	Ref. (*116*)
Modified dextran/gelatin	Emulsion	40 μm	Stromal cell-derived factor	Drug delivery	Ref. (*81*)
Hyaluronic acid	Emulsion-diffusion	350 nm	Polyhexanide	Drug delivery	Ref. (*70*)
poly(L-aspartic acid)/chitosan	LbL	~365 nm	Insulin	Drug delivery	Ref. (*84*)
Polylactic acid	Double emulsion	1-25 μm	NA	Drug delivery	Ref. (*71*)
Modified poly(L-lactic acid)	Emulsion diffusion	10-50	NA	Tissue engineering	Ref. (*120*)
Glycosaminoglycan/chitosan	Emulsion	300-2000 μm	Cells	Tissue engineering	Ref. (*86*)
Zein	Template	120 nm	NA	Wastewater treatment	Ref. (*76*)

2. Fabrication Methods

2.1. Template Method

The concept of template method to prepare hollow structure is quite simple and has been proven to be very successful. In general, there are four major steps for this method as illustrated in Scheme 1: (1) synthesis or selecting appropriate removable templates; (2) surface modification of template for anchoring of polymeric materials or polymerizing of monomers; (3) coating the templates with designed polymeric materials to form compact polymeric shells via direct polymer deposition or polymerization; and (4) degradation of the templates to obtain hollow structures. Various colloids, including polymer latexes, emulsion droplets, inorganic colloids, or even gas bubbles, could be used as templates.

There is no doubt that step 3 is considered the most challenging because it requires robust methods for polymer deposition on templates with a high yield. The most common challenge is the incompatibility between the template surface and polymeric shell material, which leads to the self-aggregation of the polymers and the unsuccessful shell formation. In such cases, step 2 is usually necessary for the successful polymer shell formation by modifying the template with special functional groups, for example vinyl groups, or physical characteristics, for example electrostatic charges. In step 4, the template is decomposed by introducing etching agent which has to be removed by repetitive and meticulous purification.

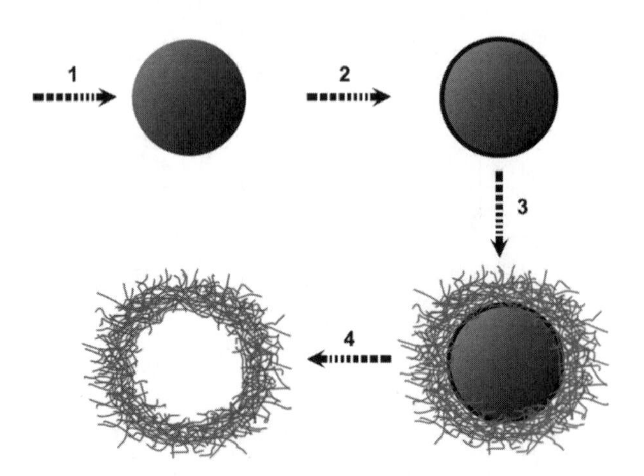

Scheme 1. Schematic illustration of a conventional template method for polymeric hollow sphere synthesis.

Biopolymers, such as polysaccharides (*27*, *65*, *85*, *86*), proteins (*22–25*, *76*, *77*, *87*, *88*) and polynucleotides (*89*), have been successfully employed to prepare hollow micro-/nano-particles by the traditional template method. Pandit group has developed a versatile method to prepare chitosan/polyglutamic acid, collagen, and elastin-like polypeptide hollow particles using sulfonated polystyrene beads as templates (*22–25*, *85*). Scheme 2 shows the graphical illustration of the preparation of hollow collagen microparticles (*25*). Positively charged collagen was first deposited on sulfonated polystyrene beads by electrostatic interaction, and then cross-linked to stabilize the polymer matrix and lock the core/shell structure. Dissolution of the polystyrene core by tetrahydrofuran led to the formation of hollow collagen particles. Morphology and structure of the hollow collagen particles were characterized by SEM and TEM as shown in Figure 1. Wang et al. developed a similar strategy to prepare hollow chitosan particles using biodegradable poly-D, L-lactide-poly(ethylene glycol) nanoparticles as templates (*65*). Chitosan was first adsorbed on template surfaces, and then cross-linked by glutaraldehyde for solidification. Further dissolving the polymer templates by acetone led to formation of chitosan hollow particles. The TEM and SEM images before and after template removal are shown in Figure 2. Zhu et al. fabricated bovine serum albumin (BSA) microcapsules using $MnCO_3$ microparticles as templates by desolvation and degradable cross-linking (*87*). BSA could be desolvated from its aqueous solution by dropwise addition of ethanol. The desolvated BSA was adsorbed onto $MnCO_3$ microparticles and further cross-linked with disulfide-containing dithiobis(succinimidylpropionate). Hollow particles were obtained after template removal at low pH. Wall thickness of the microcapsules could be controlled by the amount of ethanol.

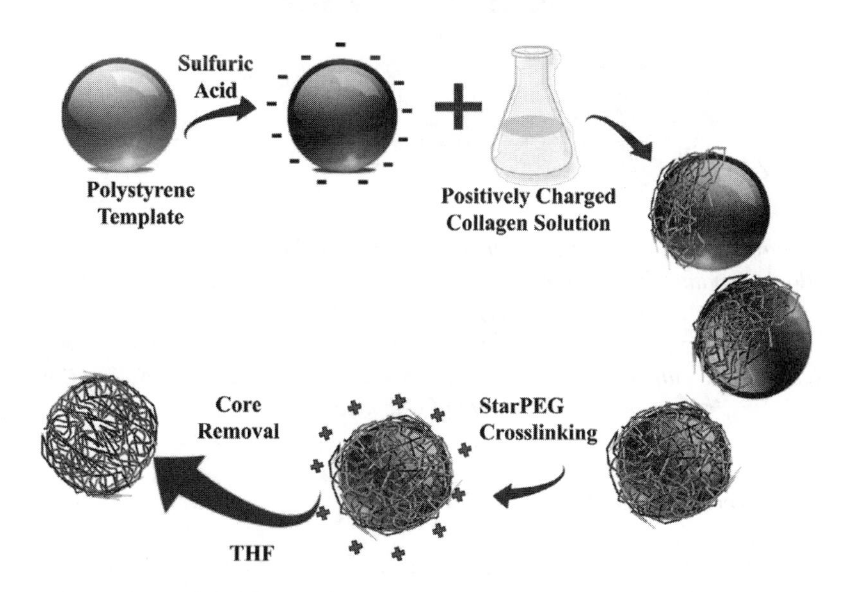

Scheme 2. Graphical representation of the fabrication of hollow collagen microspheres. Reproduced with permission from reference (25). Copyright (2012) American Chemical Society.

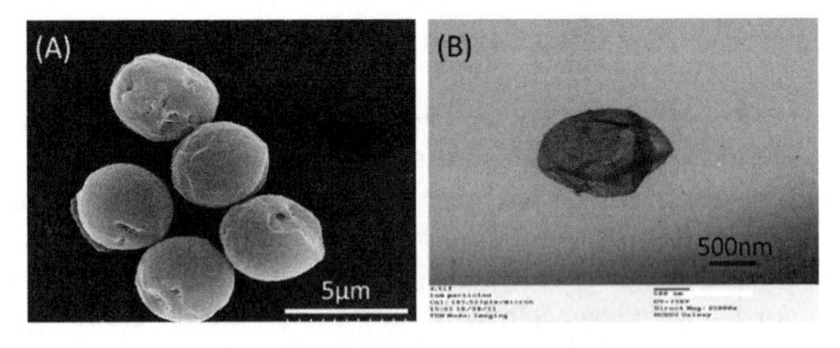

Figure 1. Isolated collagen hollow spheres. (A) SEM images of isolated 5 μm collagen hollow spheres. (B) TEM images of 1 μm collagen hollow spheres. Reproduced with permission from reference (25). Copyright (2012) American Chemical Society.

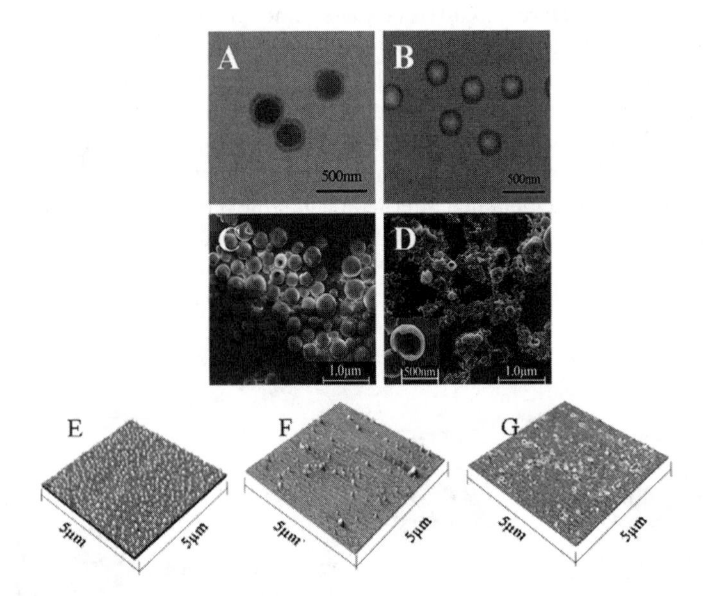

Figure 2. The morphology of the core-shell CS-PELA nanospheres (A), and hollow CS nanospheres (B) detected by TEM, the hollow nanospheres before (C) and after (D) ultrasonication detected by SEM, and PELA template (E) and hollow nanospheres before (F) and after (G) ultrasonication detected by AFM. Reproduced with permission from reference (65). Copyright (2008) The Royal Society of Chemistry.

2.1.1. Layer-by-Layer (LbL) Assembly

LbL assembly refers to the consecutive deposition of complementary/interacting polymers onto colloidal particles (templates) mediated by various interactions. Followed removal of the sacrificial colloidal templates generates the LbL capsules (Scheme 3) (*90*). Although the adsorption is mainly dependent on the electrostatic force, other interactions or reactions, such as hydrogen bonding

(*91–93*), complexation (*60*, *69*), specific recognition (*94*, *95*), and covalent reaction (*59*, *61*, *88*), have been employed as the driving forces for the LbL assembly. Template particles, such as polystyrene latex particles with surface charges, melamine formaldehyde (MF) microparticles, or inorganic particles, can be dissolved by organic solvents or acid solutions after LbL assembling. Although the LbL technique shares the same synthetic concept with the conventional template method, it provides much more advantages. The major benefit of LbL method is without doubt its versatility. The LbL capsules can be fabricated using various templates, with sizes varying from a few nanometers to hundreds of micrometers, and their chemical and mechanical properties can be precisely tailored by modulating the thickness and constitution of the shell (*5–8*, *58*). In addition, the microcapsules can be further modified with various types of compounds including polymers, nanoparticles, and biospecific motifs.

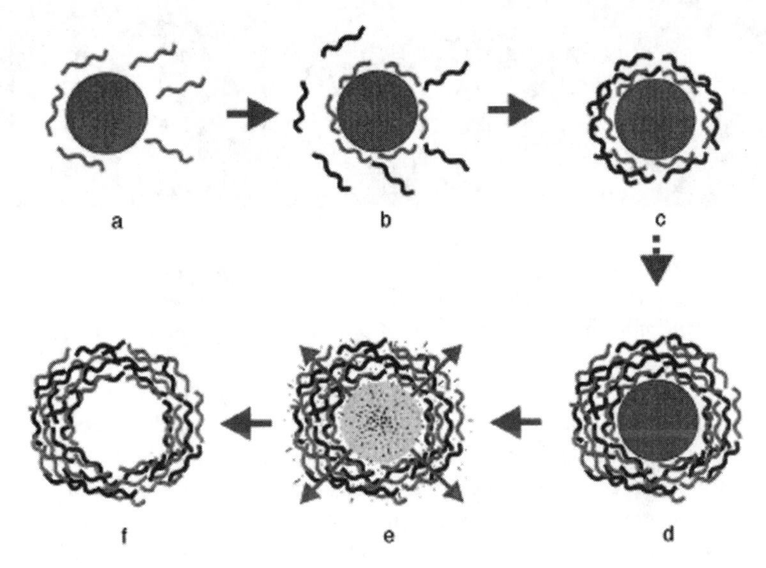

Scheme 3. Schematic illustration of the polyelectrolyte deposition process and of subsequent core decomposition. The initial steps (a-d) involve stepwise film formation by repeated exposure of the colloids to polyelectrolytes of alternating charge. The excess polyelectrolyte is removed by cycles of centrifugation and washing before the next layer is deposited. After the desired number of polyelectrolyte layers are deposited, the coated particles are exposed to etching agent (e). The core immediately decomposes, as evidenced by the fact that the initially turbid solution becomes essentially transparent within a few seconds. Finally, a suspension of free polyelectrolyte hollow shells is obtained (f). Reproduced with permission from reference (90). Copyright (1998) Wiley-VCH.

A lot of biopolymers have net charges or could be chemically modified to have net charges. This characteristic makes these biopolymers perfect building blocks for LbL microcapsules via electrostatic interaction. Schüler and Caruso synthesized decomposable hollow capsules based on deoxyribonucleic acid (DNA) and a low molecular weight organic molecule, a naturally occurring polyamine, spermidine (SP) by sequential deposition of the DNA/SP multilayers on the weakly cross-linked melamine formaldehyde (MF) particles and subsequent decomposion of MF particles by HCl (pH 1.5-1.6) (*89*). The hollow DNA/SP capsules displayed a high sensitivity to salt solutions, and could be decomposed after exposure to sodium chloride solutions. Radhakrishnan and Raichur fabricated microcapsules by LbL assembly of an arginine-rich protein protamine (PRM) and an anionic biopolymer heparin (HEP) on poly(styrene sulfonate) doped $CaCO_3$ microparticles, followed by dissolution of the cores (*96*). Figure 3 showed the morphology change of template after polymer deposition and template removal. These microcapsules exhibited high stability when stored at physiological conditions even for prolonged time periods.

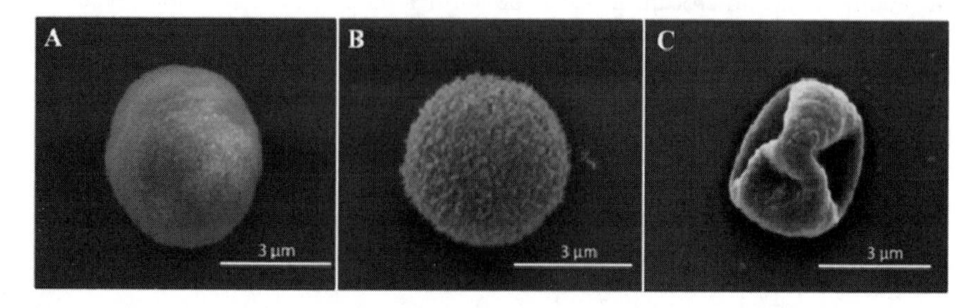

Figure 3. SEM images of a $CaCO_3$:PSS microparticle before (A) and after LbL assembly (B) and hollow air dried $(PRM/HEP)_2$ microcapsule (C). Reproduced with permission from reference (96). Copyright (2012) The Royal Society of Chemistry.

Besides electrostatic interaction, other interactions, such as complexation (*60*, *69*), specific recognition (*94*, *95*), and covalent reaction, were all employed as the driving forces to produce multilayers of microcapsules. Yu et al. utilized both electrostatic interactions and chemical complexation processes to synthesize a novel iron-heparin complexed hollow capsule. These capsules were fabricated by alternating deposition of ferric ions (Fe^{3+}) and heparin onto the surface of polystyrene latex particles with two different sizes (488 nm and 10.55 μm), followed by dissolution of the cores with tetrahydrofuran (*69*). Luo et al. first synthesized cyclodextran grafted and adamantane grafted dextrans, and then utilized the specific recognition interaction between cyclodextran groups and adamantane groups (AD) to alternately deposit the two polymers onto $CaCO_3$ particles (*94*). Hollow dextran microcapsules were formed by dissolving the $CaCO_3$ cores using EDTA aqueous solution as shown in Scheme 4. More interestingly, because the AD groups were linked with polymers or doxorubicin (DOX) by pH-cleavable hydrazone bonds, AD moieties can be removed under weak acidic condition, leading to destruction of particles and release of Dox. Zhu et al. fabricated hollow polymeric microcapsules from concanavalin A, a plant lectin, and glycogen, a polysaccharide, based on the lectin–carbohydrate interaction in a layer-by-layer fashion (*95*). These capsules have specific responses to certain carbohydrates, such as mannose, fructose, glucose and dextran at the physiological pH range. Duan et al. synthesized hemoglobin protein microcapsules through a covalent LbL technique in which multilayers of hemoglobins were covalently cross-linked by glutaraldehyde (Scheme 5) (*88*). Cyclic voltammetry and potential-controlled amperometric measurements confirm that hemoglobin microcapsules after fabrication still remain their heme electroactivity.

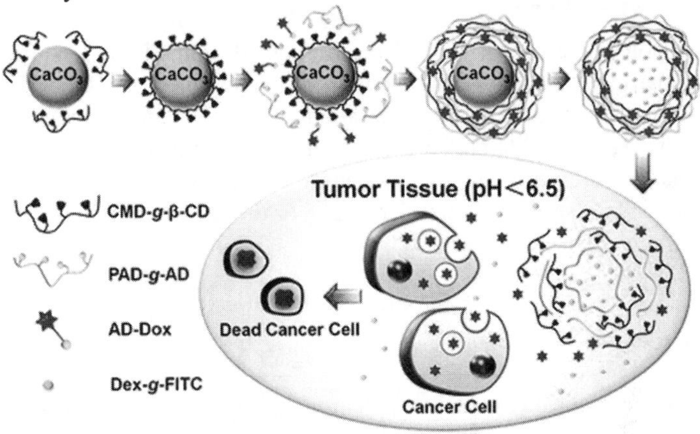

Scheme 4. Schematic diagram of the preparation of microcapsules and pH-induced drug release from a drug loaded microcapsule. Reproduced with permission from reference (94). Copyright (2012) American Chemical Society.

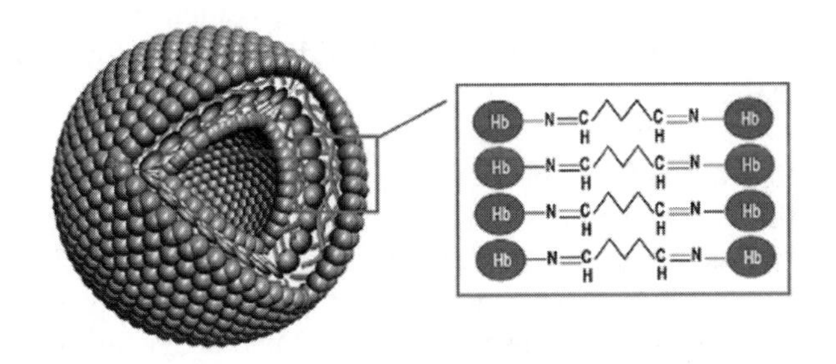

Scheme 5. Schematic representation of the assembled hemoglobin protein microcapsules via covalent layer-by-layer assembly. Reproduced with permission from reference (88). Copyright (2007) Elsevier.

Despite its versatility, the LbL method suffers from some key shortcomings. the LbL assembly procedure becomes quite tedious when many layers are required. The hollow structures prepared from this method generally lack mechanical robustness comparing to particles prepared using other approaches. As-prepared polymer capsules are also only stable when kept in solution, once dried they tend to collapse irreversibly. Another problem of the LbL method, or all the template methods, is the low efficiency and capacity of encapsulating guest molecules especially the large molecules into the capsules by the commonly used 'post-loading' strategy, which would impede the practical application as drug delivery vehicles. Although, the 'pre-loading' strategy has been developed by preloading the guest molecules in the template or directly using the drug crystal as the template (*28*, *66*, *97–101*), it is not universal and versatile, as it is restricted by the complicated design and synthesis of preloaded templates and limited choice of drug crystals.

2.2. Emulsion Method

Emulsion processing is one of the traditional methods to produce hollow polymer particles, usually combined with a subsequent solidification, such as phase separation or solvent evaporation (*37*). Emulsion method could be considered as a special template method in which emulsion droplets served as the templates either utilizing the interfaces between the emulsion droplets and the continuous phase for polymer deposition or the 3D structure of the shell-like phase in a multiphase emulsion.

2.2.1. Single Emulsion

In a single emulsion method for preparing hollow particles, it usually involves a phase separation of polymers from dissolved phase and simultaneous deposition exclusively onto the interfaces between the emulsion droplets and the continuous phase by solvent diffusion/evaporation or chemical/physical cross-linking. Obviously, the success of phase separation and interfacial deposition determine the final yield of hollow particles. This method could be further divided into two categories, emulsion-diffusion and emulsion-coacervation methods, depending on the polymers selected for shell formation dissolved in dispersed phase or continuous phase.

In the emulsion-diffusion process, polymers are dissolved in dispersed phase and phase separated to the interfaces either by solvent diffusion/evaporation or chemical/physical cross-linking. For example, Reis et al. prepared pectin hollow particles by cross-linking/polymerization of glycidyl methacrylate modified pectin in a water-in-benzyl alcohol emulsion. An aqueous phase containing glycidyl methacrylate modified pectin was mixed into benzyl alcohol by stirring to form a water-in-oil (w/o) emulsion under nitrogen atmosphere, and then the glycidyl methacrylate modified pectin was polymerized onto the interface to form a hollow structure (*42*). More interestingly, this method could not only produce the hollow particles but also *in situ* encapsulate hydrophilic cargoes inside the capsules. Baier et al. synthesized a cross-linked starch capsule containing dsDNA in a w/o miniemulsion (as shown in Scheme 6) (*102*). A starch aqueous solution was emulsified in cyclohexane to form a w/o emulsion by ultrasonication, and then the starch was cross-linked by 2,4-toluene diisocyanate and condensed onto the interfaces to form hollow particles. dsDNA was successfully encapsulated by mixing it in the initial starch solution.

In the emulsion-coacervation process, the polymers are dissolved in continuous phase and coacervated to the interfaces by physical or chemical cross-linking. As a lot of biopolymers are water soluble, the emulsion-coacervation process for preparing biopolymer based hollow particles usually involves an oil-in-water (o/w) emulsion with biopolymers dissolved in water phase. Peng and Zhang reported the preparation of biocompatible and biodegradable hollow microspheres using cyclohexane droplets as a template and the N-methylated chitosan (NMC) cross-linked with gultaraldehyde (GA) as the shell (*63*). In the preparation process, cyclohexane as the oil phase was added to a NMC aqueous solution to form an o/w emulsion, and then the NMC was cross-linked by GA and coacervated to form a polymer layer on the surface of the oil droplet. Hollow NMC particles were formed after washed with acetone to remove the oil droplets inside the polymer shells. A big advantage of this emulsion-coacervation method is that the hydrophobic drug could be encapsulated *in situ* as the hollow particles formed. Xi et al. reported the fabrication of chondroitin sulfate-methacrylate (ChSMA) nanocapsules, in which poor water-soluble drug of indomethacin (IND) could be effectively encapsulated (*103*). Chloroform containing IND was added to an ChSMA aqueous solution to form an o/w emulsion under agitation. The following polymerization of ChSMA led to formation of nanocapsules with IND encapsulated *in situ*.

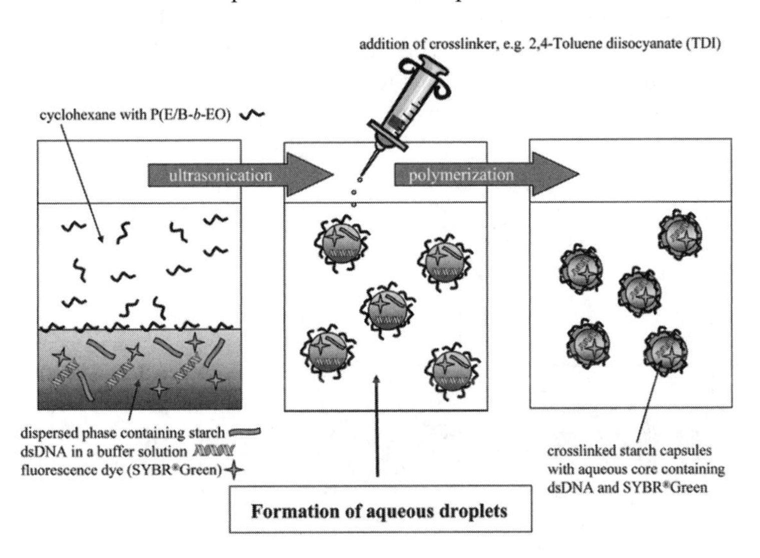

Scheme 6. Formulation process for the preparation of cross-linked starch capsules in an inverse miniemulsion. Reproduced with permission from reference (102). Copyright (2010) American Chemical Society.

A big advantage of the single emulsion method is that the drug could be encapsulated *in situ* as the hollow particles formed. However, this method is not robust and versatile. The successful formation of hollow particles highly depends on the characteristics of the polymers, the selected solvent, stabilizer and other parameters.

2.2.2. Double Emulsion Method

Double emulsion, simply defined as an emulsion in an emulsion, is a complex multiphase system, in which two liquids are separated by a third liquid which is not miscible with the first two liquids. In the case of water and oil, there are two possible cases of double emulsions: water-in-oil-in-water (w/o/w) emulsion and oil-in-water-in-oil (o/w/o) emulsion. In a w/o/w emulsion, dispersions of small water droplets within larger oil droplets are themselves dispersed in a continuous aqueous phase, while in an o/w/o emulsion, dispersions of small oil droplets within larger water droplets are themselves dispersed in a continuous oil phase. The unique structure of the middle phase provides perfect template effect to produce the hollow polymer structure after solidification of the pre-dissolved polymers by solvent evaporation or diffusion.

The w/o/w emulsion is commonly used to prepare hollow polymeric micro-/nano-particles with hydrophobic polymer materials. In a typical procedure for preparation as shown in Scheme 7, the primary w/o emulsion is formed by dispersion of a water phase into an organic phase containing the selected polymers usually with the aid of ultrasonication. The second emulsion is formed by dispersion of the primary emulsion into a aqueous phase containing a stabilizing agent, such as polyvinyl alcohol, also with ultrasonication. Finally, the solvents are removed by evaporation, leaving hardened polymeric hollow particles in an aqueous medium. As a hydrophobic biopolymer, polylactic acid (PLA) was widely employed to fabricate hollow particles (*41*, *43*, *72*, *75*).

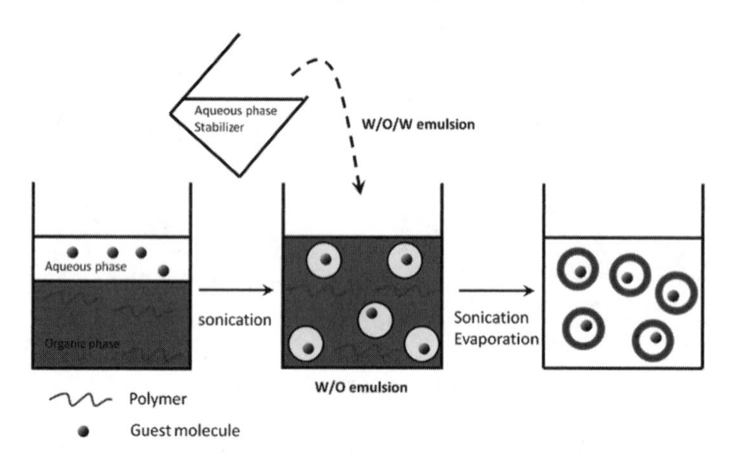

Scheme 7. Schematic illustration of W/O/W double emulsion method for polymeric hollow sphere synthesis.

Most biopolymers are water soluble, and lack of the solubility in organic solvent. For these biopolymers, the o/w/o emulsion could be used to prepare the hollow particles. Lee and Rosenberg reported the fabrication of whey protein capsules encapsulating anhydrous milk fat in a three-steps consisting of : (1) preparing a primary o/w emulsion in which anhydrous milk fat was dispersed in a whey protein aqueous solution, (2) preparing an o/w/o double emulsion in which the primary o/w emulsion was dispersed in corn oil, and (3) cross-linking the protein matrices by glutaraldehyde (*104*). Liu et al. reported the fabrication of chitosan microcapsules from o/w/o double emulsion (*45*). The strategy for preparation was to use uniform-sized o/w/o emulsions fabricated by capillary microfluidic technique as templates and convert these emulsions into core/shell microcapsules via a interfacial cross-linking reaction (Scheme 8). The cross-linking reaction occured at the inner o/w interface once the inner oil fluid containing the cross-linker contacts the middle water fluid containing chitosan in the transition tube of the microfluidic device. Hollow chitosan particles were obtained after washing.

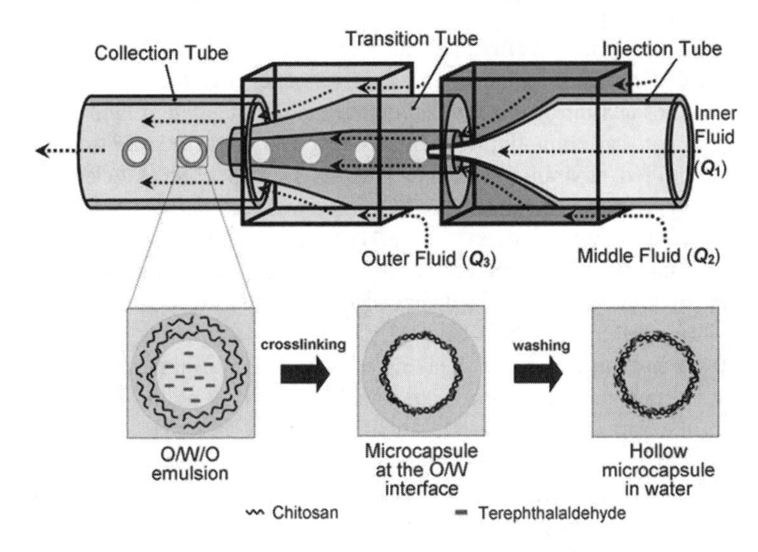

Scheme 8. Schematic illustration of the microfluidic preparation process of crosslinked chitosan hollow microcapsule. Reproduced with permission from reference (45). Copyright (2011) The Royal Society of Chemistry.

The double emulsion method is very attractive due to the high encapsulation efficiency of guest molecules *in situ* along with the hollow particle preparation. However, the double emulsion system is unstable, and usually needs complex control of each emulsion process (*105*, *106*). Although the size distribution of the particles could be improved by microfluidic technique at the micron scale (*36*, *40*, *107*, *108*), it is still polydisperse and not controllable at the submicron scale.

2.3. Self-Assembly Method

Polymer self-assembly is a type of process in which disordered polymers in a solution form organized structures or patterns as a consequence of specific, local interactions among the polymers themselves, without external direction. In the past several decades, the self-assembly behavior of polymers, especially block copolymers, in selected solvents, have been extensively investigated (*9–11*, *18*, *47–50*), and numerous supramolecular nanostructures with various morphologies, such as micelles, vesicles, rods, tubes, and other structures (*109*, *110*), have been created by polymer self-assembly. In these structures, vesicle, a special hollow particle, with a polymer bilayer enclosing a volume, have attracted great attention due to its promising application generally the same as other hollow particles. These special polymeric hollow particles are also called 'polymersomes'. Polymeric vesicles by self-assembly in aqueous solution are usually derived from amphiphilic block copolymers containing hydrophilic and hydrophobic blocks (*9*, *18*), which were usually synthesized from petroleum-derived monomers. Although most of the biopolymers do not have amphiphilic characteristic and could not form stable supramolecular structures in aqueous solution, after chemical modification with desired components, they can successfully self-assemble to hollow structures in certain conditions. For example, Chiang et al. reported the self-assembly of lipid-conjugated dextran to form vesicles with nano-scaled size by a solvent exchange method (*111*). Partial esterification of dextran with activated octadecanol-carbamate imidazole provide the dextran with an amphiphilic characteristic. Through the addition of water into the modified dextran solution in DMSO, hydrophobic association of the polymers aggregated into nano-scaled vesicles with the particle size influenced by the amount of water added. Jones reported a spontaneous formation of disulfide cross-linked nano-sized capsules based on per-thio-β-cyclodextrin (*112*). Oxidation of thiol groups by oxygen led to cross-linking of cyclodextrin and vesicles formation by self-assembly.

The sef-assembly method is a quite simple and straightforward process in which hollow particles form spontaneously without additional treatments. However, the formation of vesicle structure usually needs precise control of polymer compositions, structures and solvents selections. Moreover, it is difficult to encapsulate guest molecules into the cavity efficiently during the self-assembly process, and most of the guest molecules are still located in solution.

2.4. Other Methods

Many other fabrication methods have been developed for hollow polymer particles. Spray drying technology is a simple and efficient process to prepare hollow polymer particles. The slower diffusion rate of polymeric solute than solvent during the drying process leads to the accumulation of polymeric solute on the surface of sprayed droplet, resulting in formation of solid shell57. Acid/alkali swelling method is another fabrication method which is widely used to prepare hollow particles from petroleum-derived polymers (*3*).

3. Applications

3.1. Drug Delivery

Although biopolymers have been widely used for drug encapsulation and formulation, we will not cover all the applications of the biopolymer based hollow particles for drug delivery. In this chapter, we will focus on the application of these particles for intracellular drug delivery which has been a hot research topic in the field of 'nanomedicine' recently.

The biopolymer based hollow micro-/nano-particle is a promising candidate as a drug delivery vehicle for intracellular drug delivery for several reasons: (1) The hollow micro-/nano-particle with a large inner space is capable of loading or encapsulating large quantity of therapeutic molecules, and can protect these molecules from degradation before delivered to target sites. Their large surface areas can also allow for displaying a large number of surface functional groups such as ligands. (2) The small size of these micro/nano-particles makes them compatible with various administration routes including intravenous injection. They have a rapid absorption and release behavior provided by high abilities of their diffusion and volume change. Moreover, these small particles are especially useful for tumor-targeted drug delivery, because the angiogenic tumors tend to develop permeable vasculature and can selectively recruit circulating small particles (enhanced permeability and retention effect; EPR effect) (*113*, *114*). (3) The biocompatibility and biodegradability from the building materials, biopolymers, make them less toxic to cells and organs, and easier to release the encapsulated cargoes.

Although we all know that the biggest advantage of the biopolymer over the petroleum-derived polymer is the biodegradability, as for the application for intracellular drug delivery, the first thing we have to demonstrate is how the biodegradability of the biopolymers influences the delivery effect of the hollow particles. Rivera-Gil et al. demonstrated the internalization of degradable dextran sulfate/poly(L-arginine) and nondegradable PSS/PAH by embryonic fibroblasts for a period of 5 days (*115*). The intracellular degradability of each capsule type could be evaluated by monitoring the fluorescent change of the encapsulated fluorescent compound, DQ-ovalbumin. Due to a characteristic self-quenching mechanism, this compound exhibits a fluorescence change from red to bright green upon proteolysis. The biodegradable capsules exhibited a green fluorescence, while the PSS/PAH capsules were in their majority red after 5 days incubation (Figure 4). This result indicated that, although both types of microcapsules could be delivered to cells, the availability of DQ-ovalbumin was higher when encapsulated within dextran sulfate/poly(L-arginine) microcapsules. The result also suggests that delivery of drug into cells is just a half way of delivery, and only could be considered as accomplished after the encapsulated agent is available to the cellular machinery. The biopolymer based capsules with biodegradability could achieve the delivery much easier than the nondegradable capsules from petroleum-derived polymers.

The hollow structure of the particles provide the capability of loading and accumulating high concentration of drugs. Once these loaded particles delivered to the target sites, they can produce a high local concentration of drugs, and show more efficacy than free drugs. Zhao et al. reported loading of the antitumor drug doxorubicin in preformed chitosan/alginate microcapsules fabricated from LbL method and the application in anti-tumor treatment *in vitro* and *in vivo* (*78*). These microcapsules showed a strong ability to accumulate the positively charged DOX with the drug concentration inside the microcapsules hundreds of times higher than the feeding concentration. *In vitro* experiments showed that the encapsulated drug can effectively induce the apoptosis of HepG2 tumor cells. Moreover, the *in vivo* results showed that the encapsulated DOX had better efficacy than that of the free drug in terms of tumor inhibition in a 4-week study as shown in Figure 5.

A lot of therapeutic molecules, such as proteins, DNAs, and siRNAs, need to be protected from degradation before delivery to the target sites. Various types of biopolymer based hollow particles, fabricated by different methods, have been employed to encapsulate these molecules. These loaded hollow particles successfully delivered these cargoes to the target cells and achieved the desired therapeutic effect. Pandit group fabricated hollow collagen particles by template method, and loaded a large amount of nerve growth factor (NGF) into these hollow particles (*22*). The released NGF still showed bioactivity. A nondifferentiated neuronal PC12 Cell line, treated with NGF-loaded spheres exhibited similar morphology to NGF treated positive controls. In contrast, PC12 cells treated with unloaded spheres did not differentiate and maintained their original round shapes similar to untreated cells (Figure 6a). Significant neuronal outgrowth ($p < 0.05$) was observed in all groups treated with NGF-loaded spheres and in constant NGF treatment in comparison to single NGF-treated control (Figure 6b). More importantly, there is no cytotoxicity effect observed for cells treated with the collagen hollow particles. With the same strategy, Pandit group fabricate the hollow collagen particles and loaded them with pDNA polyplexes for gene delivery (*25*). The loaded particles showed a prolonged release of the polyplexes, and dramatically reduced the toxicity of the polyplexes to cells (Figure 7C). The polyplex loaded hollow particles transfected 3T3 fibroblasts as efficiently as the polyplex alone control (Figure 7A and 7B). Koker et al. demonstrated efficient in vitro antigen delivery to dendritic cells (DCs) using dextran sulfate/poly-L-arginine polyelectrolyte capsules as microcarriers (*101*). The capsules were internalized by DCs via a macropinocytotic route, and remained in endolysosomal vesicles, where the microcapsule shell is ruptured to release the antigen. The antigen-loaded capsules can efficiently stimulate CD4 and CD8 T-cell proliferation.

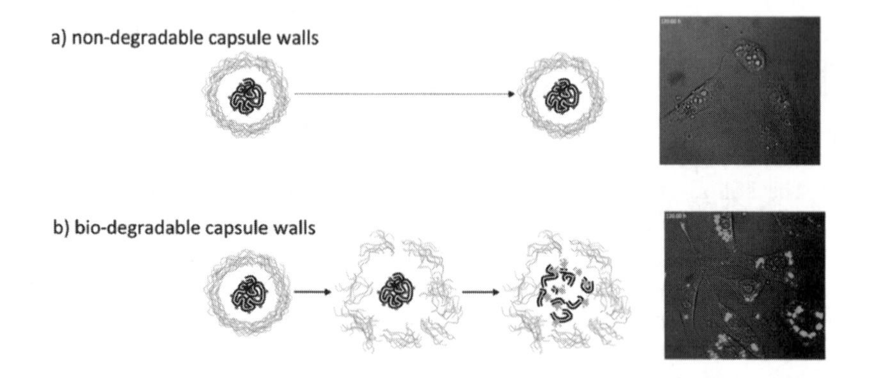

Figure 4. Enzymatic cleavage of protein cargo. Embryonic NIH/3T3 fibroblasts were incubated with (a) nondegradable PSS/PAH or (b) degradable DEXS/pARG capsules filled with the fluorogenic protein cargo, DQ-OVA. Images were taken immediately after 120 h with a confocal microscope in different channels, green, red, and transmission. An overlay of the different channels is presented in the figures. Reproduced with permission from reference (115). Copyright (2009) American Chemical Society.

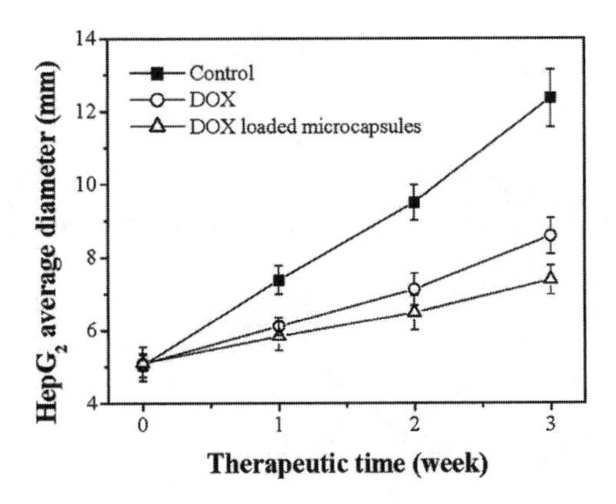

Figure 5. Diameter of the HepG2 BALB/c/nu tumor as a function of in vivo culture time. Encapsulated and free DOX with a dosage of 2 mg/kg (against the weight of mice) were injected into the tumors once a week for 3 weeks, respectively. Reproduced with permission from reference (78). Copyright (2007) Elsevier.

A lot of diseases, such as cancer and human immunodeficiency virus infection (HIV) (*116*, *117*), can not be cured by treatment with only one type of drug because of pathological complexity. The combinatorial use of multiple drugs having different pharmaceutical action mechanism might bring the synergistic or additive therapeutic effects and produce better outcomes comparing with mono-chemotherapy. Furthermore, the combination therapy may lower the dosages of drugs and concurrently lead to suppression of severe adverse effect. Therefore, the combination therapy by multiple drug treatment has been commonly used in clinical chemotherapy.The nano-/micro-particles loaded with multiple drugs are very promising for the combination therapy. The development of the fabrication methods provides the hollow particles with more functionalities. With a precise design, both the shells and cavities of the hollow particles could be loaded with drugs for dual or multi drug delivery. Wang et al. fabricated hollow deoxyribonucleic acid (DNA)/poly-l-lysine (PLL) capsules via LbL method (*118*). The capsules could be further loaded with drugs (dextran as a model) in the cavities. These dual-drug loaded capsules could be degraded by α-chymotrypsin and release the two drugs.

In most cases, release of drugs from the capsule is just controlled by the concentration difference between the cavity and surrounding environment, following a diffusion controlled manner. However, in certain cases, it would be perfect if the drug only could be released in response to certain environmental conditions, or so called stimuli responsive drug delivery.

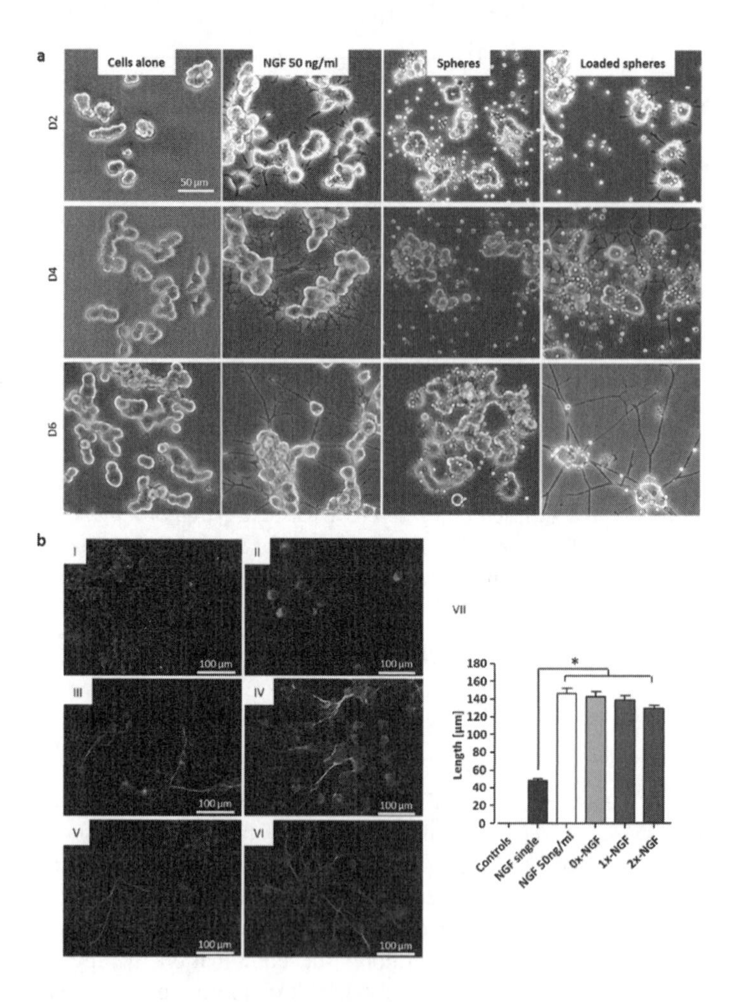

Figure 6. Bioactivity of released NGF assessed on PC12 cells. (a) PC12 cells alone, treated with 50 ng/mL NGF, with empty spheres and with NGFloaded 2× spheres at D2, D4, and D6. (b) Neuronal outgrowth in PC12 cells following 6 days culture with NGF-loaded spheres (β-III tub staining (green) of differentiated cells). (I) Representative image of nontreated cells/unloaded spheres treatment, (II) single treatment with 50 ng/mL NGF, (III) constant treatment with 50 ng/mL NGF, (IV−VI) 0×, 1×, and 2× collagen spheres treatment loaded with 250 ng of NGF. (VII) Significant ($p < 0.05$) neuronal outgrowth was observed in constant NGF treatment and in all groups treated with NGF-loaded spheres. Data represents mean ± SD, n = 50 (neurites per group). Reproduced with permission from reference (22). Copyright (2013) American Chemical Society.

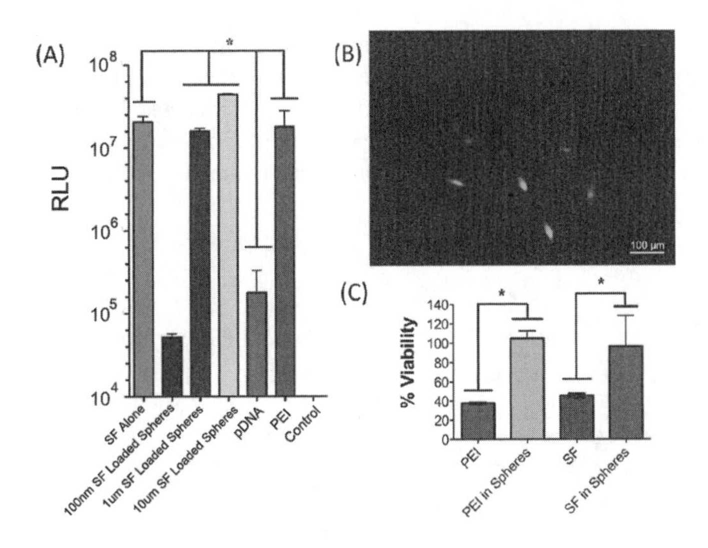

Figure 7. Transfection and polyplex toxicity. (A) Gaussia luciferase assay to assess the ability of the microspheres to release bioactive polyplexes capable of transfecting cells in vitro. In this case, the 1 and 10 µm spheres have displayed an ability to release polyplexes capable of transfecting 3T3 fibroblasts with an ability similar to that of polyplexes alone. (B) 3T3 fibroblasts expressing GFP following treatment with polyplex-loaded spheres (green for GFP and blue for DAPI). (C) Cell metabolic activity is dramatically reduced following incubation with cationic polymers. However, this effect is removed following loading of the polymers with collagen microspheres (1 µm). Data are means ± the standard deviation (n = 3). PEI stands for polyethyleneimine and SF for dPAMAM (Superfect, Qiagen). Reproduced with permission from reference (25). Copyright (2012) American Chemical Society.

3.1.1. pH-Responsive Drug Delivery

There is a pH fall between physiological conditions and intracellular endosomes and lysosomes, so the pH-responsive drug delivery is very attractive for intracellular drug delivery. Chiang et al. developed polymeric vesicles supplemented with the pH-responsive outlayered gels as a delivery system of doxorubicin (DOX) (as shown in Scheme 9) (*119*). These particles were fabricated from self-assembly of the lipid/polypeptide adduct, distearin grafted poly(c-glutamic acid), followed by sequential deposition of chitosan and poly(c-glutamic acid-*co*-c-glutamyl oxysuccinimide)-*g*-monomethoxy poly(ethylene glycol) in combination with in situ covalent cross-linking on assembly surfaces. The resultant gel-caged polymeric vesicles (GCPVs) exhibited superior capability of controlling drug release in response to the external pH change. The additional pH-responsive dual-layered gels not only effectively prevent the drug payload from premature leakage at pH 7.4 by the dense gel structure but also retain facile passage of DOX release at pH 4.7 due to the pertinent gel swelling. The GCPVs after being internalized by HeLa cells via endocytosis showed prominent antitumor ability comparable to free DOX by rapidly releasing payload in intracellular acidic organelles.

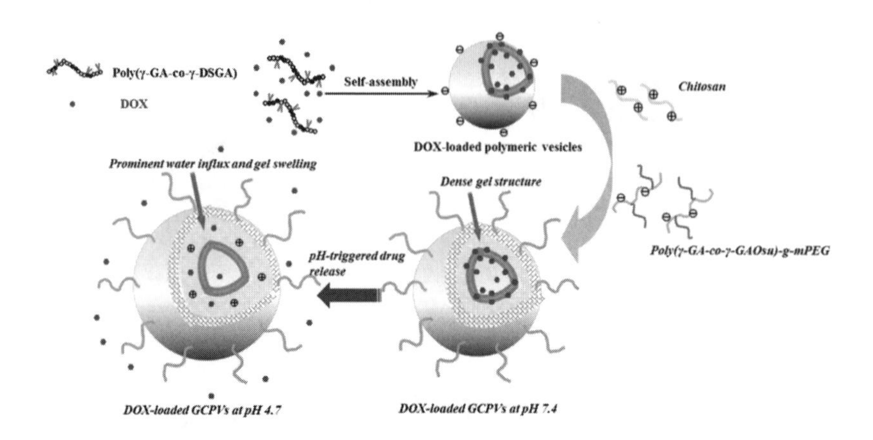

Scheme 9. Schematic illustration of the development of DOX-loaded GCPVs and their pH-triggered drug release. Reproduced with permission from reference (119). Copyright (2014) PLOS.

3.1.2. Thermo-Responsive Drug Delivery

Thermo-sensitive material, such as poly(N-isopropylacrylamide) (PNIPAAm), has been extensilvely investigated. It could be employed to biopolymer based hollow particles for controlled release of cargoes. Chen et al. designed and fabricated microcapsules consisted of polymerized glycidyl methacrylated dextran and gelatin with thermo-responsive PNIPAAm gates on their outer pore surfaces for the controlled release of stromal cell-derived factor (SDF)-1α, an important chemokine for stem cell recruitment/homing (Scheme 10) (*81*). The *in vitro* results showed that the PNIPAAm-grafted microcapsules featured thermo-responsive drug release properties due to the swollen-shrunken property of PNIPAAm gates in response to temperature changes. The SDF-1α released from this system retained its mitogenic activity, as determined by the stem cell migration assay; the migration of human periodontal ligament stem cells in response to the controlled release of SDF-1α closely followed the kinetics of SDF-1α release from the microcapsules.

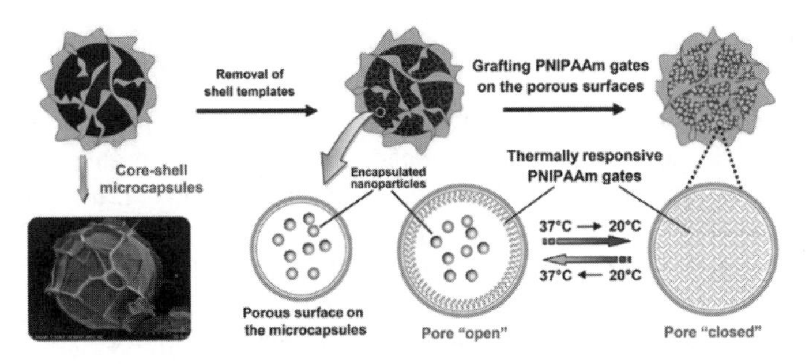

Scheme 10. Schematic illustration of the preparation process route and the design principle of the glycidyl methacrylated dextran (Dex-GMA)/gelatin microcapsules containing Dex-GMA/Gtn nanoparticles loaded with stromal cell-derived factor (SDF)-1α; the thermally controlled release of SDF-1α molecules from the resultant microcapsule platform in response to temperature change is a function of the "swollene-shrunken" property of the poly(N-isopropylacrylamide) (PNIPAAm) that are grafted into the outer porous surfaces of the microcapsules. Reproduced with permission from reference (81). Copyright (2013) Elsevier.

3.1.3. Enzyme-Responsive Drug Delivery

Biopolymers are produced by living organisms, and easy to be degraded by microorganisms with certain types of enzymes. The capsules derived from biopolymers may be sensitive to the corresponding enzyme, and release the encapsulated cargoes after enzymatic degradation. Baier demonstrated that novel hyaluronic acid (HA)-based nanocapsules containing the antimicrobial agent polyhexanide were specifically cleaved in the presence of hyaluronidase, a factor of pathogenicity and invasion for bacteria like *Staphylococcus aureus* and *Escherichia coli* (Scheme 11) (*70*). This resulted in an efficient killing of the pathogenic bacteria by the antimicrobial agent. The nanocapsules could be cleaved by hyaluronidase, and release the encapsulated polyhexanide, while control capsules formed with hydroxyethyl starch or only polyhexanide as shell material did not show any release.

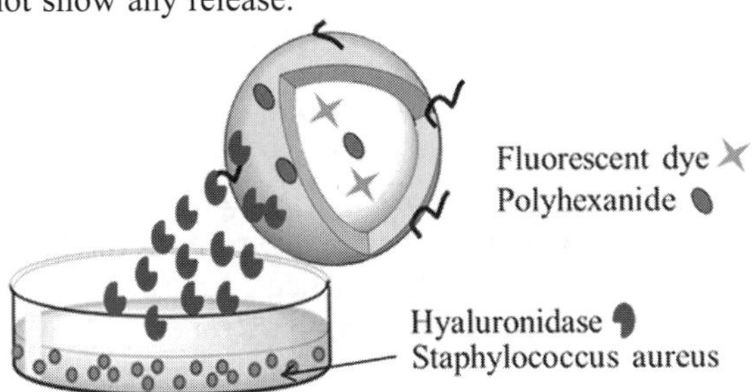

Scheme 11. Schematic illustration of enzyme responsive hyaluronic acid nanocapsules containing polyhexanide and their exposure to bacteria to prevent infection. Reproduced with permission from reference (70). Copyright (2013) American Chemical Society.

3.1.4. Redox-Responsive Drug Delivery

It is well-known that a significant difference in redox potential exists between the extracellular and intracellular compartments due to the higher Glutathione (GSH) concentrations in cytosol than the extracellular fluids (*120*). The hollow particle reducible by GSH is very attractive for controlled intracellular drug delivery, because there will be no or very little encapsulated drug released out during the delivery process in blood circulation until these particles are taken up by cells (*121*, *122*). Zheng et al. designed and fabricated novel biodegradable and redox-responsive submicron capsules through the layer-by-layer technique with poly(L-aspartic acid) and chitosan for transmucosal delivery of proteins and peptides (*84*). The cell viability test showed that all types of submicron capsules had good cytocompatibility and no cytotoxicity. Insulin could be effectively entrapped in the submicron capsules, and the release amount of insulin could be regulated by changing the GSH level.

3.1.5. Ultrasound Triggered Drug Delivery

Hollow microparticles could be destroyed by high intensity ultrasound, so the encapsulated drugs could be burst released after ultrasound irradiation. Tabata et al. described an investigation into the mechanical properties of microcapsules with a biocompatible polylactic acid (PLA) shell that can be destroyed using ultrasound irradiation (*71*). It was found that approximately 50% of capsules with a radius of 20 μm were destroyed using pulses with a pressure amplitude of 50 kPa and a frequency of 700 kHz.

3.2. Tissue Engineering

Tissue engineering can be considered as a special case of drug delivery where the goal is to accomplish controlled delivery of cells. The materials for the cell delivery should also promote cell-cell interaction, extra cellular matrix (ECM) deposition and tissue level organization (*20*). From a materials point of view, the tissue-engineering scaffolds need to be biocompatible and biodegradable. The biological functions of encapsulated cells can be dramatically enhanced by designing biomaterials with controlled organizations at the nanometer scale. The biopolymer based hollow particle has large interior space for cell encapsulation and are practically applicable for tissue engineering.

Liu et al. reported nanofibrous hollow microspheres self-assembled from star-shaped poly(l-lactic acid) (SS-PLLA) biodegradable polymers as an injectable cell carrier (*123*). The microspheres exhibited open hollow structures and nanofibrous shells (Figure 8). The injectable nanofibrous hollow microspheres self-assembled from SS-PLLA are an excellent micro-carrier for chondrocytes to facilitate high-quality hyaline cartilage regeneration. When compared with the solid-interior microspheres and the nanofibrous microspheres based on the same polymer chemistry (PLLA), the nanofibrous hollow microspheres group had a lower cell density and a substantially larger amount of new tissue matrix per cell, similar to the native rabbit cartilage (Figure 9). These findings demonstrated the significant impact of the unique physical features of the nanofibrous hollow microspheres on cell behavior and tissue regeneration. The nanofibrous hollow microspheres had been used to successfully repair a critical-size osteochondral defect in a widely used rabbit model, and had been shown to be advantageous over the chondrocytes-alone group that simulates the clinically available autologous chondrocyte implantation procedure.

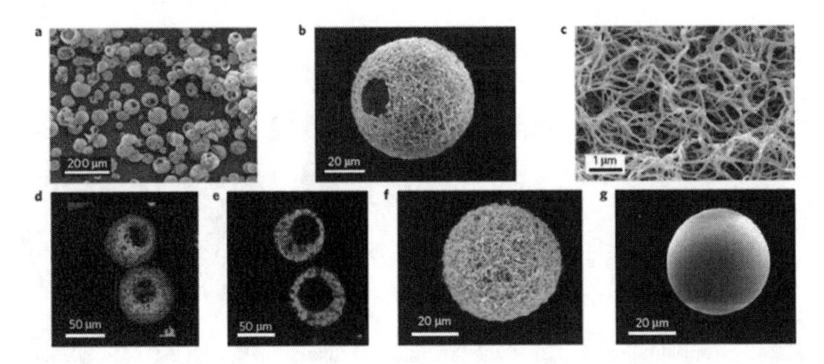

Figure 8. Characterization of nanofibrous hollow microspheres, nanofibrous microspheres and solid-interior microspheres. a, SEM image of nanofibrous hollow microspheres fabricated from SS-PLLA, showing that almost every microsphere had one or more open hole(s) on the shell. b, SEM image of a representative nanofibrous hollow microsphere, showing the nanofibrous architecture and a hole of approximately 20 m on the microsphere shell. c, A high-magnification image of the microsphere in b, showing the nanofibres, which have an average diameter of about 160 nm. d, A 3D reconstruction of nanofibrous hollow microspheres from confocal image stacks. e, A 2D cross-section confocal image of the nanofibrous hollow microspheres, confirming the hollow structure of the microsphere. f, SEM image of a representative nanofibrous microsphere, showing the nanofibrous architecture on the microsphere surface. g, SEM image of a representative solid-interior microsphere, showing the smooth surface of the microsphere. Reproduced with permission from reference (123). Copyright (2011) Nature Publishing Group.

Tiruvannamalai-Annamalai et al. reported the elements of a simple and efficient method for fabricating vascularized tissue constructs by fusing biodegradable microcapsules with tunable interior environments (*86*). Parenchymal cells of various types, (i.e. trophoblasts, vascular smooth muscle cells, hepatocytes) were suspended in glycosaminoglycan (GAG) solutions (4%/1.5% chondroitin sulfate/carboxymethyl cellulose, or 1.5 wt% hyaluronan) and encapsulated by forming chitosan-GAG polyelectrolyte complex membranes around droplets of the cell suspension as shown in Scheme 12. The interior capsule environment could be further tuned by blending collagen with or suspending microcarriers in the GAG solution These capsule modules were seeded externally with vascular endothelial cells (VEC), and subsequently fused into tissue constructs possessing VEC-lined, inter-capsule channels. The microcapsules supported high density growth achieving clinically significant cell densities. Fusion of the endothelialized, capsules generated three dimensional constructs with an embedded network of interconnected channels that enabled long-term perfusion culture of the construct.

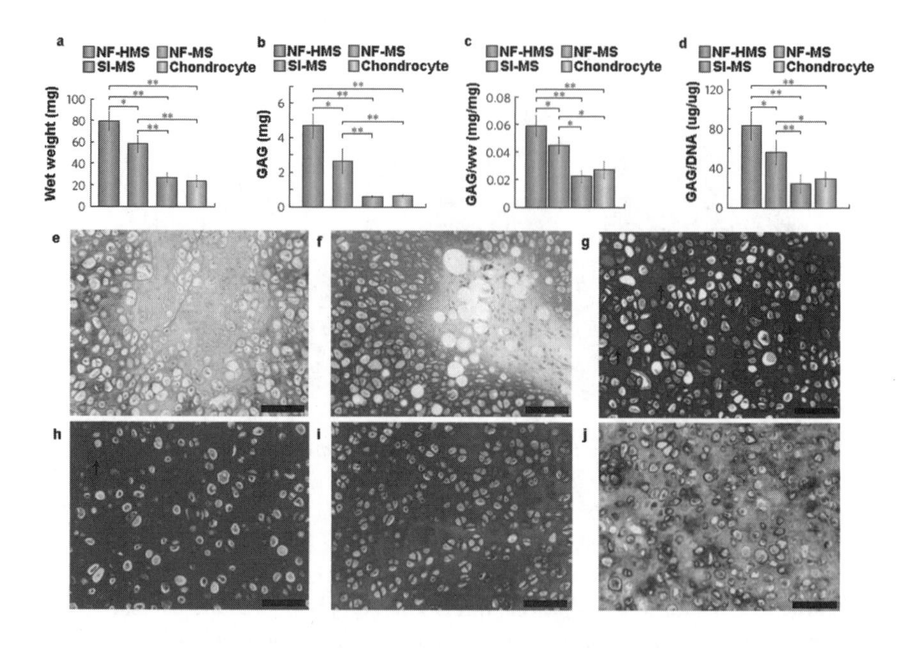

*Figure 9. Comparison of various microspheres as cell carriers for ectopic cartilage regeneration in vivo. Eight weeks after subcutaneous injection of the chondrocytes/microspheres suspension, the biochemical compositions and histological images of the ectopically engineered cartilage tissue from the same number of chondrocytes and the same mass of different types of microsphere were comparatively evaluated. a–d, The nanofibrous hollow microspheres (NF-HMS) group had a significantly higher tissue mass (a), GAG content (b), GAG/ww ratio (c) and GAG/DNA ratio (d) than the nanofibrous microspheres (NF-MS), the solid-interior microspheres (SI-MS) and the chondrocytes-alone groups. Error bars in a–d indicate standard deviation. *P<0.05;**P<0.01. e–h, Tissue sections were stained with safranin-O for proteoglycans eight weeks after subcutaneous injection: chondrocytes-alone group (e), solid-interior microspheres/chondrocytes group (f), nanofibrous microspheres/chondrocytes group (g; the arrows indicate a nanofibrous microsphere), nanofibrous hollow microspheres/chondrocytes group (h; the arrow indicates a nanofibrous hollow microsphere). i, Native rabbit knee cartilage was used as the positive control. Scale bars in e–j represent 100µ m. Reproduced with permission from reference (123). Copyright (2011) Nature Publishing Group.*

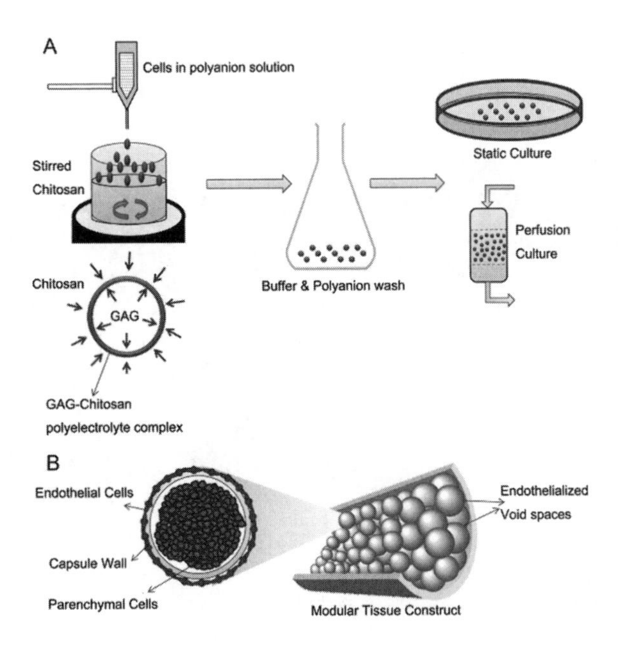

Scheme 12. Microencapsulation through complex coacervation and modular assembly. (A) Droplets of cells suspended in a polyanionic solution were dispensed into a stirred chitosan solution. Ionic interactions between the oppositely charged polymers formed an insoluble ionic complex membrane at the droplet-solution interface, thus encapsulating the suspended cells. Capsule were washed surface-stabilized with a suitable anionic polymer solution, and transferred to culture. (B) Cell laden capsules can be assembled in a packed bed fashion with interconnected endothelialized channels that may enable perfusion of fluids such as blood with limited adverse reactions. Reproduced with permission from reference (86). Copyright (2014) PLOS.

3.3. Wastewater Treatment

Biopolymers are produced by living organisms, including plants, animals and bacterias. They are always sustainable and renewable, and could become less costly as rapid technological progress. They are also biocompatible and biodegradable, could be fully degraded by microorganisms and produce no burden on environment. Using biopolymer based hollow micro-/nano-particles for the wastewater treatment is economical and environment-friendly. Moreover, the hollow particles have higher surface areas compared with the solid particles with the same mass content, and could adsorb more wastes from wastewater. Xu et al. developed biodegradable hollow zein nanoparticles with diameters less than 100 nm (TEM shown in Figure 10) to remove reactive dyes from simulated post-dyeing wastewater with remarkably high efficiency (Figure 11) (*76*). Hollow zein nanoparticles showed higher adsorption for Reactive Blue 19 than solid structures. These hollow zein particles were able to sorb up to 1016.0 mg of reactive dye per gram nanoparticles which was much higher than many other reported biodegradable adsorbents previously studied. moreover,the nanoparticles precipitated fast after adsorption and thus could be easily removed after treatment.

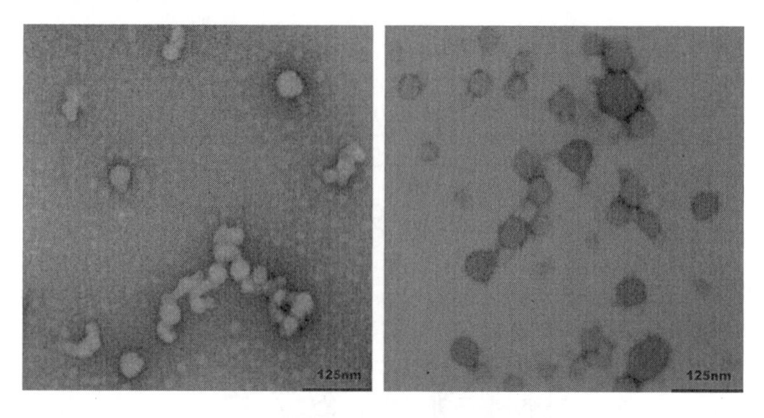

Figure 10. TEM images of hollow zein nanoparticles (left) and solid zein nanoparticles (right). Reproduced with permission from reference (76). Copyright (2013) Elsevier.

Figure 11. Adsorption of RB 19 onto hollow zein nanoparticles. Reproduced with permission from reference (76). Copyright (2013) Elsevier.

Conclusion

This review highlights the diversity of methods that have been developed to fabricate hollow micro-/nano-particles from biopolymers, and the recent applications of these particles. Hollow particles with various compositions and properties can be successfully prepared by one or combined processes. However, the commercialization of existing methods is still a big challenge. Operation cost, processing time, and complexity of fabrication process need to be improved. The application in biomedical fields makes the fabrication of these particles even more challenging, because it requires not only the facile hollow structure formation, but also the availability of multi-functionalities for biomedical utilization. Future development of the fabrication methods is still essential to expand the application of the particles.

References

1. Rösler, A.; Vandermeulen, G. W. M.; Klok, H.-A. *Adv. Drug Delivery Rev.* **2001**, *53*, 95–108.
2. Fu, G.-D.; Li, G. L.; Neoh, K. G.; Kang, E. T. *Prog. Polym. Sci.* **2011**, *36*, 127–167.
3. McDonald, C. J.; Devon, M. J. *Adv. Colloid Interface Sci.* **2002**, *99*, 181–213.
4. Narayan, P.; Marchant, D.; Wheatley, M. A. *J. Biomed. Mater. Res.* **2001**, *56*, 333–341.
5. Costa, R. R.; Mano, J. F. *Chem. Soc. Rev.* **2014**, *43*, 3453–3479.
6. De Koker, S.; Hoogenboom, R.; De Geest, B. G. *Chem. Soc. Rev.* **2012**, *41*, 2867–2884.
7. Tong, W.; Song, X.; Gao, C. *Chem. Soc. Rev.* **2012**, *41*, 6103–6124.
8. Johnston, A. P. R.; Cortez, C.; Angelatos, A. S.; Caruso, F. *Curr. Opin. Colloid Interface Sci.* **2006**, *11*, 203–209.
9. Brinkhuis, R. P.; Rutjes, F. P. J. T.; van Hest, J. C. M. *Polym. Chem.* **2011**, *2*, 1449–1462.
10. Chen, D.; Jiang, M. *Acc. Chem. Res.* **2005**, *38*, 494–502.
11. Hu, Y.; Jiang, X.; Ding, Y.; Chen, Q.; Yang, C. Z. *Adv. Mater.* **2004**, *16*, 933–937.
12. Meier, W. *Chem. Soc. Rev.* **2000**, *29*, 295–303.
13. Caruso, F.; Caruso, R. A.; Möhwald, H. *Science* **1998**, *282*, 1111–1114.
14. Delcea, M.; Möhwald, H.; Skirtach, A. G. *Adv. Drug Delivery Rev.* **2011**, *63*, 730–747.
15. Ibarz, G.; Dähne, L.; Donath, E.; Möhwald, H. *Adv. Mater.* **2001**, *13*, 1324–1327.
16. De Cock, L. J.; De Koker, S.; De Geest, B. G.; Grooten, J.; Vervaet, C.; Remon, J. P.; Sukhorukov, G. B.; Antipina, M. N. *Angew. Chem., Int. Ed.* **2010**, *49*, 6954–6973.
17. He, Q.; Cui, Y.; Li, J. *Chem. Soc. Rev.* **2009**, *38*, 2292–2303.
18. Discher, D. E.; Eisenberg, A. *Science* **2002**, *297*, 967–973.
19. Jones, O. G.; McClements, D. J. *Adv. Colloid Interface Sci.* **2011**, *167*, 49–62.
20. Nitta, S.; Numata, K. *Int. J. Mol. Sci.* **2013**, *14*, 1629–1654.
21. Oh, J. K.; Lee, D. I.; Park, J. M. *Prog. Polym. Sci.* **2009**, *34*, 1261–1282.
22. Kraskiewicz, H.; Breen, B.; Sargeant, T.; McMahon, S.; Pandit, A. *ACS Chem. Neurosci.* **2013**, *4*, 1297–1304.

23. Helary, C.; Browne, S.; Mathew, A.; Wang, W.; Pandit, A. *Acta Biomater.* **2012**, *8*, 4208–4214.
24. Dash, B. C.; Mahor, S.; Carroll, O.; Mathew, A.; Wang, W.; Woodhouse, K. A.; Pandit, A. *J. Controlled Release* **2011**, *152*, 382–392.
25. Browne, S.; Fontana, G.; Rodriguez, B. J.; Pandit, A. *Mol. Pharmaceutics* **2012**, *9*, 3099–3106.
26. Skirtach, A. G.; Yashchenok, A. M.; Mohwald, H. *Chem. Commun.* **2011**, *47*, 12736–12746.
27. Carrick, C.; Ruda, M.; Pettersson, B.; Larsson, P. T.; Wagberg, L. *RSC Adv.* **2013**, *3*, 2462–2469.
28. Luo, R.; Venkatraman, S. S.; Neu, B. *Biomacromolecules* **2013**, *14*, 2262–2271.
29. Jing, J.; Szarpak-Jankowska, A.; Guillot, R.; Pignot-Paintrand, I.; Picart, C.; Auzély-Velty, R. *Chem. Mater.* **2013**, *25*, 3867–3873.
30. Strehlow, V.; Lessig, J.; Gose, M.; Reibetanz, U. *J. Mater. Chem. B* **2013**, *1*, 3633–3643.
31. Duan, L.; Qi, W.; Yan, X.; He, Q.; Cui, Y.; Wang, K.; Li, D.; Li, J. *J. Phys. Chem. B* **2008**, *113*, 395–399.
32. Radhakrishnan, K.; Tripathy, J.; Raichur, A. M. *Chem. Commun.* **2013**, *49*, 5390–5392.
33. De Koker, S.; Naessens, T.; De Geest, B. G.; Bogaert, P.; Demeester, J.; De Smedt, S.; Grooten, J. *J. Immunol.* **2010**, *184*, 203–211.
34. Zhao, Q.; Li, B. *Nanomedicine* **2008**, *4*, 302–310.
35. Yan, S.; Zhu, J.; Wang, Z.; Yin, J.; Zheng, Y.; Chen, X. *Eur. J. Pharm. Biopharm.* **2011**, *78*, 336–345.
36. Chu, L.-Y.; Utada, A. S.; Shah, R. K.; Kim, J.-W.; Weitz, D. A. *Angew. Chem.* **2007**, *119*, 9128–9132.
37. Moinard-Chécot, D.; Chevalier, Y.; Briançon, S.; Beney, L.; Fessi, H. *J. Colloid Interface Sci.* **2008**, *317*, 458–468.
38. Petrovic, L. B.; Sovilj, V. J.; Katona, J. M.; Milanovic, J. L. *J. Colloid Interface Sci.* **2010**, *342*, 333–339.
39. Hanson, J. A.; Chang, C. B.; Graves, S. M.; Li, Z.; Mason, T. G.; Deming, T. J. *Nature* **2008**, *455*, 85–88.
40. Utada, A. S.; Lorenceau, E.; Link, D. R.; Kaplan, P. D.; Stone, H. A.; Weitz, D. A. *Science* **2005**, *308*, 537–541.
41. Na, X.-M.; Gao, F.; Zhang, L.-Y.; Su, Z.-G.; Ma, G.-H. *ACS Macro Lett.* **2012**, *1*, 697–700.
42. Reis, A. V.; Guilherme, M. R.; Paulino, A. T.; Muniz, E. C.; Mattoso, L. H. C.; Tambourgi, E. B. *Langmuir* **2009**, *25*, 2473–2478.
43. Tu, F.; Lee, D. *Langmuir* **2012**, *28*, 9944–9952.
44. Crespy, D.; Stark, M.; Hoffmann-Richter, C.; Ziener, U.; Landfester, K. *Macromolecules* **2007**, *40*, 3122–3135.
45. Liu, L.; Yang, J.-P.; Ju, X.-J.; Xie, R.; Liu, Y.-M.; Wang, W.; Zhang, J.-J.; Niu, C. H.; Chu, L.-Y. *Soft Matter* **2011**, *7*, 4821–4827.
46. Hermanson, K. D.; Harasim, M. B.; Scheibel, T.; Bausch, A. R. *Phys. Chem. Chem. Phys.* **2007**, *9*, 6442–6446.
47. Qian, J.; Wu, F. *Chem. Mater.* **2009**, *21*, 758–762.

48. Qian, J.; Wu, F. *Macromolecules* **2008**, *41*, 8921–8926.
49. Dou, H.; Jiang, M.; Peng, H.; Chen, D.; Hong, Y. *Angew. Chem. Int. Ed.* **2003**, *42*, 1516–1519.
50. Wang, J.; Jiang, M. *J. Am. Chem. Soc.* **2006**, *128*, 3703–3708.
51. Liu, G.; Jin, Q.; Liu, X.; Lv, L.; Chen, C.; Ji, J. *Soft Matter* **2011**, *7*, 662–669.
52. Lin, X.; Gao, W.; Huang, X.; Zang, P.; Chen, J.; Wu, H. *Carbohydr. Polym.* **2010**, *82*, 460–465.
53. Meng, X.-W.; Qin, J.; Liu, Y.; Fan, M.-M.; Li, B.-J.; Zhang, S.; Yu, X.-Q. *Chem. Commun.* **2010**, *46*, 643–645.
54. Yin, W.; Yates, M. Z. *J. Colloid Interface Sci.* **2009**, *336*, 155–161.
55. Zhou, X. D.; Zhang, S. C.; Huebner, W.; Ownby, P. D.; Gu, H. *J. Mater. Sci.* **2001**, *36*, 3759–3768.
56. Desai, K. G. H.; Park, H. J. *J. Microencapsulation* **2005**, *22*, 179–192.
57. Vehring, R. *Pharm. Res.* **2008**, *25*, 999–1022.
58. Tong, W.; Gao, C. *J. Mater. Chem.* **2008**, *18*, 3799–3812.
59. De Geest, B. G.; Van Camp, W.; Du Prez, F. E.; De Smedt, S. C.; Demeester, J.; Hennink, W. E. *Chem. Commun.* **2008**, 190–192.
60. Guo, S.; Zheng, J.; Dong, J.; Guo, N.; Jing, L.; Yue, X.; Yan, X.; Wang, Y.; Dai, Z. *Int. J. Biol. Macromol.* **2011**, *49*, 409–415.
61. De Geest, B. G.; Van Camp, W.; Du Prez, F. E.; De Smedt, S. C.; Demeester, J.; Hennink, W. E. *Macromol. Rapid Commun.* **2008**, *29*, 1111–1118.
62. Li, C.; Luo, G.-F.; Wang, H.-Y.; Zhang, J.; Gong, Y.-H.; Cheng, S.-X.; Zhuo, R.-X.; Zhang, X.-Z. *J. Phys. Chem. C* **2011**, *115*, 17651–17659.
63. Peng, X.; Zhang, L. *Langmuir* **2004**, *21*, 1091–1095.
64. Li, H.; Wang, M.; Song, L.; Ge, X. *Colloid Polym. Sci.* **2008**, *286*, 819–825.
65. Wang, W.; Luo, C.; Shao, S.; Zhou, S. *Eur. J. Pharm. Biopharm.* **2010**, *76*, 376–383.
66. Zhang, J.; Li, C.; Xue, Z.-Y.; Cheng, H.-W.; Huang, F.-W.; Zhuo, R.-X.; Zhang, X.-Z. *Acta Biomater.* **2011**, *7*, 1665–1673.
67. Akamatsu, K.; Chen, W.; Suzuki, Y.; Ito, T.; Nakao, A.; Sugawara, T.; Kikuchi, R.; Nakao, S.-i. *Langmuir* **2010**, *26*, 14854–14860.
68. Prego, C.; Fabre, M.; Torres, D.; Alonso, M. J. *Pharm. Res.* **2006**, *23*, 549–556.
69. Yu, L.; Gao, Y.; Yue, X.; Liu, S.; Dai, Z. *Langmuir* **2008**, *24*, 13723–13729.
70. Baier, G.; Cavallaro, A.; Vasilev, K.; Mailänder, V.; Musyanovych, A.; Landfester, K. *Biomacromolecules* **2013**, *14*, 1103–1112.
71. Tabata, H.; Kato, Y.; Suematsu, S.; Yoshida, K.; Koyama, D.; Nakamura, K.; Watanabe, Y. *Appl. Acoust.* **2014**, *78*, 89–91.
72. Sakurai, D.; Molino Cornejo, J. J.; Daiguji, H.; Takemura, F. *J. Mater. Chem. A* **2013**, *1*, 14562–14568.
73. Li, X.-D.; Liang, X.-L.; Yue, X.-L.; Wang, J.-R.; Li, C.-H.; Deng, Z.-J.; Jing, L.-J.; Lin, L.; Qu, E.-Z.; Wang, S.-M.; Wu, C.-L.; Wu, H.-X.; Dai, Z.-F. *J. Mater. Chem. B* **2014**, *2*, 217–223.
74. Pandit, S.; Cevher, E.; Zariwala, M. G.; Somavarapu, S.; Alpar, H. O. *J. Microencapsulation* **2007**, *24*, 539–552.

75. Daiguji, H.; Takada, S.; Cornejo, J. J. M.; Takemura, F. *J. Phys. Chem. B* **2009**, *113*, 15002–15009.
76. Xu, H.; Zhang, Y.; Jiang, Q.; Reddy, N.; Yang, Y. *J. Environ. Manage.* **2013**, *125*, 33–40.
77. Xu, H.; Jiang, Q.; Reddy, N.; Yang, Y. *J. Mater. Chem.* **2011**, *21*, 18227–18235.
78. Zhao, Q.; Han, B.; Wang, Z.; Gao, C.; Peng, C.; Shen, J. *Nanomedicine* **2007**, *3*, 63–74.
79. George, M.; Abraham, T. E. *J. Controlled Release* **2006**, *114*, 1–14.
80. Li, S.; Wang, X.-T.; Zhang, X.-B.; Yang, R.-J.; Zhang, H.-Z.; Zhu, L.-Z.; Hou, X.-P. *J. Controlled Release* **2002**, *84*, 87–98.
81. Chen, F.-M.; Lu, H.; Wu, L.-A.; Gao, L.-N.; An, Y.; Zhang, J. *Biomaterials* **2013**, *34*, 6515–6527.
82. Thomas, M. B.; Radhakrishnan, K.; Gnanadhas, D. P.; Chakravortty, D.; Raichur, A. M. *Int. J. Nanomed.* **2013**, *8*, 267–273.
83. Liu, Y.; Yang, J.; Zhao, Z.; Li, J.; Zhang, R.; Yao, F. *J. Colloid Interface Sci.* **2012**, *379*, 130–140.
84. Zheng, C.; Zhang, X. G.; Sun, L.; Zhang, Z. P.; Li, C. X. *J. Mater. Sci.: Mater. Med.* **2013**, *24*, 931–939.
85. Dash, B. C.; Réthoré, G.; Monaghan, M.; Fitzgerald, K.; Gallagher, W.; Pandit, A. *Biomaterials* **2010**, *31*, 8188–8197.
86. Tiruvannamalai-Annamalai, R.; Armant, D. R.; Matthew, H. W. T. *PLoS One* **2014**, *9*, e84287.
87. Zhu, Y.; Tong, W.; Gao, C.; Mohwald, H. *J. Mater. Chem.* **2008**, *18*, 1153–1158.
88. Duan, L.; He, Q.; Yan, X.; Cui, Y.; Wang, K.; Li, J. *Biochem. Biophys. Res. Commun.* **2007**, *354*, 357–362.
89. Schüler, C.; Caruso, F. *Biomacromolecules* **2001**, *2*, 921–926.
90. Donath, E.; Sukhorukov, G. B.; Caruso, F.; Davis, S. A.; Möhwald, H. *Angew. Chem., Int. Ed.* **1998**, *37*, 2201–2205.
91. Zhang, Y.; Guan, Y.; Yang, S.; Xu, J.; Han, C. C. *Adv. Mater.* **2003**, *15*, 832–835.
92. Kozlovskaya, V.; Kharlampieva, E.; Drachuk, I.; Cheng, D.; Tsukruk, V. V. *Soft Matter* **2010**, *6*, 3596–3608.
93. Sivakumar, S.; Bansal, V.; Cortez, C.; Chong, S.-F.; Zelikin, A. N.; Caruso, F. *Adv. Mater.* **2009**, *21*, 1820–1824.
94. Luo, G.-F.; Xu, X.-D.; Zhang, J.; Yang, J.; Gong, Y.-H.; Lei, Q.; Jia, H.-Z.; Li, C.; Zhuo, R.-X.; Zhang, X.-Z. *ACS Appl. Mater. Interfaces* **2012**, *4*, 5317–5324.
95. Zhu, Y.; Tong, W.; Gao, C. *Soft Matter* **2011**, *7*, 5805–5815.
96. Radhakrishnan, K.; Raichur, A. M. *Chem. Commun.* **2012**, *48*, 2307–2309.
97. Volodkin, D. V.; Larionova, N. I.; Sukhorukov, G. B. *Biomacromolecules* **2004**, *5*, 1962–1972.
98. Borodina, T.; Markvicheva, E.; Kunizhev, S.; Möhwald, H.; Sukhorukov, G. B.; Kreft, O. *Macromol. Rapid Commun.* **2007**, *28*, 1894–1899.
99. Tong, W.; Zhu, Y.; Wang, Z.; Gao, C.; Möhwald, H. *Macromol. Rapid Commun.* **2010**, *31*, 1015–1019.

100. Ai, H.; Jones, S. A.; de Villiers, M. M.; Lvov, Y. M. *J. Controlled Release* **2003**, *86*, 59–68.
101. De Koker, S.; De Geest, B. G.; Singh, S. K.; De Rycke, R.; Naessens, T.; Van Kooyk, Y.; Demeester, J.; De Smedt, S. C.; Grooten, J. *Angew. Chem., Int. Ed.* **2009**, *48*, 8485–8489.
102. Baier, G.; Musyanovych, A.; Dass, M.; Theisinger, S.; Landfester, K. *Biomacromolecules* **2010**, *11*, 960–968.
103. Xi, J.; Zhou, L.; Fei, Y. *Int. J. Biol. Macromol.* **2012**, *50*, 157–163.
104. Lee, S. J.; Rosenberg, M. *J. Microencapsulation* **2000**, *17*, 29–44.
105. Ficheux, M. F.; Bonakdar, L.; Leal-Calderon, F.; Bibette, J. *Langmuir* **1998**, *14*, 2702–2706.
106. Yafei, W.; Tao, Z.; Gang, H. *Langmuir* **2005**, *22*, 67–73.
107. Liu, L.; Yang, J.-P.; Ju, X.-J.; Xie, R.; Yang, L.; Liang, B.; Chu, L.-Y. *J. Colloid Interface Sci.* **2009**, *336*, 100–106.
108. Zhang, H.; Tumarkin, E.; Peerani, R.; Nie, Z.; Sullan, R. M. A.; Walker, G. C.; Kumacheva, E. *J. Am. Chem. Soc.* **2006**, *128*, 12205–12210.
109. Gohy, J.-F. Block Copolymer Micelles. In *Block Copolymers II*; Abetz, V., Ed.; Springer: Berlin, 2005; Vol. 190, pp 65−-136.
110. Harada, A.; Kataoka, K. *Prog. Polym. Sci.* **2006**, *31*, 949–982.
111. Chiang, W.-H.; Lan, Y.-J.; Huang, Y.-C.; Chen, Y.-W.; Huang, Y.-F.; Lin, S.-C.; Chern, C.-S.; Chiu, H.-C. *Polymer* **2012**, *53*, 2233–2244.
112. Jones, L. C.; Lackowski, W. M.; Vasilyeva, Y.; Wilson, K.; Chechik, V. *Chem. Commun.* **2009**, 1377–1379.
113. Fang, J.; Nakamura, H.; Maeda, H. *Adv. Drug Delivery Rev.* **2011**, *63*, 136–151.
114. Wang, A. Z.; Langer, R.; Farokhzad, O. C. *Annu. Rev. Med.* **2012**, *63*, 185–198.
115. Rivera-Gil, P.; De Koker, S.; De Geest, B. G.; Parak, W. J. *Nano Lett.* **2009**, *9*, 4398–4402.
116. Al-Lazikani, B.; Banerji, U.; Workman, P. *Nat. Biotechnol.* **2012**, *30*, 679–692.
117. Levy, V.; Grant, R. M. *Clin. Infect. Dis.* **2006**, *43*, 904–910.
118. Wang, Z.; Qian, L.; Wang, X.; Zhu, H.; Yang, F.; Yang, X. *Colloids Surf., A* **2009**, *332*, 164–171.
119. Chiang, W.-H.; Huang, W.-C.; Shen, M.-Y.; Wang, C.-H.; Huang, Y.-F.; Lin, S.-C.; Chern, C.-S.; Chiu, H.-C. *PLoS One* **2014**, *9*, e92268.
120. Schafer, F. Q.; Buettner, G. R. *Free Radical Biol. Med.* **2001**, *30*, 1191–1212.
121. Saito, G.; Swanson, J. A.; Lee, K.-D. *Adv. Drug Delivery Rev.* **2003**, *55*, 199–215.
122. Qian, J.; Wu, F. *J. Mater. Chem. B* **2013**, *1*, 3464–3469.
123. Liu, X.; Jin, X.; Ma, P. X. *Nat. Mater.* **2011**, *10*, 398–406.

Editors' Biographies

Yiqi Yang

Dr. Yiqi Yang received his Ph.D. in 1991 from Purdue University after receiving his undergraduate and master's degrees from Donghua University. As a textile chemical engineer, he is a distinguished professor at the University of Nebraska-Lincoln, a state specially recruited expert of China, and a Changjiang Scholar Lecture Professor at Jiangnan University. Dr. Yang's research interests are in green polymers and materials, biotextile engineering, and fiber and textile chemistry. Examples of his research include development of new lignocellulosic, protein and synthetic fibers from agricultural wastes and byproducts, and application of green materials in textile, composite, and medical industries. Dr. Yang has extensive experience in fiber and textile productions, and has close relationships with international fiber and textile industries.

Helan Xu

Dr. Helan Xu received her Ph.D. in 2014 from University of Nebraska-Lincoln after receiving her undergraduate and master's degrees in textile science and engineering from Donghua University. She has been working on textile and medical applications of agricultural byproducts and wastes. Her research projects include developing textile fibers and 3D ultrafine fibrous tissue engineering scaffolds from proteins derived from agricultural byproducts and wastes, protein nanoparticles for biomedical and industrial applications, and biodegradable lightweight composites from renewable agricultural byproducts. Dr. Xu has published 25 SCI journal papers, including nine first-authored papers, and has written one ACS book chapter.

Xin Yu

Ms. Xin Yu, a double master's degree holder, graduated from Ecole Nationale Supérieure des Arts et Industries Textiles (ENSAIT) in France and Donghua University in China, and majored in textile engineering, textile technology, and advanced materials. She has worked in the Australian nationwide research organization Commonwealth Scientific and Industrial Research Organisation (CSIRO) and the division of Materials Science and Engineering (CMSE). Currently, Ms. Yu is a lecturer at the International School of Zhejiang Fashion Institute of Technology in China. Her research interests include textile technology, textile materials science, bioengineering, and functional textiles.

Indexes

Author Index

I

L

S

T

U

V

W

Z